Deep Homology?

Uncanny Similarities of Humans and Flies Uncovered by Evo-Devo

Humans and fruit flies look nothing alike, yet their genetic circuits are remarkably similar. Here, Lewis I. Held, Jr. compares the genetics and development of the two to review the evidence for deep homology, the biggest discovery from the emerging field of evolutionary developmental biology (evo-devo). Remnants of the operating system of our hypothetical common ancestor 600 million years ago are compared in chapters arranged by region of the body, from the nervous system, limbs, and heart, to vision, hearing, and smell. Concept maps provide a clear understanding of the complex subjects addressed, while encyclopedic tables offer comprehensive inventories of genetic information. Written in an engaging style with a reference section listing thousands of crucial publications, this is a vital resource for scientific researchers, as well as for graduate and undergraduate students.

Lewis I. Held, Jr. is a fly geneticist who has taught human embryology for 30 years. He studied molecular biology at MIT (BS 1973), investigated bristle patterning under John Gerhart at UC Berkeley (PhD 1977), and conducted postdoctoral research with Peter Bryant and Howard Schneiderman at UC Irvine (1977–1986). *Deep Homology?* is his fifth scholarly monograph, following *Models for Embryonic Periodicity* (Karger 1992), *Imaginal Discs* (Cambridge University Press 2002), *Quirks of Human Anatomy* (Cambridge University Press 2009), and *How the Snake Lost Its Legs* (Cambridge University Press 2014).

Deep Homology?

Uncanny Similarities of Humans and Flies Uncovered by Evo-Devo

LEWIS I. HELD, JR.

Texas Tech University, Texas, USA

CAMBRIDGE
UNIVERSITY PRESS

University Printing House, Cambridge CB2 8BS, United Kingdom

One Liberty Plaza, 20th Floor, New York, NY 10006, USA

477 Williamstown Road, Port Melbourne, VIC 3207, Australia

4843/24, 2nd Floor, Ansari Road, Daryaganj, Delhi - 110002, India

79 Anson Road, #06-04/06, Singapore 079906

Cambridge University Press is part of the University of Cambridge.

It furthers the University's mission by disseminating knowledge in the pursuit of education, learning and research at the highest international levels of excellence.

www.cambridge.org
Information on this title: www.cambridge.org/9781316601211

First published 2017

A catalogue record for this publication is available from the British Library

Library of Congress Cataloging in Publication data
Names: Held, Lewis I., Jr., 1951- , author
Title: Deep homology : uncanny similarities of humans and flies uncovered by
    evo-devo / Lewis I. Held, Jr., Texas Tech University, Texas, USA.
Description: New York : Cambridge University Press, 2017. | Includes
    bibliographical references and index.
Identifiers: LCCN 2016045814 | ISBN 9781107147188 (hardback)
Subjects: LCSH: Flies—Anatomy. | Human anatomy. | Anatomy, Comparative. |
    Homology (Biology) | Evolutionary developmental biology.
Classification: LCC QL538 .H45 2017 | DDC 595.7/35—dc23 LC record available at
    https://lccn.loc.gov/2016045814

ISBN 978-1-107-14718-8 Hardback
ISBN 978-1-316-60121-1 Paperback

Contents

Preface

Given the immensity of the Universe, it is hard to imagine that Earth is the only planet teeming with living things. Some day, therefore, we will likely encounter extraterrestrial beings, provided that our own species doesn't extinguish itself first.

Fortunately, our yearning for an encounter with alien life can be sated here and now by the expedient equivalent of *Gulliver's Travels*. Geneticists have been exploring a sort of Lilliput for over 100 years, but their exploits are not well known.

Ever since Thomas Hunt Morgan started experimenting with the fruit fly *Drosophila melanogaster*, *c.* 1908 [474,1180], this tiny insect has beguiled those of us who have followed in his footsteps. Its anatomy (exoskeleton, six legs, compound eyes) looks "alien" to our own (endoskeleton, two legs, simple eyes) [363], but its behaviors are eerily humanoid. Three examples were featured in Jonathan Weiner's lyrical biography of Seymour Benzer, *Time, Love, Memory* [2404] – namely, its sleep–wake cycle, its courtship rituals, and its ability to learn.

Still, the greatest similarities between humans and flies are completely invisible to the naked eye, or even to a high-powered microscope. They are genetic. The ways in which the human and fly genomes encode anatomy are amazingly alike. What we've learned about those codes is impressive, but it is even more fascinating to contemplate what we have yet to learn in this regard. This book surveys both the Known and the Unknown and seeks to trace a boundary between them.

The terrain charted here lies in a fertile hybrid zone between the realms of evolutionary and developmental biology – an area that has come to be called "evo-devo" [1251]. Evo-devo attained notoriety in 1988 with the discovery that vertebrates share clusters of *Hox* genes with flies [128,2021]. Additional "toolkit" genes [357,2435] have since been found that are likewise involved in constructing anatomies across the animal kingdom [482,494,678,1927,2310]. The history of these epiphanies is recounted in the Introduction, *Hox* complexes are analyzed in Chapter 1, and the remaining chapters cover a variety of organs or organ systems where vertebrates and flies share deep similarities.

The time seems right for a synthesis of this kind. Evo-devo is entering its Golden Age [1505,1522] and is ripening nicely: 2017 marks the fortieth anniversary of Gould's *Ontogeny and Phylogeny* [785] and the twentieth anniversary of the term "deep homology" [2086].

To the extent that biology textbooks mention evo-devo at all, they usually include diagrams of *Hox* homologies, but those oversimplified renditions omit many of the intriguing nuances [551]. Indeed, the entire field of evo-devo is brimming with Gordian Knots that beckon the next crop of would-be Alexanders. Some of those brain teasers are offered here as "puzzle boxes" with enough citations (numbers in square brackets) to allow novices to conduct their own customized investigations. Those neo-evo-devo devotees are the primary audience for whom this book was written, though other fans of the field might enjoy the narrative also. The corpus of text, puzzles, and diagrams (Chapters 1–7) is just long enough to serve as a staging area for launching readers into the ocean of literature (References section). In this way I am trying to "pay forward" the priceless gift that Martin Gardner gave me and the other budding child-scientists of my generation with his marvelous "Mathematical Games" column for *Scientific American*.

The citations that infest the text are admittedly annoying, but all of them are pertinent, and some of them are gems. For example, there is one reference buried in the "neuron identity" row of Table 2.1 with the lackluster title "Opposing intrinsic temporal gradients guide neural stem cell production of varied neuronal fates" [1336]. It solves a riddle posed some 50 years earlier by the British theoretician John Maynard Smith about how cells tell time [1452]. The dueling clock genes that were identified were given the cute names *Imp* and *Syp*. The deep thinker who would be most amused by such deep homologies is Aristotle, who was fascinated by the natural world but bewildered by its inner workings.

Evo-devo terms are deftly explained in Hall and Olson's *Keywords and Concepts in Evolutionary Developmental Biology* [848], and most of the other technical terms used here should be defined in any college genetics textbook. Nevertheless, readers may still find the jargon daunting unless they are at least acquainted with the basic tenets of evo-devo. Those principles are accessible in the popular books by Sean Carroll (*Endless Forms Most Beautiful* [327] and *From DNA to Diversity* [330]), and the more advanced treatises by John Gerhart and Marc Kirschner (*Cells, Embryos, and Evolution* [742] and *The Plausibility of Life* [1164]).

How can a book that focuses on only two species offer any wider insights? Because this comparison is only a means to an end. The rationale is to use whatever clues are unearthed by the comparison to propel broader surveys of animal phyla. *Homo sapiens* and *Drosophila melanogaster* occupy such distant twigs on the tree of animal life that any similarities must indicate either (1) enduring conservation, which could allow us to deduce aspects of our last common ancestor [90,953], or (2) convergent evolution, which could allow us to discern constraints on the range of available outcomes [633,2360]. In either case this "compare and contrast" exercise should prove worthwhile.

In 1997 an exercise of exactly this sort was conducted for vertebrate versus arthropod limbs by Neil Shubin, Cliff Tabin, and Sean Carroll [2086]. They showed that vertebrate and arthropod limbs are built by similar algorithms, despite manifesting starkly different anatomies, and they argued that the simplest explanation for the genetic similarities was conservation rather than convergence. These authors coined the term "deep homology" to denote this phenomenon, and they illustrated the meaning of this term by a familiar example of wing evolution (boldface added):

> **Determination of whether two structures are homologous depends on the hierarchical level at which they are compared.** For example, bird wings and bat wings are analogous as wings, having evolved independently for flight in each lineage. However, at a deeper hierarchical level that includes all tetrapods, they are homologous as forelimbs, being derived from a corresponding appendage of a common ancestor. Similarly, we suggest that whereas vertebrate and insect wings are analogous as appendages, the genetic mechanisms that pattern them may be homologous at a level including most protostomes and deuterostomes. Furthermore, **we propose that the regulatory systems that pattern extant arthropod and vertebrate appendages patterned an ancestral outgrowth** and that these circuits were later modified during the evolution of different types of animal appendages. Animal limbs would be, in a sense, developmental "paralogues" of one another; modification and redeployment of this ancient genetic system in different contexts produced the variety of appendages seen in Recent and fossil animals. [2086]

In 2009 those same authors reprised their theme in an essay entitled "Deep homology and the origins of evolutionary novelty" [2087]. In that paper they argued that vertebrate and insect eyes are also deeply homologous. They admitted surprise at the accumulating evidence consistent with their thesis, and they offered a firmer definition of the deep homology concept (boldface added):

> One of the most important, and **entirely unanticipated**, insights of the past 15 years was the recognition of an ancient similarity of patterning mechanisms in diverse organisms, often among structures not thought to be homologous on morphological or phylogenetic grounds . . . Homology, as classically defined, refers to a historical continuity in which morphological features in related species are similar in pattern or form because they evolved from a corresponding structure in a common ancestor. **Deep homology also implies a historical continuity, but in this case the continuity may not be so evident in particular morphologies; it lies in the complex regulatory circuitry inherited from a common ancestor.** [2087]

The present book grew from the seeds sown by those two seminal papers. Although the goal here is to compare humans and fruit flies, experimentation on humans is unethical, so virtually all of the data assigned to the human side of the ledger come from mice instead; mice are close enough to us genetically to serve as a proxy. Moreover, many comparisons will involve broader taxonomic levels (e.g., mammal–insect, vertebrate–arthropod, deuterostome–protostome) as we trace the similarities back through geologic time, and some contrasts will be lopsided (e.g., vertebrate–fly) due to the nature of the relevant data available in the literature.

The history of science is punctuated with epiphanies where two things which had been thought to be qualitatively different turned out to be fundamentally alike [316]. Thus, Newton compared the moon to a falling apple, Faraday united electricity with magnetism, and Einstein wedded space with time and matter with energy [261]. Now, thanks to decades of work by legions of evo-devotees, we realize that *H. sapiens* and *D. melanogaster* are not as different as we'd imagined. Rather, humans and fruit flies seem more like twins separated at birth half a billion years ago [927].

Evo-devo thus finds itself at the same sort of threshold as the one crossed by Galileo ~400 years ago when he gazed at Jupiter through a state-of-the-art, but still rather crude, telescope. There he saw what he first thought were nearby stars but soon realized were orbiting moons [2424]. That discovery astounded him and revolutionized astronomy. Even so, his telescope could not resolve Saturn's rings, which he mistook for lateral bulges [154]. Evo-devo is giving us a first glimpse of our metazoan forebears, and we are bound to draw some wrong conclusions about them (until we uncover their fossil remains), but even ghostly images are arguably better than none at all [2035]!

Discerning whether a given similarity is due to common descent (homology) or to independent evolution (convergence or parallelism) can be quite tricky [847,1350,2033], and partisans have clashed over the origins of the brain [2198], the eye [1629,1666,1787], and other organs [601]. One especially instructive debate concerns whether the chordate notochord is homologous to the annelid axochord [274].

Guidance on the homology–convergence controversy can be found in Günter Wagner's *Homology, Genes, and Evolutionary Innovation* [2363] and George McGhee's *Convergent Evolution* [1459], and in the practical checklists devised by Ehab Abouheif *et al.* [10] and Cliff Tabin *et al.* [2221]. Readers attuned to the philosophical side of science should relish the Talmudic literature that has grown up around this topic, including incisive treatises by Antonio Fontdevila [653], Stephen Jay Gould [788], Brian Hall [846], Jason Hodin [953], Gerd Müller [1572], David Stern [2184], Pat Wilmer [2439], Greg Wray and Ehab Abouheif [9,2463], and others [128,753,1667,2198,2366].

By stressing similarities over differences, this book runs the risk of being misinterpreted as endorsing the notion of deep homology as a default assumption. For that reason a question mark was included in the title to serve as a disclaimer. Even for those features where homology is implied by a preponderance of available evidence, any surmise to that effect will be subjective and must remain tentative. Any author foolish enough to enter this tangled swamp – and I do so reluctantly – must heed the warning signs. Seth Blair, for example, has pointed out the risks of unintentional "cherry-picking" (boldface added):

> Finding similarities is something of a self-fulfilling prophecy. More detailed, mechanistic information from more taxa could certainly help in this debate, especially information based on unbiased screens instead of candidate genes. If mechanisms vary greatly, however, **it will still come down to an argument about plausibility, and one scientist's homolog is often another's convergence** . . . Can some level of mechanistic similarity ever rule out convergence, or is that wishful thinking? The identity of ancestral organisms has been the subject of intense debate since the 1800s and it is interesting to think about what kind of data it would take to settle that debate. On the other hand, what fun would that be? [215]

So why even try, given all the uncertainties, ambiguities, and complications in attempting to envision the so-called "urbilaterian" ancestor [484,486,601] of humans, flies, and other bilaterally symmetric animals [90,1158,2035]? The saving grace of this otherwise quixotic quest is that it does not aspire to reach any lasting conclusions, but merely to stimulate further research. Dan Nilsson has offered wise advice on how to proceed with regard to the study of eye evolution in particular (boldface added):

> For understanding eye evolution we are left with a number of cues from morphology and ontogenetic paths, from developmental genes and their interactions, and from physiology and effector genes. **All of these cues can be deceptive, and none is principally more important than any other.** Hypotheses on eye evolution will also have to agree with phylogenetic trees, datings of molecular divergence, and the fossil record. The best we can do is to aim for a synthesis. [1629]

This undertaking may also prove worthwhile for a more practical reason: the fly genome is a gold mine of genes whose homologs cause cancer [937,1948,2361], aging [1616,1775,1797], neurodegeneration [368,672,1466,2316], and a host of other human maladies [1719,2380,2392,2486], so some of the similarities revealed here could have broad clinical applications [204,1378,1775,1935,2444]. Indeed, ~77% of human disease-causing genes have a fly homolog [187,1889,1905]. The relevance of fly circuitry to human pathology [526,1935] even extends to neurological [1618,1679] and psychiatric [2321,2494] disorders, and detailed assays have recently been published for how to measure the behavioral parameters [1525,2312].

Utility is surely a virtue, but so is beauty, and evo-devo has dramatic and aesthetic dimensions that merit our attention as well. For example, few phyla have ever ventured onto land; two that have done so are chordates and arthropods, and both taxa have even taken to the skies [759]. We humans are rightly proud of our history, but flies can boast a heroic odyssey too. The impetus for this book is thus partly Homeric: to extol the fly's genomic exploits vis-à-vis our own. Tom Brody's *Interactive Fly* website is a terrific resource for delving into the comparative genetics of flies versus vertebrates, including the human diseases that are currently being investigated using flies as a model system. Andreas Schmidt-Rhaesa's *The Evolution of Organ Systems* [2003] treats some of the topics covered here in a more encyclopedic way in terms of animal phylogeny.

For the sake of terminological consistency, the nomenclature of fly genetics is followed. Protein names are set in roman type and capitalized (e.g., Spalt), while gene names are italicized (e.g., *spalt*) and capitalized only if the first mutation described was dominant. Abbreviations: ATP (adenosine triphosphate), bp (base pair), cAMP (cyclic adenosine monophosphate), kb (kilobases), LOF (loss of function), GOF (gain of function), mb (megabases), and MY (millions of years). Fly/vertebrate homologs are denoted by a slash mark (e.g., *eyeless/Pax6*), whereas the order is reversed (vertebrate/fly) in Figures 6.1 and 6.2 to conform with schematics.

Ancillary material has been exiled to tables and figure legends so as not to disrupt the flow of the narrative. The deeply conserved role of *doublesex/Dmrt1* in sex determination [167,1324,1822,2420,2538] is omitted entirely because I have covered this subject in a previous book [925] (cf. gonad and germ cell evolution [611,612,613]). Other topics have been excluded because the evidence for deep homology, although enticing, is preliminary: gut [282,288,885,1073,2228], liver [2312], kidney [934,2312,2400], integument [1052,2271], skeleton [1579,2242], muscle [164,700,1687], tendons [1692,2032,2125], and the recently discovered deep homology of cartilage [2245]. Finally, I have refrained from discussing the universal "machine code" of the genome unless it is germane to anatomy [2272]: microRNAs [8,386,834,1841,1941], introns [1943], enhancers [394,1383,1569,1874,2336], epigenetic tags [1294], transcription factors [567,1636], genomic ontology [2365], gene orthology [1458], and polyploidy [668].

Every effort has been made to ensure literal accuracy by soliciting expert scrutiny. David Hillis verified my template for bilaterian phylogeny. Jason Hodin checked Figure 1.3 and shared his vast PDF collection to expand it. Volker Hartenstein approved Figure 2.3. Dan Nilsson vetted Figure 3.3 and furnished additional unpublished data. Other scholars generously critiqued sundry drafts of the Preface and Introduction (Michael Akam, John Gerhart, and Volker Hartenstein), Chapter 1 (Michael Akam, Richard Campbell, Joseph Frankel, John Gerhart, Jason Hodin, Thurston Lacalli, and Arnaud Martin), Chapter 2 (Alain Ghysen and Thurston Lacalli), Chapter 3 (Gordon Fain, Markus Friedrich, Roger Hardie, Sönke Johnsen, Thurston Lacalli, Ivan Schwab, and Jeff Thomas), Chapter 4 (Daniel Eberl, Gordon Fain, Alain Ghysen, and Martin Göpfert), Chapter 5 (John Carlson and Alain Ghysen), Chapter 6 (Peter Bryant, Susan Bryant, and Vernon French), and Chapter 7 (Richard Cripps, Volker Hartenstein, and Eric Olson). To all of these luminaries I am grateful. I owe an even greater debt of gratitude to Ellen Larsen and Marc Srour, who read the entire manuscript in serial installments. Despite this feedback, errors may remain for which I take responsibility.

A few colleagues asked why all the drawings of humans in this book are male. They are all versions of da Vinci's Vitruvian Man, but no gender bias entered into my decision to use this icon to represent humans, mammals, vertebrates, or chordates, depending on the context. I had to redraw Leonardo's original because he sketched *two* pairs of arms and legs to show the metrics called for by the Roman architect Vitruvius [2340], while I needed only *one* pair of each to make my points. Also, he rendered a *frontal* view, but I needed *side* views in some cases, so I ordered a sculptural facsimile of Vitruvian Man from a collectibles catalog, took a picture of it from the side, imported the image into Adobe Illustrator, and traced it for that purpose. The "Vitruvian Fly" took me a week to trace from a montage of scanning electron micrographs. One last disclaimer: in a previous book [925] (Figure 6.3f), I attributed a drawing of a human–fly chimera to William Blake, based on a caption in Claudio Stern's 1990 paper [2182], and I planned to use it again here. When I wrote to inquire about copyright permission, Claudio told me that he had drawn the man with a fly's head in the style of Blake as a prank.

Marc Srour and Ruth Serra-Moreno graciously translated quotes from Geoffroy St.-Hilaire and Ramón y Cajal, respectively, in the Introduction. Several anonymous reviewers critiqued key aspects of my initial plan during the proposal process and helped refine the eventual focus of the book. Dominic Lewis, life sciences editor at Cambridge University Press, backed this (admittedly offbeat) project from its conception, and Jade Scard, content manager, nursed the manuscript through the various stages of production. I also thank Hugh Brazier, the copy-editor who deftly edited my last monograph, for magically polishing the rough edges of this book's prose as well.

Many undergraduates unwittingly served as guinea pigs to "beta-test" this material as a source for term-paper topics in an evo-devo seminar I've taught for a few years. One of the assignments is to pick a favorite gene from *D. melanogaster* and explain what's interesting about it from a clinical standpoint. If I were forced to play this game I'd probably choose *Dscam*, which encodes a cell adhesion molecule associated with Down syndrome (cf. Puzzles 2.7–2.10). *Dscam* is a lovely example of the gadgetry that evolution has cobbled together [258], and the experiments that have been conducted to dissect its modes of action (e.g., [2554]) are not only masterpieces of detective work; they are also great didactic vehicles for showcasing the subtle power of the genetic approach [849].

Robert Frost once mused that his goal in life was "to unite my avocation and my vocation as my two eyes make one in sight." My exploits as a fly geneticist over the past ~40 years have been so enjoyable that they seemed like a hobby, while my teaching of human embryology to premedical students over the past ~30 years was more of an assigned job, though not really a chore. Imagine my delight, therefore, when these two realms started merging at the dawn of the genomic era. Never in my wildest dreams did I think that humans and flies would turn out to be so much alike at the genetic level. My aim here is to share some of that shock and awe.

This is the final book of an evo-devo trilogy [925,928] that has entertained me, in the writing thereof, for more than a decade. The journey has at times felt as daunting as Frodo's trek to Mordor, but, like Frodo, I've been assisted along the way by wise wizards and faithful friends, some of whom are listed above. With all due respect to Professor Tolkien and his mastery of mythology, the field of evo-devo offers at least as much adventure for the hobbit-child in all of us as his fabulous world of Middle Earth.

Lewis I. Held, Jr.
Lubbock, Texas
May 2016

Foreword

Deep Homology? is the final installment in Lewis Held's grand trilogy concerning the field of evolutionary developmental biology (evo-devo). This book follows his *Quirks of Human Anatomy* and *How the Snake Lost its Legs*. Overall, this trilogy makes the excitement and promise of evo-devo accessible to a broad readership. Let me first put Held's latest book into context. As Held puts it, "evo-devo has come into its Golden Age;" there are four specialized journals and many books dedicated to evo-devo, and just in the last few years a new evo-devo society – the Pan-American Society for Evolutionary Developmental Biology (www.evodevopanam.org) – held its inaugural meeting at the University of California, Berkeley, and its sister the European Society for Evolutionary Developmental Biology (http://evodevo.eu) held sold-out meetings in Vienna and Uppsala. These societies, meetings, journals, and publications show that evo-devo is on track to living up to its promise not only to understand how developmental processes and organismal body plans originate and evolve, but also to enrich evolutionary theory and enhance its predictive power. Indeed, evo-devo has already begun to provide critical insights into medicine, biodiversity conservation, agriculture, and animal breeding.

What sparked this rejuvenation and success was the discovery that animals as diverse and distantly related as flies and humans share a similar set of highly conserved genes and interactive gene networks that regulate their development. In *From DNA to Diversity*, Sean Carroll and colleagues called these highly conserved regulatory genes the "genetic toolkit" for animal development [5]. An early example of this toolkit comes from Walter Gehring and colleagues, who stunned our field with their discovery that the gene *eyeless*, which regulates eye development in the fruit fly *Drosophila melanogaster*, has a counterpart in mice called *Pax6*. Misexpressing mouse *Pax6* in developing fruit fly tissues leads to growth of fruit fly eyes on wings, legs, antennae, and head [11]. Conversely, misexpressing fruit fly *eyeless* in developing frog tissues leads to the development of frog eyes on different parts of the frog body [6].

This discovery was immediately met with both excitement and controversy, because it touched the core of the most important concept in comparative biology – "homology." The homology concept, which is most often defined as a trait in two species that evolved from the same trait in their most recent common ancestor, is notoriously difficult to define [3]. An analysis of eye anatomy across a wide range of animals by Salvini-Plawen and Mayr led them to conclude that the compound eyes of insects and the camera eyes of vertebrates are not homologous as functional image-forming eyes because eyes were absent in the most recent common ancestor of insects and vertebrates [14]. Yet, Gehring and colleagues had shown that these independently evolved or "convergent" eyes are regulated by the same homologous gene: the most recent common ancestor of flies and vertebrates possessed the *Pax6* gene. What triggered controversy was their interpretation of this remarkable discovery: because *Pax6* is homologous, the compound eyes of insects and camera eyes of vertebrates are also homologous.

This controversy gave way to broader discussion about how to reconcile the discovery of homologous toolkit genes and their networks regulating non-homologous (novel) morphologies. In a famous essay, Sir Gavin de Beer had presciently recognized that homology at the level of genes does not equate homology at the level of morphology and vice versa [7]. Bolker and Raff (1996), Abouheif *et al.* (1997), Abouheif (1997), and Wray and Abouheif (1998) further proposed that homology at the level of genes should be considered separately from homology at the level of morphology, and that one should also always specify the level at which they are assessing homology [1,2,4,16]. Abouheif [1] and Wray and Abouheif [16] took this one step further and demonstrated that a "hierarchical approach to homology," where all levels of biological organization (genes, gene expression, embryological origin, and morphology) are considered separately yet simultaneously in the framework of a phylogenetic tree, can uncover important features of the evolutionary process. In the case of eyes, it reveals that *Pax6* was likely "co-opted" to facilitate the independent evolution of image-forming eyes in insects and vertebrates. This hierarchical approach emphasizes homology at distinct levels of biological organization and co-option of developmental regulatory genes.

Around the same period, Bolker and Raff (1996) and Shubin *et al.* (1997) introduced another term, "deep homology," to describe the scenario in which homologous toolkit genes and their networks regulate non-homologous morphology [4,15]. In contrast to the hierarchical approach, which clearly separates levels of homology, deep homology blurs these levels, and by doing so draws greater attention to the roles these highly conserved genes and their networks played in the ancestor of two distantly related species. In the case of eyes, deep homology suggests that *eyeless/Pax6* may have played a role in specifying a "light-sensing" organ in the ancestor of insects and vertebrates, and that image-forming eyes would have been independently elaborated in insects and vertebrates from this light-sensing organ. However, the concept of deep homology is open to the possibility that insect and vertebrate eyes are actually homologous because they evolved from the same cell types regulated by *eyeless/Pax6* in the ancestor. Deep homology is also open to the possibility that *eyeless/Pax6* did not regulate a light-sensing organ in the ancestor, but was co-opted to facilitate the *de novo* evolution of *eyeless/Pax6*. Deep homology and the hierarchical approach highlight the potentiating role of the conserved genetic toolkit and its interactive networks in the evolution of novel morphologies between closely and distantly related species.

It is hard to imagine the field of evo-devo before Nüsslein-Volhard and Wieschaus [13] discovered the genetic toolkit in fruit flies and McGinnis *et al.* [12] provided the first evidence that this toolkit was highly conserved – they showed that the binding motif in developmental regulatory genes called the homeobox was conserved in a broad range of animals. The Modern Synthesis, which integrated genetics with Darwinian evolution, has been the predominant paradigm within evolutionary biology for much of the twentieth century [8]. Architects of the modern synthesis predicted that natural selection as the primary engine of evolution should erase any homology at the genetic level between the 35 animal phyla, including arthropods and chordates [10]. At this scale of evolutionary time, natural selection would have "recrafted" the nucleotide sequences of each gene in the genome through the constant accumulation of beneficial and neutral mutations during the process of adaptation of each organism to a constantly changing local environment [10]. During this same period, experimental embryologists were searching and struggling to find "grand homologies" or conserved embryological features between animals, such as the three germ layers shared by all bilaterian animals or the cleavage patterns shared by flatworms, annelids, and molluscs [9]. Therefore, the discovery of deep homology was a major surprise for evolutionary biology and a major triumph for experimental embryologists [9,10].

Yet, the deep homology concept has not been fully grasped and incorporated into evolutionary theory, or into other areas of biology, including ecology, medicine, and agriculture. Like its predecessors, this latest book in Lewis Held's trilogy will help lay the foundation for such a synthesis. This book is the first to provide an updated and detailed analysis of several cases of deep homology in one place, allowing the reader to vividly see the concept applied to different developmental processes and different organ systems. By doing this, the reader begins to realize that deep homology may be the rule, not the exception. Held's book is timely, as evo-devo is rapidly changing and incorporating approaches from many fields of study, including genomics, ecological and quantitative genetics, developmental plasticity, ecology, paleontology, cell and systems biology, theoretical biology, behavior, and population genetics. While integrating this long list of subdisciplines is a recipe for success, the central tenet of evo-devo, deep homology, sometimes gets lost in the flood of data from so many different directions. Therefore, his enjoyable writing style, which makes deep homology accessible to a broad range of scientists, is crucial for proper integration of these subdisciplines into evo-devo.

Finally, and I never understood why, but many great book and movie trilogies seem to be written or produced in reverse order. To understand *how the snake lost its legs* and why there are so many *quirks of human anatomy*, we should probably first try to fathom the depths of *deep homology*. So read all three books, but start here and work your way backwards!

Ehab Abouheif
Department of Biology, McGill University, Canada

References

1. Abouheif, E. (1997). Developmental genetics and homology: a hierarchical approach. *Trends Ecol. Evol.* 12, 405–408.
2. Abouheif, E., Akam, M., Dickinson, W.J., Holland, P.W.H., Meyer, A., Patel, N.H., Raff, R.A., Roth, V.L., and Wray, G.A. (1997). Homology and developmental genes. *Trends Genet.* 13, 432–433.
3. Bock, G.K., and Cardew, G. (eds.) (2007). *Homology*. Wiley & Sons, New York, NY.
4. Bolker, J.A., and Raff, R.A. (1996). Developmental genetics and traditional homology. *BioEssays* 18, 489–494.
5. Carroll, S.B., Grenier, J.K., and Weatherbee, S.D. (2001). *From DNA to Diversity: Molecular Genetics and the Evolution of Animal Design*. Blackwell Science, Malden, MA.
6. Chow, R.L., Altmann, C.R., Lang, R.A., and Hemmati-Brivanlou, A. (1999). Pax6 induces ectopic eyes in a vertebrate. *Development* 126, 4213–4222.
7. de Beer, G.R. (1971). *Homology, an Unsolved Problem*. Oxford Biology Readers 11. Oxford University Press, London.
8. Futuyma, D.J. (2013). *Evolution*, 3rd edn. Sinauer, Sunderland, MA.
9. Gilbert, S.F., Opitz, J.M., and Raff, R.A. (1996). Resynthesizing evolutionary and developmental biology. *Dev. Biol.* 173, 357–372.
10. Gould, S.J. (2002). *The Structure of Evolutionary Theory*. Harvard University Press, Cambridge, MA.
11. Halder, G., Callaerts, P., and Gehring, W.J. (1995). Induction of ectopic eyes by targeted expression of the *eyeless* gene in *Drosophila*. *Science* 267, 1788–1792.
12. McGinnis, W., Garber, R.L., Wirz, J., Kuroiwa, A., and Gehring, W.J. (1984). A homologous protein-coding sequence in *Drosophila* homeotic genes and its conservation in other metazoans. *Cell* 37, 403–408.
13. Nüsslein-Volhard, C., and Wieschaus, E. (1980). Mutations affecting segment number and polarity in *Drosophila*. *Nature* 287, 795–801.
14. Salvini-Plawen, L., and Mayr, E. (1977). On the evolution of photoreceptors and eyes. *Evol. Biol.* 10, 207–263.
15. Shubin, N., Tabin, C., and Carroll, S. (1997). Fossils, genes and the evolution of animal limbs. *Nature* 388, 639–648.
16. Wray, G.A., and Abouheif, E. (1998). When is homology not homology? *Curr. Opin. Genet. Dev.* 8, 675–680.

Introduction

Ever since Aristotle called insects "bloodless" [149] and Linnaeus banished them to a separate class that we now call the phylum Arthropoda [1329], we have had little reason to view insects as anything but alien lifeforms. After all, they have compound (vs. simple) eyes, six (vs. two) legs, and a chitinous (vs. calcified) exo- (vs. endo-) skeleton . . . plus they're so tiny that many escape our notice completely.

Our age-old chauvinism ended in 2001 when the genomes of humans and fruit flies became available for comparison [105,143,1946], and we could fathom the large (~50%) overlap of the respective gene repertoires [180,2486,2549]. Even our smug predictions of having many more genes were dashed [1905,2199]: we have ~22,000; they have ~15,000 [1771], though we can still cling to the slight excess to salvage some shred of superiority. As Gerry Rubin put it, flies turned out to be just "little humans with wings" [71].

The egotistical wall separating us from insects may have crumbled in 2001, but it had been badly cracked by unsettling findings on several earlier occasions. Those episodes are worth recounting, at least briefly, because they convey the thrill that scientists often feel upon discovering heretical facts that challenge a prevailing paradigm before it finally topples. To allow readers to experience these epiphanies in their purest form, the leading pioneers are quoted in their own words.

In 1822 the first hint of a similarity between vertebrate and arthropod body plans came to light when Étienne Geoffroy St.-Hilaire described his dissection of a crayfish ($\approx$ lobster) [486,787]. He was startled to find that crayfish – and, by inference, insects as well – have the same strata of organs as vertebrates, though the layers are inverted. That is, the back of a crayfish corresponds to the belly of a vertebrate, and vice versa (boldface added):

> Je viens de trouver que tous les organes mous, c'est-à-dire, que les organes principaux de la vie sont reproduits chez les crustacés, et par conséquent chez les insectes, **dans le même ordre, dans les même relations et avec le même arrangement que leurs analogues chez les hauts animaux vertébrés** . . . **Quelle fut ma surprise**, et j'ajoute, de quelle admiration ne fus-je pas saisi, en apercevant une ordonnance qui plaçoit sous mes yeux tous les systèmes organiques de ce homard dans l'ordre où ils sont rangés chez les animaux mammifères? Ainsi sur les côtés de la moëlle épinière, je vis tous et chacun des muscles dorsaux; au-dessous étoient les appareils de la digestion et les organes thoraciques, plus bas encore, le coeur et tout

le système sanguin, et plus bas enfin formant la dernière couche, tous et chacun des muscles abdominaux. [740][1]

The idea that a lobster is essentially an armored, upside-down mouse sounds as absurd today as it did then, but Geoffroy defended his theory by pointing out that the flatfish embodies an equally absurd, but undeniably real, instance of postural shift relative to other fish – albeit a rotation of 90°, versus the 180° he envisioned in the crayfish case. His analogy, clever though it may have been, failed to convince his colleague Georges Cuvier, and these titans of the French academic elite clashed in a historic debate (1830, Paris) that was widely deemed to have been won by Cuvier [74]. It took 172 years for Geoffroy's inference of an inversion to be vindicated by evo-devo research [83,1654]. The story of that redemption will be recounted in Chapter 1.

In 1915 the neuroanatomist Santiago Ramón y Cajal charted the wiring of neurons in the optic lobes of the fly brain [303]. He had wanted to understand the vertebrate brain but was daunted by its complexity, so he had sought a simpler nervous system that he could analyze in order to figure out the basics of the circuitry, with the intention of going back and applying what he learned to vertebrates. Surely, he thought, the fly must have a simpler visual system. But he found that, on the contrary, it is just as intricate, and it even operates in a similar way [1976]. Some of those similarties are covered in Chapter 3, and the chiasms to which he refers ("cruces intra-retinianos") are treated in Chapter 2 (boldface added):

Confrontando esta retina ideal del insecto con la de los vertebrados (figuras 82 y 83), todas las dudas se disipan *in continenti*, imponiéndose imperiosamente las homologias esenciales sugeridas por Kenyon, Cajal y Zawarzin. Y todavia se acentuaria el parecido si se hubiera prescindido de los dos cruces intra-retinianos, que constituyen una de las originalidades más notables de la retina de los articulados . . . **Con todo lo cual no pretendemos afirmar que la retina de los insectos deje de ofrecer algunos rasgos de organización originales, especificos ó poco ó nada representados en los vertebrados.** [303][2]

In 1917 the embryologist Ross Harrison described extra legs that sprouted after he transplanted leg rudiments to ectopic sites on the opposite flanks of host salamanders [880]. Occasionally, the operations produced triplicated legs, and those

---

[1] I just discovered that all the soft organs, as in the main organs of life, are found in crustaceans and consequently in insects **in the same order, with the same relations, and with the same arrangements as their analogs in the higher animals, the vertebrates** . . . **What surprise** and, might I add, what admiration came over me when, before my eyes, I saw a prescription that placed all the organic systems of this lobster in the same order in which they are arranged in mammals! On the sides of the dorsal cord, I saw each and every dorsal muscle; beneath them were the digestive and thoracic organs, under them the heart and the entire circulatory system, and finally beneath those, the abdominal muscles formed the last layer.

[2] All doubt is dispelled when comparing this ideal retina of the insect with the one of vertebrates (figures 82 and 83). The essential homologies, already suggested by Kenyon, Cajal and Zawarzin, become evident. If they lacked the two intra-retinal crossings, which is one of the most notable originalities of articulated organisms, their resemblance would be even more obvious . . . **With this being said, we do not try to claim that the retina of insects no longer offers original organizational characteristics, which may or may not be represented in vertebrates.**

legs obeyed the same rules of symmetry that William Bateson had formulated in 1894 for triplicated legs of abnormal arthropods (insects and crustaceans) [160]:

> There may be further reduplication [of salamander legs], so that more or less complete triple limbs may result. The three limbs then have approximately the same relations as found by Bateson, especially in arthropods. [880]

In 1976 the notion that arthropods and amphibians regenerate their append-ages by similar rules resurfaced [278], when the developmental biologists Vernon French, Peter Bryant, and Susan Bryant proposed a unifying model to explain how triplications occur at the cellular level (boldface added):

> The results of the contralateral grafts [of salamander leg blastemas] are **exactly analogous to those obtained with cockroaches** and can be explained in the same way. The handedness and axial orientation of the supernumerary limbs are the same as those of the stump. [673]

Their model has since been supplanted by alternative explanations [300,023], but the similarity of the underlying mechanisms upon which it was based has been con-firmed genetically. The evidence for a common appendage algorithm in arthropods and vertebrates is presented in Chapter 6.

In 1978 the fly geneticist Edward Lewis echoed another theme (aside from limb trip-lications) from Bateson's 1894 classic *Materials for the Study of Variation* [160,1297]. In Bateson's vast collection of aberrant animal specimens there were many cases where one organ had been converted into the likeness of another – e.g., an antenna that devel-oped as a leg instead. Bateson coined the term "homeosis" for such transformations.

Lewis studied homeotic mutations – the most famous of which, *bithorax*, con-verts the tiny third thoracic segment into a larger second thoracic segment, yielding a four-winged fly with two thoraxes [911]. His 1978 paper summarized his genetic dissection of this part of the fly's third chromosome, and it set forth a model for how the "Bithorax-Complex" (BX-C) dictates identities of body segments [1296]. Lewis's model was wrong in the number of genes expected for the BX-C (9 vs. 3) [1974], but it was right in the colinearity predicted between chromosomal loci and the segments they control [1218,1460,1461].

In 1984 a 180-base-pair motif was identified in the homeotic genes of both the BX-C and its sister "Antennapedia-Complex" (ANT-C) by Matt Scott's lab in the USA and Walter Gehring's lab in Switzerland [1463,1899,2037,2039]. It was named the "homeobox" [731] and the genes in these complexes, which specify consecu-tive groups of body segments, came to be called *Hox* genes [298]. The homeobox encodes a DNA-binding domain that allows proteins to regulate target genes by binding their *cis*-enhancers [1751]. Homeoboxes were subsequently found through-out the animal kingdom [509,732,972] (boldface added):

> The discovery of the homeobox . . . enforced the idea that evolutionarily distant organisms might share common developmental pathways and common genetic circuits. This idea is now taken for granted in all current genomic approaches, and today it seems strange that **it was completely unanticipated in 1980 at the beginning of the cloning era.**
>
> — Eric Wieschaus [2431]

It is difficult to understand now how surprised people were by the finding of vertebrate cognates of *Drosophila* homeotic genes; it was quite amazing.

— Denis Duboule [1899]

In 1988 mice were shown to have *Hox* complexes homologous to the combined BX-C and ANT-C of flies [2021], and in 1989 the order of genes in those clusters was found to be colinear with the order of body zones along the head–tail axis [552,790]. Why should flies and mice be using the same system of "area codes" to subdivide bodies that have overtly different metameric units – ectodermal segments versus mesodermal somites, respectively [430]? Evidently, *Hox* complexes act like abstract yardsticks to mark locations [31,1462] without regard to embryological origin (ectoderm or mesoderm) or histological character (segments or somites) [2114]. This ancient system of axial zonation is discussed in Chapter 1.

Today we take the colinearity and clustering of vertebrate *Hox* genes for granted, and everyone thinks that it was a logical step after cloning *Drosophila* homeobox sequences in 1984. But it wasn't, mostly because our minds were not prepared for this.

— Denis Duboule [1899]

In 1994 Geoffroy's old hypothesis of dorsal–ventral inversion was confirmed at the molecular level [83]. Vertebrates were shown to use the same signaling molecules as insects along their dorsal–ventral axis but to do so with inverted polarity. The inversion event itself was later traced to the base of the chordate phylum [1356]. The evidence that led to this conclusion is considered in Chapter 1.

In 1995 the fly's *eyeless* gene was shown to be capable of inducing extra eyes when it is artificially misexpressed at ectopic locations [844]. The spectacle of flies with eyes on their legs, etc., was shocking enough, but what startled the research community even more was the ability of the mouse's orthologous *Pax6* gene to elicit extra fly eyes in the same way. Could the same "master gene" be regulating the assembly of a compound eye in an insect and a camera eye in a vertebrate [732]? Evidence for and against this idea is presented in Chapter 3, along with an assessment of the many other parallels between the eyes of insects and those of vertebrates.

In 1996 this fast-moving field was codified by Rudy Raff, a tireless champion for evo-devo in the USA, in his monograph *The Shape of Life* [1847]. A sampling of his chapter titles conveys the scope of his synthesis: "Deep time and metazoan origins, The developmental basis of body plans, Building similar animals in different ways, Developmental constraints, Modularity, dissociation, and co-option." In the following passage from his preface, Rudy alludes to the progress that had been made since he and Thom Kaufman first laid the foundations for evo-devo with their seminal 1983 book *Embryos, Genes, and Evolution* [1848]:

Over a decade has passed since *Embryos, Genes, and Evolution*, and an experimental discipline that integrates developmental and evolutionary biology has begun to coalesce. Most important, the whole emphasis of work on development and evolution has shifted to new ground due to the transformation of our understanding of the genes that regulate development. [1847]

In 1999 Rudy founded the journal *Evolution and Development*, together with other leading lights (Wallace Arthur, Sean Carroll, Michael Coates, and Greg Wray). It quickly became the premier outlet for evo-devo researchers to publish their findings and has remained so ever since.

In 2000 the DNA sequence of the fruit fly genome was published [105], with the human genome following shortly thereafter. The expectation had been that the number of human genes would dwarf the number of fly genes [1373], but this prediction proved to be misguided [1771]. Indeed, the more that these two genomes have been compared, the more similar our "operating systems" appear to be, despite the differences in the anatomical "apps" that they control (boldface added):

> The remarkable similarity of the genetic regulation of development in distant organisms has heralded a new conception of evolution. It was a big surprise when evolutionary conservation of the Krebs cycle, the genetic code, and classes of structural proteins was extended to regulation of development. **The diversity of organisms had fooled everyone into thinking that the evolution of completely different regulatory processes, or at least completely different uses of the same genes, was likely to be responsible for evolutionary change.**
> — Matt Scott [2036]

The primary aim of the present book is to survey those unexpected similarities. Admittedly, these revelations are still so fresh that it is hard to know what to make of them. For that reason this book will inevitably be disappointing, since it cannot reach any firm conclusions based upon the piecemeal nature of the evidence. Even so, for those of us who happen to like Swiss cheese with our ham, we can still savor the taste in spite of the holes.

Aristotle was the first author ever to mention the fruit fly *Drosophila melanogaster* [801,1780,2204] (though it had no such name then) when he described a "gnat" emerging from vinegar slime in *History of Animals*, c. 350 BCE (Book 5, part 19, line 552b5) [149]. If he were alive today, he would surely be a card-carrying evo-devotee – surfing the internet databases to study the history of animals from a modern genomic perspective. There is also little doubt that he would be as amazed as the rest of us to see how much we resemble the little vinegar gnat when, like the mythical Orpheus – or better yet like Lewis Carroll's Alice – we descend into the hidden world beneath our superficial differences.

1   Body Axes

Humans, flies, and other bilaterally symmetric animals have three axes (Figure 1.1): an anterior–posterior (A–P) or head–tail axis, a dorsal–ventral (D–V) or back–belly axis [1410], and, at least nominally, a left–right (L–R) axis, though this L–R "axis" [268,327] differs from the other two in its binary, quantal nature, and some authors do not consider it to be a bona fide axis at all [1160] but just left–right differences. Humans and flies turn out to use the same sets of conserved genes along their A–P and D–V axes, but they rely on different genes for the L–R axis [925]. Each of these axes will be examined in turn. Some of the topics were broached in the Introduction, but only superficially.

The anterior–posterior axis is subdivided by *Hox* genes

Our fascination with *Hox* genes began with gene mutations that transform parts of the fly's body [911,1297]. Cloning of these homeotic genes revealed a shared 180-bp sequence [1899]. The 60 amino acids encoded by this homeobox form a DNA-binding homeodomain that allows the resulting proteins to bind enhancer sites near target genes so as to affect their transcription [1751,1894]. The fame of the *Hox* genes is certainly warranted [357], but much of the "*Hox* mania" hype that has grown up around them is not [1914,2462].

First, there is nothing special about the helix-turn-helix configuration of the homeodomain. Evolution has spawned a zoo of transcription factors from zinc fingers to leucine zippers [2406], and homeoproteins comprise just one class [725]. All that is needed for a protein to act as a transcription factor are (1) a few apt amino acids whose R groups can bind DNA bases, (2) a scaffold of α-helices or β-sheets to orient those R groups, (3) a domain to dovetail with transcriptional or chromatin-remodeling machinery, and (4) a tag for entry into the nucleus [1293]. What sets homeodomain proteins apart is that a subset of them was recruited to assign "area codes" to body regions. Any of the other ~20 transcription-factor families existing at the dawn of animals could presumably have fulfilled this role instead [482,494].

A second aspect of the homeobox story that needs demystifying is clustering. Long before cloning unveiled the homeobox, the fly's major homeotic genes were mapped to two loci on the third chromosome [1327], so it came as no surprise when the genes in these Antennapedia and Bithorax complexes (ANT-C and BX-C) turned

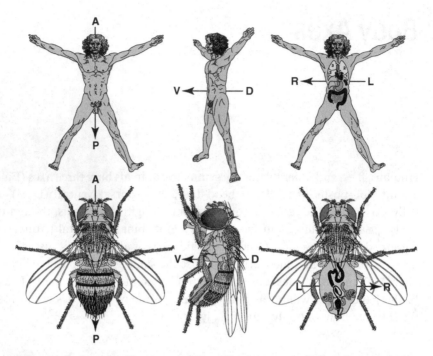

Figure 1.1 Body axes of humans and flies. The three axes of bilaterian anatomy are anterior–posterior (A–P), dorsal–ventral (D–V), and left–right (L–R). We tend to envision three-dimensional objects in this way (i.e., from a Cartesian standpoint), and it turns out that animal genomes actually do construct the body using different genetic gadgetry along each of the three axes. The term "axis" is not as fitting for L–R as it is for A–P and D–V, since what we really mean is that the two sides differ anatomically (mainly in the viscera). Perhaps we should instead call it a "medial–lateral" axis, whose mirror symmetry is broken by an asymmetric overlay of binary states (M. Akam, personal communication)?

out to be paralogs. Paralog clusters are not unusual [497,768,769,1457]. For example, humans have two nests of hemoglobin genes [878]: an array of three α-globin genes on chromosome 16 and an array of five β-globin genes on chromosome 11 [1174], with the β cluster having been founded by an escapee from the α cluster [877,2159]. Clusters arise whenever founder genes undergo tandem duplications (via unequal crossing over) [2532], followed by functional divergence of the copies [2252].

Thus, what was so intriguing about the ANT-C and BX-C genes was neither their homeoboxes nor their clustering per se, but rather the fact that they govern the A–P axis in the same order as their 3′-to-5′ sequence along the chromosome [714,1461]. This property of "spatial colinearity" was reported for the BX-C by Ed Lewis in 1978 [1296] and for the ANT-C by Kaufman *et al.* in 1980 [1122]. Their evidence entailed the mapping of mutant phenotypes onto genetic loci.

Molecular studies later confirmed that colinearity is also manifested in the transcription zones of ANT-C and BX-C genes along the A–P axis of the fly embryo [30], but, oddly, each zone trails off posteriorly, overlapping with the zones of genes that lie more 5′ in the array (Figure 1.2) [1544]. Such overlaps suggested a combinatorial code of some sort, but that turned out not to be true in most cases [32] (e.g., see [1183] for an exception). When two genes overlap in their expression, only one usually assigns identity – a rule that was termed "posterior prevalence" because the posteriorly transcribed gene tends to overrule the anterior one in the BX-C, though the opposite (i.e., anterior overruling posterior) was later found to characterize the ANT-C [565].

These findings were impactful, but they paled in comparison to the 1989 discovery that galvanized evo-devo. In that year mice were shown to have clusters of homeobox genes orthologous to those in ANT-C and BX-C [552,790]. The murine clusters exhibited spatial colinearity and were later found to obey posterior prevalence. To distinguish those clustered homeobox genes that partition the A–P axis from differently functioning homeobox genes elsewhere in the genome, the term "*Hox*" was coined – a contraction of "homeobox" [298]. Mice have four *Hox* complexes, each of which corresponds to a fusion of ANT-C and BX-C. Indeed, all vertebrates have four clusters due to two rounds of genome duplication that occurred near the base of the clade [716,772,1113,2324].

When the *Hox* clusters of *Homo sapiens* are aligned alongside one another, "Swiss cheese" gaps become apparent where particular paralogs are clearly missing (Figure 1.2) [1593,2139,2541]. These losses are attributable to the redundancy that results whenever extra copies of a gene arise via either genome duplication or unequal recombination [1373]. Unless the nascent paralogs adopt different roles after they arise, all but one of them will typically erode due to accumulation of deleterious mutations [1021]. Those mutations are tantamount to "noise" that degrades the base-sequence "signal" [444]. All of our 13 paralog groups retain at least two members, so some functional divergence must have occurred within each of them in enough time to preserve these remnants [114]. Those changes must have occurred before the radiation of placental mammals, because mice have the same 39 *Hox* genes as humans [511].

A different kind of "noise" affects DNA on a larger scale. Gene order gets scrambled whenever a chromosome breaks at two sites and the intervening piece flips 180° (like a one-dimensional Rubik's cube). For example, the gene sequence ABCDEFGHIJ might become ABC(HGFED)IJ and then A(FGHCB)EDIJ, where parentheses indicate breakpoints [1327]. Inversions occur quite frequently in *Drosophila* [770], and in one herculean project researchers managed to decipher the history of virtually all inversions among Hawaiian drosophilids [331]. Genes can also rearrange when chromosomes crack in half or fuse together [2530]. In the face of all this chaos, *Hox* clusters have somehow managed to remain more intact than we would have expected by chance alone.

The "Conserved Cluster" epiphany of 1989 was as earth-shaking for geneticists as the 1922 discovery of King Tut's tomb was for archaeologists, because it revealed a relic from a long-lost ancestor who was 200,000 times older than Tut (~600 MY old vs. ~3000 years old) [538,2259]. How incredible that a stretch of linked genes could endure intact for such a long period of time [999,1384]! Thus, it was the

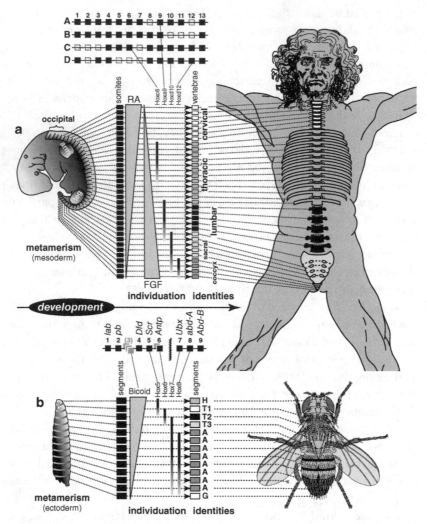

Figure 1.2 Colinearity of *Hox* gene expression along the A–P axis. Developmental processes are cartooned from left to right in **a** and **b**. In both cases *Hox* genes impose "tagmosis" [66] – regional identities (e.g., thoracic vs. lumbar in humans or thoracic vs. abdominal in flies) in groups of metameres (vertebrae or segments) along the body column (cf. [1430,2121]). Adapted from [925,928]. Overlapping domains of *Hox* expression in both humans and flies imply a combinatorial code of some sort, but only one *Hox* gene typically governs any particular site [333,996] due to the rule of "posterior prevalence" (see text for details) [551,2106,2498].

a. Colinearity of *Hox* transcription in humans. We have four *Hox* complexes (A–D at top; 3′-to-5′ from left to right) that evolved via two genome doublings in our fish ancestors [312]. Some paralogs were later lost (dotted outlines) in basal mammals [1593,2139,2541], leaving 39 (out of 52) residual genes in mice and humans [511]. Our 33 vertebrae develop from mesodermal somites (see 31-day-old embryo at left with limb buds labeled) [2010]. The first four (occipital) somites form the base of our skull. Initially, all somites look alike (black rectangles) [323], but they later acquire regional identities (arbitrary shades for different tagmata) by "individuation" (the creation of differences among iterated

antiquity of the clustering, not the clustering per se, that was so amazing . . . and ultimately so perplexing [671,1954]. The phyletic extent of this synteny could not be assessed until metazoan genomes began to be sequenced [460,1860,2095], starting with the nematode in 1998 [1,1658].

Our clearest view of the archetypal urbilaterian *Hox* complex came in 2014, when the relatively less scrambled genome of the centipede *Strigamia maritima* (a protostome) was compared to the similarly "low-noise" genome of amphioxus (a deuterostome) [377,716,1838]. This juxtaposition allowed us to span the protostome–deuterostome divide [999] at a level much closer to the trunk branchpoint in terms of gene sequence

---

←

Figure 1.2 (*cont.*)

units) in response to the relative dosages of RA (retinoic acid) [2075] and FGF (Fibroblast Growth Factor) [554]. Thus, each *Hox* gene is activated by a specific RA/FGF ratio [439,1156,1240]. Those ratios are established by reciprocal concentration gradients (gray triangles) [137,763]. Four examples are given to illustrate the trend [291,324,1007]: the sequence of expression of *Hox* genes *c6*, *a9*, *d10*, and *d12* (vertical bars) along the body's A–P axis matches their 3′-to-5′ order within the clusters. The level of transcription (shading) in each bar fades from high (dark) to low (light). Based on the area codes [424,652,2408] that are specified by these (and other) *Hox* genes [807,2010], postcranial somites adopt one of five tagma-specific fates: cervical (7 vertebrae), thoracic (12), lumbar (5), sacral (5), or coccygeal (4) [989]. Similar (RA/FGF-dependent) *Hox* circuitry mediates segmentation of the vertebrate hindbrain (not shown) [564,1733,1996] and spinal cord [465,1092] (cf. Figure 2.1). NB: The RA and FGF gradients as drawn are misleading. At no time does either morphogen span the entire A–P axis. Rather these gradients act *locally* in a discrete growth zone as the somites form [763].

b. Colinearity of *Hox* transcription in flies. For clarity a larva is drawn (left) instead of an embryo. The *Hox* cluster of *D. melanogaster* is split into two parts between genes *6* and *7*. Fly orthologs share corresponding numbers with human genes, but they are usually called by the names given above (*lab*, labial; *pb*, proboscipedia; *Dfd*, Deformed; *Scr*, Sex combs reduced; *Antp*, Antennapedia; *Ubx*, Ultrabithorax; *abd-A*, abdominal-A; *Abd-B*, Abdominal-B). Gray squares along the horizontal array are rogue genes that have adopted non-*Hox* functions (see text) [734,749]. Ironically, one of them – the *Hox3* paralog *bicoid* [911,1464] – now controls expression of all the other *Hox* genes (cf. [273]). The Bicoid gradient [808,1811] turns ON gap genes (not shown) at different thresholds [876,1025,1659], and they, in turn, activate *Hox* genes in the same sequence as their 3′-to-5′ order along the chromosome [1726,2433]. Four examples of *Hox* expression (vertical bars, as in **a**) are shown (*Hox5*, *Hox6*, *Hox7*, and *Hox8*) [329,333,1544,1761]. Unlike in humans [649], each bar ends in midsegment because *Hox* expression in flies is *para*segmental [512,1218,1248] (cf. annelids [2173]). The "area codes" specified by these (and other) *Hox* genes dictate the tagmata typical of insects (cf. [998]): "head" (H); thorax (T), whose segments differ radically (T1, T2, and T3); and abdomen (A), whose eight segments are relatively uniform [257]. The tiny "x" to the left of the abdomen means that males lack the seventh abdominal segment (a weird dimorphism) [655]. NB: The top block marked "H" for "head" actually entails six segments [881]. Nor is the bottom block, marked "G" for "genitalia," meant to denote a single segment: male and female genitalia are made by a genital disc composed of tissue from several segments [1972]. Finally, posterior *Hox* genes of protostomes may not be orthologous to those of deuterostomes, owing to independent duplication events, so the *Hox7–8–9* numbering should be taken with a grain of salt (M. Akam, personal communication).

(albeit not in terms of time) than the more highly derived "twigs" represented by flies [450,2530] and mice [1586]. The authors of the study put it this way (boldface added):

> With the exception of some conserved local gene clusters, the location of genes on the chromosomes of *Drosophila* and other Diptera retains no obvious trace of the ancestral bilaterian gene linkage. Other holometabolous insects such as *Bombyx mori* and *Tribolium castaneum* do show significant conservation of large-scale gene linkage with other phyla, for example, in the chordate *Branchiostoma floridae* (amphioxus) and the cnidarian *Nematostella vectensis*. The last common ancestor of these two lineages pre-dated the ancestor of all bilaterian animals, and yet the genomes of these species retain detectable conserved synteny: orthologous genes are found together on the same chromosomes, or chromosome fragments, far more often than would be expected by chance.
>
> **We find the S. maritima genome also retains significant traces of the large-scale genome organisation that was present in the bilaterian ancestor.** Although the assignment of scaffolds to chromosomes is not determined in *S. maritima*, there are **sufficient gene linkage data within scaffolds to reveal clear retained synteny between amphioxus and S. maritima**, at a higher level than any of the Insecta or Pancrustacea we have examined.
>
> . . . By implication, **the last common ancestor of the arthropods retained significant synteny with the last common ancestor of bilaterians as well as the last common ancestors of other phyla, such as the Chordata.** This conserved synteny is more complete with this *S. maritima* genome sequence, due to the relative scrambling of the genomes of those other arthropods that have been sequenced previously. [377]

The most plausible explanation for the durability of the *Hox* complexes (and other syntenic blocks elsewhere in the genome) is that the *cis*-enhancers of the clustered genes are shared or interspersed [1022,1024,2030]. Under such circumstances it would be difficult, if not impossible, to split an array without detaching genes from their regulators [1278,1535,2079,2159] and thereby killing any mutant individual experiencing such a split.

If so, then what then are we to make of *D. melanogaster*? The inversion that severed ANT-C from BX-C [1298] somehow allowed both of the flanking genes (*Antp* and *Ubx*) to keep their *cis*-enhancers [1607], and similar "lucky breaks" evidently occurred in other *Drosophila* lineages between *Hox1* and *Hox2* or between *Hox7* and *Hox8* [1609].

The fact that evolution allowed three distinct break sites in this one genus raised the concern that there might be something distinctive about *Drosophila*, and that is precisely the conclusion reached by the researchers who analyzed these *Hox* splits. Negre *et al.* became suspicious that something peculiar was happening because of a further heresy associated with the split between *Hox1* and *Hox2* – namely, a relocation of *Hox1* to the opposite end of the array [1608]. In their 2005 paper they stressed five cardinal differences between *Drosophila* and vertebrate *Hox* clusters [1607] (cf. [2151]):

1. (ANT+BX)-C is six times larger than in a typical vertebrate (665 vs. 110 kb).
2. ANT-C and BX-C have transposons; vertebrate *Hox* clusters usually don't.
3. ANT-C contains non-*Hox* genes; vertebrate *Hox* complexes typically don't.
4. Some fly *Hox* genes have an inverted 3′–5′ polarity; not so for vertebrates.
5. The fly cluster was split yet still works; vertebrates haven't managed this.

Based on these differences, the authors proposed an interesting hypothesis: the normal constraints on *Hox* complex integrity were exceptionally loosened for *Drosophila* because these flies develop quickly and express *Hox* genes all at once [1609]. In contrast, vertebrates develop slowly and express *Hox* genes not only with spatial colinearity but also with temporal colinearity [1175,2345].

In 2007 Denis Duboule endorsed this argument in a seminal review entitled "The rise and fall of *Hox* gene clusters" [551]. He decried the sanitized versions of the *Hox* story that were then pervasive in textbooks and instead offered a more nuanced scheme which classifies *Hox* clusters as organized, disorganized, split, or atomized. The four classes are not mutually exclusive, however. Fly clusters can be both disorganized and split [1609]. Vertebrates have the neatest clusters for the reasons listed above [1698], as well as because their *Hox* genes contain small introns. Intron size matters when speed of transcription is critical [940].

Duboule made a preliminary attempt to apply his scheme to phyla across the animal kingdom, but additional relevant information has been published since then. Figure 1.3 is based on his diagram, but it also incorporates abundant data from those recent studies, including Jason Hodin's extensive survey for the 2014 textbook *Evolutionary Analysis* [939]. The term "nested zones" denotes colinearity for bilaterian species, but it is broad enough to include cnidarians like hydra, where nesting is observed despite the absence of a head–tail axis sensu stricto [223].

Based on the limited information at his disposal, Duboule inferred that the *Hox* cluster of urbilaterians had to be disorganized, but the centipede genome challenges that conclusion. The *Hox* cluster of *S. maritima* is moderately large (457 kb) for an arthropod [1708], but all of the key genes are present and oriented with the same 3′–5′ polarity [377]. Likewise, the *Hox* cluster of amphioxus is big (~400 kb) [57] but relatively tidy in gene order, polarity, and colinearity [1738]. Based on these "primitive" species, the urbilaterian *Hox* cluster was likely larger than that of a vertebrate but neater than that of a fly.

What remains unclear is why some taxa that have uniform metameres along the trunk portion of their body should need to maintain either (1) a dapper *Hox* complex or (2) spatial colinearity of *Hox* gene expression [513]. Enigmatic animals in this category include centipedes [252,995] and amphioxus [1738], as well as onychophorans [1058] – the "living fossil" velvet worms [657] that comprise a sister phylum of Arthropoda [219].

To solve this paradox we may need to look beneath the surface. The integument of an animal may lack tagmata, but the underlying nervous system could be using *Hox* area codes in subtler ways that are not discernible at the level of gross anatomy [252,333]. Indeed, there is a growing consensus that the primordial role of the urbilaterian *Hox* complex was to partition the nervous system into groups of modules (see Chapter 2) – computer chips, if you will – rather than to balkanize the body into blocks of metameres [513,746,939,1971].

In Duboule's view it was an irony of history that our first taste of *Hox* clusters came in the flavors of flies and mice, because both of these taxa turned out to

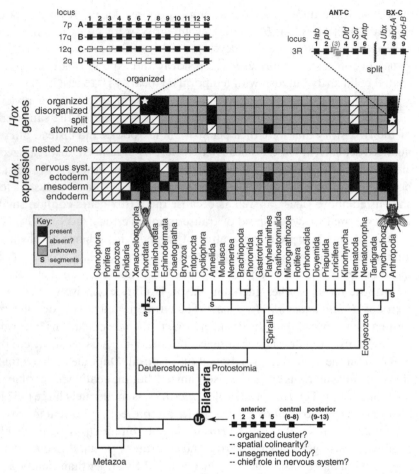

Figure 1.3 Organization and expression of *Hox* genes among animal phyla. In the panels marked "*Hox* genes" the black squares indicate presence of the trait in at least some members of the phylum (but not universal presence throughout the phylum), diagonal lines denote absence in all species examined thus far (though future analyses of more species could show otherwise), and gray squares signify insufficient data to draw any conclusion at present (see key). The same applies to panels marked "*Hox* expression," but some of those categories are not mutually exclusive at the level of individual species. For example, expression in the nervous system is implicitly coupled with expression in the ectoderm, though the reverse is not true. The term "nested zones" denotes sequentially overlapping expression domains, without implying a Russian-doll arrangement per se.

Hox arrays of humans and flies (starred) give chromosome number and arm (p or q for humans; L or R for flies). The four sets (A–D) in humans (upper left) arose from genome doublings in basal vertebrates (4×) [312], with 13 paralogs having later been lost (dotted outlines) [2139,2541] due to redundancy [807]. (Teleosts underwent an extra doubling [1824], and separate doublings apparently occurred in lampreys [1476], scorpions [2061], and horse-shoe crabs [1652].) The *D. melanogaster* complex (upper right) is split into two parts (ANT-C and BX-C) [1327], and the gray boxes along the ANT-C denote loss of *Hox* function [734,749,1788,2168] in three *Hox3* paralogs (*zen*, *zen2*, and *bicoid*) and one *Hox6* paralog (*ftz*). (Curiously, lepidopterans have added *zen*-like copies [631] as freely as chordates have added *Hox9*-like copies [1740].)

be unusual, though for different reasons. Vertebrates are eccentric insofar as we appear to use our *Hox* arrays as timing devices. These "*Hox* clocks" work best if the clusters are tightly packed (cf. the freaky *Rhox* array [1380,2541]), apparently because (1) the genes are spooled ($\approx$ rosary beads) into a transcriptionally active compartment within the nucleus [1598,1646,2159,2296], and (2) the clocks must mesh in perfect synchrony with the wave that makes successive somites [566,1157,2183]. Thus, selection pressure has evidently been compacting vertebrate *Hox* genes into tighter and tighter clusters ever since temporal colinearity arose ~500 MY ago (before the genome quadrupled) [551,1549,2306,2358].

Meanwhile, flies have been drifting in the opposite direction. Flies don't need a *Hox* clock because they make all of their segments simultaneously [566,1932]. This "long-germ" mode of development differs from the "short-germ" style of

←_____

Figure 1.3 (*cont.*)

This diagram was compiled from surveys by Jason Hodin [939] (his Figure 19.5) and Denis Duboule [551] with primary source data added for Ctenophora [1560,1722], Placozoa [1536], Cnidaria [555,1419,1877,1954,2171], Chordata [1012,2139,2358], Hemichordata [99,671,2317], Echinodermata [96,162,306,866,1539], Xenacoelomorpha [917,1547,1548], Chaetognatha [1729], Annelida [133,690,1029,1207,2173], Mollusca [680], Platyhelminthes [163,447,1693,1790], Nematoda [7], Onychophora [1058], and Arthropoda [377,1430,1708,2121].

In Duboule's scheme, "organized" means the cluster is compact (sizes range from 0.3 to 13.6 mb in arthropods [1708]) with uniform polarity, while "disorganized" means that genes are widely spaced or scrambled with no uniform polarity. In Echinodermata, for example, the sea star cluster is organized [162], whereas the sea urchin one is disorganized [300,1539]. *Hox* genes are "atomized" (unlinked) but still expressed in nested zones in *Oikopleura dioica* (Chordata) [2057], *Caenorhabditis elegans* (Nematoda) [7], and *Nematostella vectensis* (Cnidaria) [1954].

The structure of the ancestral *Ur-Hox* complex (below right) is based on minimum estimates [44,298,487,714,1278], with "6–8" and "9–13" marking as-yet unduplicated genes, though it is more likely that ~10 genes were present [141,963,1788], with "6–8" already split into *Hox6, 7,* and *8*. A total of ~88 homeobox genes (not shown) existed in urbilaterians [1593], some of which occupied *Hox*-related clusters [970,1247]. Based on the tissues where *Hox* expression is most common (lowest panel with four rows of squares), *Hox* genes may have originally acted in the nervous system (cf. Chapter 2) [99,958,1971,1995,2358] – an idea proposed by Alain Ghysen [746] and championed by Jean Deutsch [513].

Urbilaterians were probably unsegmented, since segments only exist in a few phyla [376,562,1755], depending on how narrowly you define "segment" [284,862,1054,2011] relative to "metamere" [430,789,1506]. Only three phyla that are classically deemed to be segmented are labeled as having segments (S) [1755,2040], though Tardigrada and Onychophora are widely considered to be segmented as part of the greater "panarthropod" clade (M. Akam, personal communication), and Kinorhyncha might also qualify under looser criteria [215,562,597,1055,1451]. Even nematodes can't be dismissed entirely [2023]!

For supplementary information on gene polarity see [993,1278], for conserved microRNAs see [484,913,1409,1747,2498], and for *Hox* cluster antecedents see [555,2171]. Spiralian *Hox* genes are inventoried in [155]. The tree of extant phyla is based on [314,562,2259], but a provocative recent essay by Andreas Hejnol and Casey Dunn challenges the way we've traditionally handled the phylum concept itself [915]. For numbers of species per phylum see [405].

basal insects [953], where segments arise sequentially as they do in vertebrates [472,1157,1743]. Once the *Hox* array was released from the constraints associated with the rhythmic cadence of segmentation [2307], its integrity became moot. Fly clusters have been disintegrating ever since. The cluster cracked in half between ANT-C and BX-C in the lineage leading to *D. melanogaster*, but an even greater rupture occurred in the *D. repleta* clade, where *Hox1* was ripped away from *Hox2* and exiled to the hinterlands beyond *Hox9* [1607].

The epitome of disintegration is atomization, which destroys colinearity completely. No fly has reached that point yet, but the tunicate (chordate) *Oikopleura dioica* has. Incredibly, it nevertheless managed to retain a nested pattern of *Hox* gene expression [2057]. How? We don't know. Presumably, the atomized genes took their enhancers with them when they dispersed into the diaspora, but we have yet to figure out how they did so [2057]. The deeper question is: how were *cis*-enhancers linked to the A–P axis in the first place?

Here we encounter a second irony of history. The most rigorous analyses of colinearity have been done with *D. melanogaster*, but that circuitry is so derived that it is virtually useless as a standard for Bilateria in general [1395,1645,1757]. *D. melanogaster* organizes its A–P axis via a morphogen called Bicoid [1659]. A morphogen is a signaling molecule whose concentration is graded along an axis [257]. The scalar levels of the gradient allow cells (or syncytial nuclei) along that axis to gauge their distance from the poles and hence to adopt fates based upon this positional information [1923]. Bicoid was actually the first molecule that was proven to function as a morphogen in any animal [543–545,1811].

As shown in Figure 1.2, a concentration gradient of Bicoid spans the A–P axis, and *Hox* genes are turned ON at different levels of that gradient in a colinear manner [2038]. This cartoon makes it look as if Bicoid controls *Hox* genes directly, but that is not how the circuitry really works. Bicoid actually regulates *Hox* genes *indirectly* via intervening tiers of "gap" [876,1025,2423] and "pair-rule" [1020,1309] genes that interact as a hierarchical network of transcription factors [30,1047,2018].

The *Drosophila* network is typical of advanced (cyclorrhaphan) dipterans [2131], but only a subset of gap genes govern body segmentation in other insects [1157,1332,1422]. The oldest parts of the hierarchy are the lowest two echelons containing the pair-rule and "segment-polarity" genes [459,800,1054]. Curiously, one of the fly's primary pair-rule genes – *fushi tarazu* or *ftz* for short [2018] – is a rogue paralog of *Hox6*, more commonly known as *Antennapedia* (*Antp*; see Puzzle 1.1).

Amazingly, *bicoid* is another rogue paralog that, like *ftz*, never left its chromosomal birthplace [749,997]. It arose as a *Hox3* duplicate ~150 MY ago, near the dawn of the cyclorrhaphan clade [808,1275]. Its ensuing rise to power was as dramatic as a Shakespeare play and as enigmatic as a Sherlock Holmes mystery (see Puzzle 1.2).

The 1988 proof that Bicoid functions as a morphogen corroborated the "French flag" model that Lewis Wolpert proposed in 1968 [1923]. According to this model, positions along an axis are specified by the concentration of a morphogen diffusing from one pole [2456,2457]. In this case, *bicoid* mRNA is stored at the anterior end

---

Puzzle 1.1   Why did *ftz* leave its *Hox* career for a pair-rule job?

The *ftz* gene arose as a duplicate of *Antp* (*Hox6*) before the dawn of arthropods ~550 MY ago [912,2256] and has stayed beside its twin ever since. In flies, *ftz* has a homeobox but no longer serves as a *Hox* gene. Nevertheless, *ftz* still helps pattern the embryo.

Indeed, it was the zebra-esque photo of *ftz*'s in-situ transcription pattern that first revealed the two-segment periodicity of the seven pair-rule stripes that presage the 14 segments of the fly body [326]. That iconic picture was published in 1984 [841] – the same year as the discovery of the homeobox: an *annus mirabilis* indeed!

For decades *ftz* was classified as a *secondary* pair-rule gene (i.e., a gene controlled by other pair-rule genes), but recent findings have upgraded it to *primary* status (i.e., a gene controlled by gap genes) at a higher level of the segmentation hierarchy [2018].

How did a gene that started with *one* stripe of expression in its original *Hox6* role acquire *six more* stripes to become what we now call *ftz*? One clue to this mystery comes from the "upstream element" (UE) enhancer near *ftz*. When UE is stitched to a *lacZ* reporter in one orientation, it mimics the stripe pattern of *ftz*, but when inverted, it yields a single stripe coincident with *Antp*'s expression domain [947]!

The Ftz protein itself strayed from the orthodox Hox tradition in two motifs that are each only ~5 amino acids long, so these transitions could have been mutationally trivial to achieve [910,1342]. Surprisingly, the homeodomain is not needed for Ftz's pair-rule duties [418], which arose in stem insects [911], but it *is* needed for Ftz's chores in the nervous system [458,912,1565], which are evolutionarily more ancient [1056].

The shifting fortunes of *ftz*'s roles over the eons may seem capricious [912], but they offer useful lessons in how evolution can use a spare protein to rewire old circuits for new functions when selective constraints shift over time due to the rerouting of developmental pathways [911,996,1756,1788] (cf. the complex resumé of *engrailed* [273,1086,1554,1744] and its paralogs [201,752,923]).

---

of the egg. After translation, Bicoid protein diffuses to form a gradient. Once the gradient stabilizes [1048], particular gap genes are turned ON at different concentration thresholds – like the broad stripes of a French flag [2202]. The gap genes then proceed to interact with one another and cooperate with pair-rule genes so as to activate *Hox* genes (colinearly) along the A–P axis.

Fewer than half of the fly's eight *Hox* genes have their *cis*-enhancers attuned to Bicoid directly [1668], rather than to gap or pair-rule genes. Therefore, we can rule out the simplest conceivable French flag scenario, where the Bicoid-binding affinity (or number) of *cis*-enhancers varies linearly from gene to gene along ANT-C and BX-C [1669,1727], with consecutive genes being switched ON in 3′-to-5′ sequence by successively lower Bicoid levels [1726]. Indeed, Bicoid turns out to be less autocratic than anyone had initially imagined [362,1811,1937,2445].

Puzzle 1.2   How did *bicoid* become the ruler of its *Hox* siblings?

At the site where a *Hox3* gene should be, *D. melanogaster* has three genes: *bicoid*, *zen*, and *zen2* [996,1607]. Each of these paralogs has a homeobox, but none serves a *Hox* role any more.

These three genes arose from two duplications. First, *Hox3* begat *bicoid* and *zen* near the dawn of advanced dipterans ~150 MY ago [808,911,2167]. Then *zen* begat a *zen2* twin before the *Drosophila* genus radiated ~50 MY ago [1607]. Even before the first event, insect *Hox3* had abdicated its *Hox* role for some unknown reason [631,996] as it started to be expressed (maternally and zygotically) in rostral tissues that were outside the embryo proper [1370,2016]. The paralogs then divided their labors [504,911]: *zen* kept zygotic duties while *bicoid* kept maternal ones [2168]. Eventually, Bicoid formed a concentration gradient that acts like a scalar ruler – bestowing A–P coordinates upon nuclei throughout the syncytial blastoderm [911] (cf. morphogen roles of other homeoproteins [273,1085,1554,1825]).

How did *bicoid* become king of the segmentation hierarchy [492]? Apparently, a single base change in its homeobox was all that was necessary for this usurper to impersonate the old king, *orthodenticle* (see text) [1332,1756,2016]. However, the stage had been set long before the final coup by several events, including (1) a passive shift in *bicoid* mRNA positioning via germ band elongation [1274,1370], (2) the intervention of devices for mRNA localization [281,1993], and (3) a newly acquired ability to quash *caudal* translation [2014].

Once the embryo had two coequal regents (*orthodenticle* and *bicoid*), they must have jockeyed for power as mutations hobbled one or the other. Henceforth, *bicoid*'s fate probably hinged on speed: selection would have favored any gene that could build a larva sooner [1464].

Surprisingly, *Hox3* may not be the only apostate to erect a gradient that enslaves its brethren. *Hox7* (*Ubx*) of Onychophora (the sister phylum of Arthropoda) is expressed in an A–P gradient that seems to specify area codes for caudal segments [1058]. Nor has *bicoid* reigned unopposed since its ascension: in a few species it was dethroned and replaced [1173,1274] (cf. the murder of a morphogen in chordates [313]). As with *ftz*, the story of *bicoid* offers insights into how gene networks rise and fall [996,1756].

This simplistic scenario (cf. Ed Lewis's original proposal [1296]) may not apply to flies, but it does have validity in chordates [1185]. The main A–P morphogen in vertebrates is retinoic acid (RA) [1623,1895], a derivative of vitamin A. RA has been shown to (1) form a concentration gradient, (2) bind *Hox* enhancers at "RAREs" (retinoic acid response elements) via nuclear receptor proteins [28], and (3) assign regional identities to somites [554] and thereby to the final vertebrae [1146,1147]. Interestingly, RA plays a comparable role in the hindbrain [564,1996].

Amphioxus is the most basally branching chordate alive today [959,1355,1838, 2019], so, in theory, it can afford us a glimpse of how the *Hox* cluster of early

chordates may have operated before the vertebrate genome underwent two doublings [716,1443]. As in vertebrates, the *cis*-enhancers of amphioxus' *Hox* genes
have conserved RARE motifs [1601,2357], and its genome contains a sufficient
(albeit reduced) set of genes for RA metabolism [41].

Intriguingly, treating amphioxus embryos with excess RA causes A–P shifts in *Hox*
gene expression [964,1738,2020] that are eerily reminiscent of the deviations seen in fly
embryos at higher dosages of Bicoid [1811]. These RA-mediated shifts occur even when
protein synthesis is artificially blocked [1185] – indicating that the action of RA on *Hox*
transcription is direct. Hence, it is possible that an RA gradient was involved in establishing *Hox* area codes along the A–P body axis of urbilaterians. However, as discussed
in the next section, the preponderance of evidence points to a culprit other than RA.

otd/Otx may have ruled the front as *caudal/Cdx* ruled the rear

In vertebrates the RA signal is, to some extent, conveyed to the *cis*-enhancers of
Hox genes indirectly by Cdx homeoproteins [352,980,1341], which act in a dose-
dependent manner [727,904,2352]. *Cdx* genes of vertebrates are homologous to the
caudal gene of flies. In both taxa the Caudal/Cdx proteins form a gradient whose
slope is opposed to that of the Bicoid or RA gradient [257,1377]. In flies this reciproc-
ity is due to Bicoid's ability (unusual for a homeoprotein) to bind *caudal* mRNA and
prevent its translation [1811]. The fly's Caudal gradient specifies posterior identities
[839], just as the Cdx gradient does in vertebrates [593,1990], though it may not do
so via positional information per se [419] (cf. [1055]). Consistent with this idea, the
caudal/Cdx ortholog in nematodes (*pal-1*) is needed for tail development [582].

It would seem, therefore, that urbilaterians employed *caudal/Cdx* to provide
area codes for their rear end [419,488,1683,2014], possibly under Wnt control
[376,1396,1410,1431,1624,1952]. This inference is supported by (1) the expression of
caudal/Cdx in posterior tissues of amphioxus [262], (2) the antiquity of the gene
itself, and (3) its membership in the ancient *ParaHox* cluster of three *Hox*-related
genes [639,970]. The *ParaHox* cluster manifests *Hox*-like A–P colinearity [724] and
is thought to have arisen (along with the *Hox* cluster) from a *ProtoHox* cluster
[262,638,715,1843] before the urbilaterian–cnidarian split [298,382,659,1468,1698].

The question remains: What morphogen was originally employed *anteriorly*?
Bicoid evolved too recently to have served this role [1932], and there is no evidence
that RA functions as an A–P signal in protostomes [39,510,1388,1416], though we
cannot rule out a secondary loss in ecdysozoans [310,836]. Our best clue comes
from short-germ insects, which typically use the gap genes *orthodenticle* (*otd*) and
hunchback (*hb*) in a similar way to how flies use *bicoid* [1371,1465,2016,2440], though
otd's A–P role in the beetle *Tribolium* may have been lost [1190,1997], and hemimetabolous insects rely more on the posterior pole than the anterior one as a reference
point for segmental patterning [1231,1756].

Both Otd and Hb are transcription factors, and this fact alone would seem to
disqualify them as candidates for the role of a morphogen, because such proteins

are widely thought to be incapable of diffusing in a *cellularized* (vs. syncytial) milieu like the blastoderm of a short-germ embryo [1514,1752]. However, transcription factors – including homeoproteins like Otd – can actually be secreted, taken up by cells, and imported into nuclei, even in the presence of cell membranes [355,1389], so we cannot rule out diffusion gradients in this case [1085,1825,2249]. Moreover, much of the fly's segmentation gene network operates similarly in short- and intermediate-germ insects [472,1756,2250], so cell boundaries may not be impeding transport as much as we might suspect [385,1574,2200].

Compared with *otd* and its vertebrate ortholog *Otx*, the role of *hb* in anterior patterning evolved recently, dating back only to the dawn of insects or perhaps of arthropods [1795]. In contrast, the usage of *Otx* as a cephalizing agent may go back all the way to urbilaterians [1313], because it helps specify rostral neural identities in chordates [840,1861,2054,2096], hemichordates [958,1357], and platyhelminths [2315]. Hence, *otd/Otx* may have reigned over the anterior of the urbilaterian embryo [2101] or at least the rostral nervous system [950,2173], just as *caudal/Cdx* governed the posterior. Other clues, however, point to *six3* having played this role instead [1003,1418], and the issue is far from settled. The primordial targets of the A–P morphogens (whatever they were!) appear to have included the gap gene *teashirt* (as well as the *Hox* genes, of course) [1399].

Segmentation is so rare among phyla that it must be convergent

A third great irony of *Hox* history is that the first clusters to be studied were in "segmented" animals, though this term can only loosely be applied to vertebrates, whose somite blocks (mesoderm) and hindbrain rhombomeres (ectoderm) are not truly body segments in the classical terminology of comparative morphology [284,430]. Even if we do include chordates [2249], only three of the 28 bilaterian phyla (Chordata, Annelida, and Arthropoda) would qualify as "segmented" (Figure 1.3) [284,562,1506,1757]. The simplest interpretation of this disjointed distribution is that segmentation evolved separately three times [376,1755] – long after urbilaterians split into deuterostomes and protostomes.

In contrast, all of the bilaterian phyla for which data are currently available (11 out of 28 total) express their *Hox* genes in nested (colinear) zones, so urbilaterians probably did likewise [155]. These contrasting facts imply, as Michael Akam surmised in 1989, that "the evolution of metameric segmentation has most probably been superimposed independently onto a regionalized embryo" [31]. Assuming that he is right, how did evolution overlay metameres onto a pre-existing "French flag" of *Hox* expression zones so neatly that the edges of the *Hox* zones coincide perfectly with the boundaries between certain metameres? This coincidence is actually less perfect than we have imagined, depending upon *when* during development we measure it.

The registration in the fly embryo is initially quite precise, with *Hox* zones terminating at "*para*segment" boundaries. (Parasegments are primary metameres that have the same periodicity as segments but are out of phase with them by half a wavelength [512,1248].) At later stages, however, this congruence fades [32,1796], and the mismatches become blatant in the imaginal discs that form the fly's append-ages. For example, *Ubx* expression is very uneven in both the haltere and third-leg discs that belong to the third thoracic segment (T3). Strangely, the intensity of *Ubx* RNA is highest in T3-specific transverse rows and sensilla nests [923].

Why is a gene that is supposed to be specifying segmental identity in a broad-brush manner being used on such a fine scale, like "touch-up paint" for a subset of structures within a segment? A clever answer was proposed by Michael Akam [32,33] and his colleague James Castelli-Gair [333] in 1998. They argued that the *Hox* paradigm, as then formulated, had misled us into thinking that *Hox* genes act like digital switches. Hox proteins, they stressed, are merely transcription fac-tors [1894,2111], so it should come as no surprise that they are only used when and where they're needed in an analog way. To wit, if T3 differs from T2 in only a few structures, then it makes sense that *Ubx* is only expressed at those sites within the segment. In short, *Hox* genes are "micromanagers" [33].

> *Hox* genes are not controlling segment identity, but cellular behavior that will result in a certain segment morphology. This is what *Hox* genes do in unsegmented organisms like *C. elegans* and is probably what they did in the common ancestor of all metazoans. Segment identity is a subjective concept that originates from the observation that in a particular species, a number of cell characteristics are always associated in a given segment. [333]

> By accepting a role for the regulation of *Hox* genes within compartments, we demote them from their privileged status as stable binary switches. The revised model allows the activity of the *Hox* genes to be modulated in a complex way throughout development, by local signals, hormone receptors or any of the other stimuli that commonly mediate gene regulation. In this regard, it makes the *Hox* genes like any other genes. It predicts that small changes, particularly in the structure of their promoter modules, will change the phenotype of segments. [32]

> In each of these cell types, expression of a *Hox* gene means something different – to divide or not to divide, to make or not make a bristle, to die or not to die. In any given lineage, that mean-ing probably changes several times during development . . . If you want to make that bristle or bump specific to a particular segment, you add a Hox transcription factor to the code that controls the relevant enhancer module. [33]

If we were to abandon the hypothesis of three segmentation origins and assume instead that segmentation arose only once in urbilaterians, then we would have to account for why such a seemingly useful trait was lost in so many descendant clades [471,1755,2258]. No convincing explanation for such a pervasive forfeiture has ever been put forward [375], nor has any persuasive vestige of segmentation ever been adduced for any unsegmented phylum [140,215,2381], though some annelid taxa are equivocal in the latter regard (M. Akam, personal communication). We would also have to explain why segmentation is implemented so differently at the cellular and tissue levels in the three phyla that manifest it [2040].

Despite its flaws, the *Single*-origin Scenario has had its share of advocates [140, 485,957], going back to Geoffroy St.-Hilaire [1158]. It has been bolstered recently by a litany of genes that have been shown to be used similarly by vertebrates, annelids, and arthropods during the segmentation process [2040]. Those clues are reviewed below in the form of a fanciful debate (à la Galileo's *Dialogue*) between two straw men: Homer contends that all segmentation is homologous due to its having been inherited from urbilaterians, while Parker argues for "parallel" evolution. Parallel evolution occurs whenever two clades evolve similar structures independently via the *same* genetic devices. It differs from "convergent" evolution, which entails the independent adoption of *different* devices [953].

HOMER   Most arthropods use a short-germ style of segmentation [459] that resembles how vertebrates make somites [1042]. In both cases, the metameres emerge cyclically [1938,1981] from a posterior growth zone [140,1472,1504], and the same is true for annelids [139] (but cf. [1450]).

PARKER   Yes, but lots of seriated structures aside from segments arise with rhythmic periodicity [922], like the chambers of a nautilus . . . or even the leaves of a tree! Coupling an oscillator clock with a growth zone seems like a trivial contrivance that could evolve quite easily in any clade [1900]. Moreover, it seems silly to base your argument on somites, because vertebrates use *Hox* genes not only for area codes in their paraxial (somite-making) mesoderm, but also in their lateral (non-somite) mesoderm, which never gets chopped into metameres [789].

HOMER   The same homeobox gene – *caudal/Cdx* – is integral to the operation of the posterior growth zone in both arthropods and vertebrates. Doesn't that suggest deep homology for the segmentation process?

PARKER   Perhaps. But *caudal/Cdx* also specifies posterior identities in nematodes, which are not segmented [582], so there is no firm link to segmentation per se. The same objection applies to *nanos*, which may also be a universal posterior determinant [376].

HOMER   The way that the *Notch* signaling pathway defines intersegment boundaries in annelids [1912] and some arthropods (albeit not flies) [1515] resembles how vertebrates use *Notch* to delineate somites [485,1754,1816,1962,2249].

PARKER   So what? Evolution often uses *Notch* to draw lines and subdivide territories [126,2467], even in the unsegmented hindgut [1034,2229], wing [452], and eye [1026] of the fly (cf. [126,2383,2553]). Moreover, recent data question *Notch*'s role as a "salami slicer" (segment-making device) in arthropods [450,1094] and vertebrates [1706], not to mention onychophorans, where it has nothing to do with segmentation [1055]. Indeed, the segmentation machinery varies so wildly at the cell and tissue levels [215,1506] that it will take more than a few conserved cogs and gears at the gene level to prove homology of the overall clockwork [2363].

HOMER   So the comparable usage of *engrailed* in the segments of annelids [1831], arthropods [473,1148], and amphioxus [967] won't sway you either?

PARKER   Hardly! Vertebrates were not in your list. They don't use *engrailed* to achieve segmentation [967], though they do express it later in somites [857]. The original function of *engrailed* seems to have been in the nervous system, rather than in segmentation [273,1086,1554,1744]. This example, and the others you cite, are more plausibly explicable as instances of independent co-option [215,376,939,953,2249].

Based on the evidence presented above, Parker must be declared the victor, though we can't bury the Single-origin Scenario just yet [969,1989]. There are simply too many phyla whose genomes have not yet been analyzed in sufficient depth for us to draw a firm conclusion right now [484,563].

The ultimate manifesto for bilaterian parallelism *in general* was set forth in 2002 by Douglas Erwin and Eric Davidson in a review entitled "The last common bilaterian ancestor" [601]. They not only refuted the hypothesis that urbilaterians were segmented, but they also rejected most other claims for ancestral complexity that were based on trans-phylum genetic comparisons. Indeed, the only conserved features they were willing to concede were a two-ended gut, *Hox*-based A–P subdivisions, mesodermal derivatives, and a nervous system. Homologous genes, in their view, do not build homologous *organs*, but rather execute homologous *cellular* functions that were separately recruited into *analogous* organs by different clades (boldface added):

> Although the heads, hearts, eyes, etc., of insects, vertebrates and other creatures carry out analogous functions, neither their developmental morphogenesis, nor their functional anatomies are actually very similar if considered in any detail. However, in each of the body parts, respectively, the same differentiated cell types are employed across the Bilateria, and it is this fact that underlies their analogous functions: **heads all require various types of neurons and their ganglionic associations; hearts necessitate certain kinds of slow contractile cells; eyes require photoreceptor cell types, guts require digestive and secretory cell types; and so forth** . . . The regulatory processes that underlie development of specialized differentiated cells are indeed very old, conserved, plesiomorphic features. In contrast, the morphogenetic pattern formation programs by which the body parts develop their form are clade-specific within phyla or classes. [601]

> So a possible solution to our paradox is that the regulatory genes which we find employed in patterning analogous body parts originally ran the differentiation gene batteries required for their respective functions, and **since these genes were expressed in the right place they could be coopted during evolution to produce successively more elaborate pattern formation functions, differently in each clade** . . . A corollary is that some regulatory genes will often be found operating at several different levels of the developmental process leading to body part formation, from initial specification to terminal installation of particular cell types. [468]

Whether Erwin and Davidson's perspective proves as convincing for the other shared traits of bilaterians as it has been for segmentation remains to be seen. The truth for at least some traits may lie somewhere between the two extremes [1666], but readers can judge for themselves as the story unfolds.

The dorsal–ventral axis is established by *BMP* and *Chordin*

In 1994 Detlev Arendt and Katharina Nübler-Jung presented molecular data which revived Geoffroy St.-Hilaire's old idea that arthropods are essentially upside-down vertebrates [83,956,1654,2297]. In both animal groups the same morphogen is used to encode positional information along the dorsal–ventral (D–V) axis [973]. That morphogen is known as Dpp (Decapentaplegic) in flies and BMP4 (Bone Morphogentic Protein 4) in vertebrates [894,1307,1857], and its antagonist is called

Sog (Short gastrulation) in flies and Chordin in vertebrates [35,728,1527,2525]. Dpp/ BMP4 is expressed dorsally in fly embryos but ventrally in vertebrate embryos, while Sog/Chordin comes from the side opposite to Dpp/BMP4 in both cases [203,1479,2000].

Given this new evidence that the D–V axis of insects is inverted relative to that of vertebrates, the question became: How was the axis originally oriented in urbilaterians, assuming that they used the same morphogen in a similar manner? Based on the BMP–Chordin polarity of the taxa that have been surveyed thus far (Figure 1.4), vertebrates are clearly the exception, rather than the rule [207]. The available clues imply that the inversion occurred in a primitive chordate [1246,1356] that may have resembled amphioxus – the most basal chordate alive today [716,729].

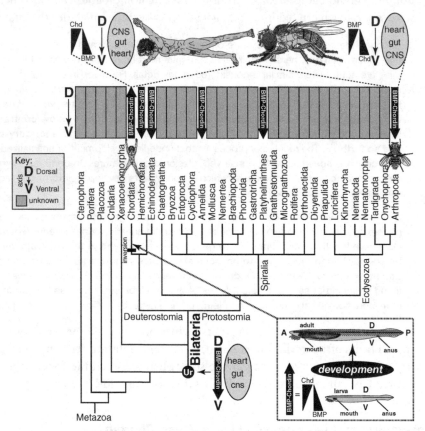

Figure 1.4 Orientation of BMP–Chordin polarity along the D–V axis among animal phyla.
Cellular positions along the D–V (dorsal–ventral) axis of bilaterian embryos are specified by a concentration gradient of the morphogen BMP (Bone Morphogenetic Protein = Dpp in flies), depicted as a solid triangle in the schematics at the top. The BMP gradient is shaped by a reciprocal gradient (antiparallel triangle) of Chordin (Chd) or other antagonists [1492]. In the survey of animal phyla below, BMP–Chordin polarities are denoted by thick arrows from high BMP to high Chordin.

In flies the BMP–Chordin vector matches the D–V vector, whereas in chordates the two vectors point in opposite directions. Why? Based on the D–V polarities in other phyla

If so, then the D–V axis of urbilaterians would have had the same orientation as that of flies (i.e., D = BMP; V = Chordin), not that of vertebrates.

The BMP–Chordin polarity of amphioxus matches that of vertebrates [1199,2020], but it has asymmetries that may be relics of the inversion event [231,299,2142]. D–V reversal is like a swimmer flipping from breaststroke to backstroke – a maneuver that reorients the mouth pointing up. Curiously, the amphioxus mouth starts out on its left flank and migrates down to the ventral midline during metamorphosis [507,716,1730] – as if it were re-enacting a compensatory "retrofit" inherited from its inverted ancestor [743,1222,1223,1226,1356]. Re-enactments of this kind were codified in Haeckel's famous dictum that "ontogeny recapitulates phylogeny" [785], though many exceptions to this principle exist [1354], and the phrase has lost its claim to dogmatic status [101,247].

Despite its fish-like appearance, amphioxus is a filter feeder that only swims intermittently [716]. Most of the time it inhabits vertical burrows with only its head sticking out to harvest tiny plankton [192,303,2190] (cf. hemichordates [743,1366]). In such a vertical posture, "dorsal" and "ventral" have little meaning. Hence, this niche could have provided an ideal inflection point of relaxed selection where body

←──

Figure 1.4 (*cont.*)

that have been studied thus far, the inversion event seems to have occurred at a point just before the stage when amphioxus diverged from its fellow chordates (box at lower right) [1356]. Flipping of the formerly dorsal (back) side to the ventral (belly) side presumably placed the mouth at an awkward angle for feeding. In amphioxus the mouth migrates from the left flank to a ventral site during metamorphosis [507,716,1730] – suggesting that this species was "caught in the act" of adjusting its mouth location to suit its new posture [299,646,1222,1223,1226]. If so, then the mouth might have been expected to start dorsally instead of laterally. Indeed, recent evidence appears to refute the notion of recapitulation [2141].

Because of this inversion (regardless of its cause), the sequence of organs along the D–V axis in Chordata – CNS (central nervous system), gut, heart – is exactly opposite to that in Arthropoda – heart, gut, CNS – and other bilaterian phyla – just as Geoffroy St.-Hilaire reported in 1822 [740]. In vertebrates the mouth is pushed ventrally by the growing brain during development [384].

This diagram is based on Figure 6 of Bier and De Robertis (2015) [207], with additional data from the following references: Chordata [2000] (including amphioxus [966,1199,1688,2515]), Hemichordata [743,1355,1356,1940], Echinodermata [549,942,1246,1526], Annelida [505,1214], Platyhelminthes [728,1528,1876], and Arthropoda [35,767]. Data are omitted for Cnidaria because the usage of BMP and Chordin, albeit along a body axis, is neither simple nor homologous to bilaterians [739,1262].

In some phyla Chordin is augmented or replaced by other BMP antagonists [205,1492], and the circuitry often differs from the vertebrate norm [1372,1707], so the situation is not as clean as implied here [207,1232]. In particular, sporadically distributed taxa that rely more upon stereotyped cell lineages than upon morphogens have experienced secondary losses of the D–V gradient mechanism [1492] by either (1) degrading their BMP–Chordin circuit (leech) [1214] or (2) assigning it a lesser role (tunicates) [462] or (3) abandoning it entirely in favor of cytoplasmic determinants (nematodes) [1793,2105]. Tree of extant phyla as in Figure 1.3.

> ### Puzzle 1.3   How did the mouth relocate after the body inverted?
>
> If Geoffroy was right, and vertebrates are basically upside-down arthropods, then we are faced with a geometrical dilemma exemplified by a cockroach flipped on its back: if its mouth points up, then the insect cannot feed in the normal way. Assuming that the mouth of chordate ancestors pointed down as in modern arthropods (and in our hemichordate cousins [743]), then an inversion would have caused the mouth to point up [1655]. Interestingly, that is what we see in the larvae of tunicates [384,396], which form the sister clade of vertebrates [500,962]. Because tunicates are filter feeders, this orientation does not pose a problem because the mouth works fine at any angle [1467]. However, vertebrates typically feed on a substratum with their mouths pointing down, so at some point the mouth's location must have changed [1492]. One possibility, discussed in the text, is that it migrated ventrally – tracing the same path down the flank as we see during amphioxus metamorphosis [1223]. However, recent evidence strongly contradicts this scenario [2141]. Alternatively, the mouth may have been pushed ventrally as the forebrain enlarged evolutionarily – a possibility suggested by what actually happens during vertebrate embryonic development [384]. Still another idea is that the mouth arose independently from the original oral opening (cf. [1224,1435]), and that the old (dorsal) mouth disappeared, leaving only the new (ventral) one [83,84,1654].

tilt and mouth site could have been altered without jeopardizing feeding ability [742]. The mouth may have eventually relocated in response to a change in body posture (see Puzzle 1.3).

Left–right asymmetry emerges from cytoskeletal chirality

Humans and flies are essentially symmetric on the outside but asymmetric on the inside (Figure 1.5) [925]. Our heart and spleen, for example, are shifted to the left, while our liver and appendix are displaced to the right [1187]. In contrast to humans, the fly's heart is symmetric [438], but its kidneys – or the equivalents thereof – are not [374], nor are its testes [738]. One asymmetric organ that humans and flies do share in common is the gastrointestinal tract. In both cases the digestive tube meanders through the abdominal cavity in flagrant disregard for the body's midline [433,2010], though its species-specific shape in both humans and flies is somehow programmed to remain relatively constant from one individual to the next [1171,1279,1285,2410].

The human brain looks remarkably symmetric at the level of gross anatomy, but it was obvious to most of us as children (except those who are ambidextrous) that something was amiss when we tried to use our left and right hands with equal grace in one sports activity or another, but couldn't. Why our brain should be so asymmetric in its control of motor function is a profound mystery (see Puzzle 1.4).

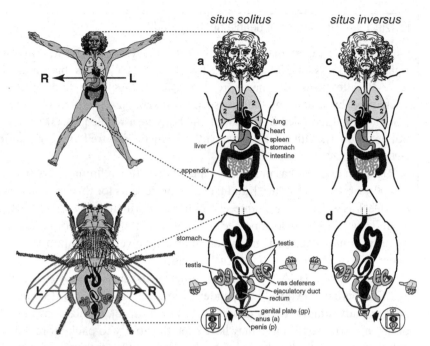

Figure 1.5 Visceral asymmetry in humans and flies.

a. Asymmetric organs in *H. sapiens* include the lungs, which have two lobes on the left but three on the right [239], the heart, spleen, and stomach [1187], all of which are displaced to the left [283,294], the liver, which is shifted to the right [2010], and the large intestine [1171,2410], which contorts itself into the shape of a question mark with the appendix protruding where it joins the small intestine [1614,2437]. Except for the lungs, all these organs start out symmetric during development and then become asymmetric [332].

b. Asymmetric organs in *D. melanogaster* include the stomach, which forms an S-shaped loop [433], the testes, both of which coil into left-handed spirals around the vas deferens (denoted by the curl of fingers in the outlying hands relative to thumb direction) [1500,2180], and the genital plate (cartooned as a square containing anus and penis), which rotates 360° clockwise, hence twisting the ejaculatory duct into a loop around the rectum [985]. Testis coiling may have evolved to accommodate long sperm whose tails grow to absurd lengths in some flies [1798]. (Does coiling match tail rotation direction?) Genital rotation is a quirky legacy of the twists and turns of dipteran mating postures throughout evolutionary history [928,2215]. The asymmetry of Malpighian (renal) tubules (not shown) may have evolved as a way of tailoring their services to the needs of visceral organs that were already asymmetric [374].

c, d. Reversals of left–right asymmetry, which are termed "*situs inversus*" to contrast this condition with the normal "*situs solitus*" situation (**a, b**) [1257,2291]). Such reversals are caused by *dynein*-affecting mutations in humans [118,828,1691] but by *myosin*-affecting mutations in flies [738,985,2147]. The regulatory genes in these two taxa appear to be non-homologous (see text).

Puzzle 1.4   Why are most people right-handed?

One of the strangest manifestations of L–R asymmetry is the prevalence of right-handedness in *Homo sapiens* [322,422,423,2118]. This bias must be due to L–R differences in gene expression [246,421,2209], but hand preference is unchanged by mutations that cause *situs inversus* [24,1471], so its genetic control must be uncoupled from that of the body as a whole [548,1470,2210]. We do not yet know the full extent to which the hemispheres are hard-wired differently [398,933,1155,2422].

Right-handedness has prevailed at a level of ~90% in humans [2499] for at least 10,000 years [120,243,426,627], and there are precedents for this trend in other great apes [832], as well as hints of antecedents in other primates [144,420,855,1546]. Brain lateralization has been documented anatomically [628] or behaviorally [1490,2109] in other mammals [756,1236] and in non-mammalian vertebrates [210,410,1322,1381], and it even exists in insects [175,1284], including the fruit fly [1737].

No one knows (1) when brain asymmetry first arose [351,416,667,1924,1925], nor (2) whether it evolved independently in different taxa, nor (3) what, if any, adaptive advantages it may confer [421,548,1713,2332]. Conceivably, one benefit of asymmetric partitioning may have been the ability to pack more sensory and behavioral "apps" into the brain by installing each of them on one side only [416,417,1413,1552,1570].

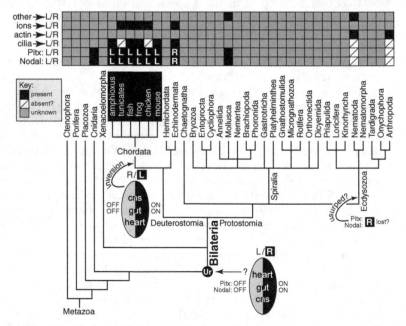

Figure 1.6 Mechanisms of symmetry breakage along the L–R axis among animal phyla. Black squares signify usage of the indicated mechanism in at least some members of the phylum (or clade within Chordata) but not universal presence throughout the taxon. Diagonal lines denote its absence in all species examined thus far, though future analyses of more species could show otherwise. Gray squares mean that conclusive facts are lacking (see key).

What causes asymmetries – be they subtle or dramatic – to arise during development? In 1990 Nigel Brown and Lewis Wolpert proposed an explanation [268]. They reasoned that asymmetries at the *anatomical* level must rely on some sort of chirality at the *molecular* level. Over the ensuing years their hypothesis has proven to be essentially correct [1287,2458] for both humans and flies [221]. However, humans employ the chirality of tubulin–dynein motors, whereas flies use the chirality of actin–myosin motors [738,771]. This difference may reflect an ancient divergence between the deuterostome and ecdysozoan clades to which humans and flies belong (Figure 1.6), though it is

←

Figure 1.6 (*cont.*)

Asymmetric expression of the morphogen Nodal and its effector Pitx (lowest two rows) had been thought to occur only in deuterostomes [356], but in 2009 Nodal was shown to control handedness of shell coiling in snails (Spiralia; phylum Mollusca) [793], and in 2014 Nodal and Pitx were detected asymmetrically in Hydra (phylum Cnidaria) along the mother–bud ($\neq$ L–R) axis [2395]. These results suggest that lateralized expression of Nodal and Pitx predated the cnidarian–bilaterian split [2395] but was later lost from Ecdysozoa (and usurped by a myosin-based mechanism) when they split from Spiralia [220,1595,2510]. Sinistral (L) expression of Nodal and Pitx in chordates is attributable to their D–V inversion, which turned right into left as it flipped the order of organs (heart, gut, and CNS = central nervous system) along the D–V axis (Figure 1.4) [2142]. Dextral (R) expression of Nodal and Pitx in sea urchins is consistent with chordate body inversion [550], but has recently been questioned based on larval topology [221]. Hemichordates fail to use *nodal* along the L–R axis (though they do so along the A–P and D–V axes) [963]. Even so, their polarity opposes chordates in terms of mouth position relative to BMP–Chordin orientation [1356]. Flies and nematodes lack *nodal* homologs, and although *Pitx* homologs exist, they are not expressed asymmetrically [231], so the "L/R" squares in both cases are marked "absent."

Several types of mechanisms have been implicated in the breaking of symmetry along the L–R axis [407,475], based on a few model chordates [1590]. Those hypothetical devices are inventoried in the upper four rows of squares, where the arrow means "causes" and symmetry breaking is denoted by "L/R." "Cilia→L/R" (cf. Figure 1.6a–d) [1649] refers to cilia-propelled flow of an unknown morphogen in the ventral node of the mouse [467] or in the node's homologs in fish (Kupffer's vesicle) and frogs (gastrocoel roof) [2031,2283]. "Actin→L/R" indicates the involvement of actin–myosin motors [813,1704], which, like cilia, have an inherent chirality that imparts handedness to their rotational movements [1523,1588,1805,1839,2254]. "Ion→L/R" designates ion fluxes initiated by asymmetric protein pumps in cleavage-stage embryos [2332], though the causal link to visceral asymmetry has been questioned [221]. "Other→L/R" is a catchall category [193,388,977,1592] that includes agents aside from cilia, actin, or ion fluxes [2332,2333], such as cytoplasmic determinants for shell coiling in snails [475,669,2070], microRNAs [1083], intercellular junctions, extracellular-matrix components [407], and unknown factors that dictate neural asymmetry in nematodes [604,1697,1978].

The ciliary gadgetry is thought to be a derived trait of vertebrates that was later lost in chickens and pigs [220,816,2177], but there is no consensus as to which, if any, of the other symmetry-breaking devices dates back to urbilaterians [416,853,2332,2458]. For evidence that cilia dictate asymmetry in amphioxus and sea urchins see [220], and for puzzles about cilia in general see [624].

Tree of extant phyla as in Figure 1.3. The "checkerboard" entries are from previous phyletic surveys [220,432,1288,1595,1714], with data added from primary sources for Cnidaria [2395], amphioxus [2142,2514], tunicates [232,1558,2072,2273,2284], fish [19], frog [1290], chicken [1290], mouse [2232], Echinodermata [549,550,941,942,2386], Mollusca [475,793,1382], and Nematoda [305,1588,2013].

premature to attempt to draw such a conclusion before more phyla have been analyzed. The history of insights in this field is briefly reviewed below.

One of the most useful clues unearthed thus far was the 1995 discovery of genes that are expressed on the left – but not the right – side of the chick embryo at a certain stage of development [1037,1159,1286]. One of those genes – a homolog of *nodal* – turned out to be the key to our current understanding of how bilaterians tell their left from their right [432]. Like BMP, Nodal is a member of the TGF-β family of signaling molecules [1994,2065].

Additional clues have come from anomalies in conjoined twins [1289,2331] and from mutations that reverse the asymmetry of the viscera in humans, mice, and flies, resulting in what is termed a "*situs inversus*" phenotype (Figure 1.5) [2155]. The nature of these latter mutations will be considered first.

In our species, *situs inversus* occurs in 50% of men whose sperm are immotile due to the absence of dynein from their flagellar tails [23]. Dynein is needed not only for the motility of sperm but also for the mobility of cilia throughout the body [1177], and those cilia rely on the same microtubule architecture as sperm axonemes [828]. It turned out that defective cilia were responsible for the reversals of visceral laterality in these men, rather than any side effects of the mutations on their sperm [22,1653,1758,2507].

In 1959 a loss-of-function mutation in the mouse gene *inversus viscerum* (*iv*) was found to cause *situs inversus* in 50% of the homozygous mutant individuals [267,1000]. The reversed mice express *nodal* on the right – instead of the (normal) left – side of the body [1358]. Based on this 50:50 ratio, mouse embryos would seem to have no initial preference to express *nodal* on one side or the other [1257] until the wild-type *iv* gene tips the balance toward left-sided *nodal* expression [410]. In the absence of *iv* function, the embryo essentially flips a coin to see whether it will express *nodal* on the left or right. As was the case in humans, the mouse's *iv* gene was later (in 1997) shown to encode a ciliary dynein [2213].

In 1993, this scenario was challenged by the discovery of a recessive mutation in the mouse gene *inv* (*inversion of embryonic turning*), which was reported to cause ~100% of homozygous mice to show L–R visceral reversals and right-sided *nodal* expression [1358], instead of the 50% typical for *iv* [266,610]. In contrast to what had been concluded previously, this phenotype implies that the default state of the mouse embryo is not neutral after all, but rather is biased to express *nodal* on the right. The implication is that (1) symmetry is broken at an early stage, and (2) the wild-type *inv* gene overrides this bias in the opposite direction when it is turned ON [267]. Like the product of the *iv* gene, the Inv protein [1550] is an integral component of the cilium [1314,2394], but surprisingly, unlike *iv* mutant cilia, *inv* mutant cilia are not paralyzed [1675].

In 1998 mobile cilia were shown to be necessary for normal L–R asymmetry in mice [1643]. Mammal embryos appear to use cilia like miniature propellers (Figure 1.7b) to waft an unknown molecule "X" [2240] to the left side of the body, where it turns ON the *nodal* gene (Figure 1.7d) [1643,1649] (but cf. [2222,2511]). When *iv* is disabled, this leftward flow fails to occur [1675] – a result that is consistent

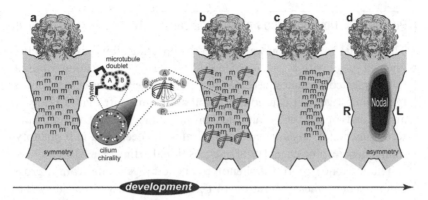

Figure 1.7 Development of L–R asymmetry in mammals. Adapted from [928]. L–R asymmetry in mammals has mainly been explained by two alternative models [2507], both of which involve unidirectional fluid flow generated by rotating cilia. In mouse embryos this one-way flow occurs in the ventral node [467] on the eighth day of development [1649], though, for the sake of clarity, the events are rendered here as if they were taking place on a man's torso instead of the embryo's node. According to the first model (illustrated here), the flow propels an unknown morphogen molecule ("m") toward the left side of the body [2240], causing its concentration to become increasingly asymmetric, culminating in the unilateral expression of *nodal* [432]. According to the second model (not shown), there is no morphogen [854]: the flow itself is sensed mechanically (not chemically) by immotile cilia at the rim of the node, prompting the affected cells to express *nodal* [2507].

a. A morphogen ("m") of uncertain identity [2240] is initially distributed symmetrically about the ventral midline.

b. Nodal cilia (enlarged in inset cross section) lack the central pair of microtubules that defines orthodox "9 + 2" cilia [2542]. This unusual "9 + 0" structure enables them to twirl (clockwise) like a propeller instead of rowing back and forth like an oar [2077,2230]. Moreover, their posterior tilt prevents them from moving the surrounding fluid on the lower half of their transit (recovery stroke) due to viscosity near the surface, so they can only move fluid on the upper half (effective stroke) – a constraint that produces a net flow of the morphogen to the left [1643].

c. Eventually, this flow raises the morphogen concentration on the node's left side sufficiently to activate the *nodal* gene [221].

d. The Nodal signal, which is thus produced, is conveyed to the lateral plate mesoderm [1129] and to organs that develop later from this tissue (e.g., the heart and spleen) [406]. This Nodal-ON identity allows these organs to develop differently from the Nodal-OFF organs on the right (default) side of the body [1868]. NB: Nodal is acting here in a digital (binary) mode, not in the sort of analog (graded) mode that we usually associate with a (TGF-β) morphogen. In this regard, the left–right "axis" differs from the other two body axes [1160].

with the *iv* phenotype of 50:50 randomization [1714]. However, following this logic, researchers expected the *inv* mutation to be causing *right*ward flow that would reliably produce *situs inversus*. Instead, the flow in that case was found to be *left*ward, albeit somewhat reduced in intensity [1675]! This paradox remains to be resolved (see Puzzle 1.5) [416].

Puzzle 1.5   How does the *inv* mutation cause ~100% reversals in mice?

The original *inv* mutation was induced by injecting albino mouse embryos (at the one-cell stage) with a *tyrosinase*-tagged transgene and creating inbred lines from the pigmented neonates [2505]. When the transgene inserted itself into the wild-type *inv* host gene, it caused a deletion and frameshift that left only the first 90 amino acids of the 1062-amino acid protein intact [1551]. This remnant is unlikely to be functional, so *situs inversus* can safely be deemed a null phenotype, rather than a neomorphic effect [1573] or a positional artifact [1551].

The inference of 100% phenotypic penetrance came from a cross between heterozygotes where 17 out of 74 F_1 offspring had *situs inversus*. The number 17 roughly equals the expected 18.5 homozygous recessives (= ¼ of 74) based on Mendelian segregation, whereas this number is far from the 9.25 (= ⅛ of 74) predicted for 50:50 randomization ($p < 0.05$).

Disconcertingly, however, other experiments point to a 50:50 (*iv*-like) effect for *inv*. In the most contrary study, 6 out of 16 *inv* mutant mice failed to show reversed stomachs [2233], and knocking down the *inv* ortholog in zebrafish results in randomized *situs* [1703]. Penetrance for the human ortholog is uncertain [2012].

If, despite the contradictory data, the purported 100% effect is valid, then how might the *inv*-null be acting if it does not reverse nodal flow? The domains of the Inv protein resemble those of the fly protein Diego, which regulates planar cell polarity [1314] (cf. Diego's role in fly asymmetry [771]), so Inv might be affecting asymmetry via the cell membrane, rather than via the cilium per se [116] (but cf. [72,2133]). Consistent with this conjecture, Inv does interact with junctional cadherins and catenins [1656].

The situation is complicated, however, by Inv's promiscuity and pleiotropy: its splicing isoforms bind calmodulin [1550], modulate the actin cytoskeleton [2412], and bind microtubules of the spindle, centriole, midbody, cilium, and cortex [1657], and for all we know, any of these roles could be instrumental. More asymmetry puzzles can be found in [115].

The cilium itself has an inherent handedness. Its outer ring of microtubule doublets and dynein arms endows it with a chirality that forces it to spin clockwise (Figure 1.7b). This spin, in turn, dictates the leftward flow of the fluid in the ventral node of the embryo. Hence, the gross L–R asymmetry of visceral organs in the mammalian body ultimately seems to stem from the cytoskeletal chirality of tubulin–dynein linkages within the cilium organelle [158].

Flies exhibit a number of L–R asymmetries (Figure 1.5). The gut is starkly lopsided [738], as are the left and right renal tubules, which function differently on the two sides of the body [374]. Strangest of all, perhaps, are the left and right testes of *D. melanogaster*, both of which develop as left-handed spirals, instead of becoming mirror images of one another (see Puzzle 1.6) [1500,2180].

As with mice, the best clues about L–R asymmetry in flies come from mutations that cause *situs inversus*. The most penetrant such mutations recovered thus far affect

---

Puzzle 1.6   Why are both of *D. melanogaster*'s testes left-handed?

Unlike most flies, where the testes are mirror images of one another, both of the testes of *D. melanogaster* coil in a left-handed spiral as they revolve around the vas deferens (Figure 1.5) [1500,2180]. This heretical anatomy is as strange as a man with two left hands (or feet)! The only other animal with such a paradoxical geometry is a species of arctic whale called the narwhal [1620]. Male narwhals typically have a single tusk (like the mythical unicorn) that they appear to use for jousting [2094] and for sensing salt concentrations [1660]. The tusk is a modified left canine tooth [1661] with a twisted groove down its length that makes it looks like a giant cork-screw [1161]. The groove twists as a left-handed helix down the tusk. Rarely, males have two tusks, and in those individuals both of the tusks have sinistral grooves [1882], despite the fact that the extra tusk grows out from a *right* canine precursor.

Puzzle 1.7   How do *myoID* mutations cause ~100% reversals in flies?

The ~100% penetrance of the fly's *myoID* mutations poses the same dilemma as the putative ~100% penetrance of the mouse's *inv* mutation (see Puzzle 1.5), because it implies that the wild-type gene is overturning a pre-existing asymmetry, rather than breaking symmetry in the first place. What possible selection pressure could there be for such a reversal [266]? The inference of default states can be tricky, since it depends upon the mutations that are being evaluated. Although *myoID* suggests that the baseline is sinistral asymmetry, the affected organs are visibly symmet-ric before they become asymmetric [738], and the mutant phenotypes of other genes imply a symmetric (or randomized) default state [902,2243]. The *Hox* gene *Abdominal-B* in particular is required for either dextral or sinistral development of the male genitalia [431]. At one time it was thought that the presumptive sin-istral default state might be enforced by *myoIC* [1779], but recent research refutes that notion, and shows that, on the contrary, *myoIC* acts redundantly with *myoID* [1681]. Indeed, it is hard to imagine any myosin defying the chirality of its class to counteract a fellow member [1523,2254]. Thus, the nature of the implied sinistral default state remains a mystery.

---

the *myoID* gene, which rivals the mouse's *inv* in its ~100% potency [1681]. This gene encodes a class I myosin of type D. Class I myosins resemble the familiar myosin of muscles in terms of their structural domains and mode of action [986], but they do not form filaments [1203]. Rather, they ratchet along actin filaments as monomers. To effect net movement of actin filaments, myosins must be anchored to a substrate, and that substrate here appears to be the cell membrane. MyoID binds β-Catenin in adherens junctions [2146], and it requires both DE-Cadherin [1779] and Dachsous, an atypical cadherin, to influence the asymmetry of organs in the abdominal cavity [771].

Mutations in the fly's *myoID* gene cause a pervasive *situs inversus* syndrome of L–R reversals throughout the body (see Puzzle 1.7), with the curious exception of a tiny neural module that functions in long-term memory [738,1737]. However, unlike the situation in mammals, this widespread pleiotropy is not due to a single primary site of action like the mouse's ventral node. Rather, each asymmetric organ appears to have its own dedicated command center [738] (cf. [2181]). Separate L–R "organizers" have been mapped for the midgut [1216,1680,2243], hindgut [771], and genitalia [431,2146].

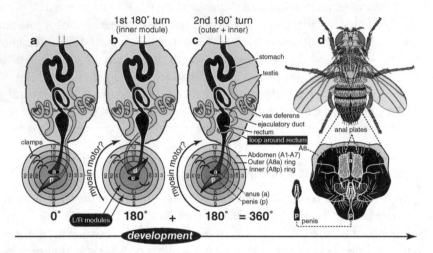

Figure 1.8 Development of L–R asymmetry in flies. Adapted from [928,2215]. Among the various L–R asymmetries in flies [738] the most striking is arguably the 360° rotation of male terminalia (anus and penis), which occurs during the pupal period [2215]. Rotation is clockwise in wild-type pupae but *counter*clockwise in *myoID* mutants (Figure 1.5d) [1681,2146]. Terminalia are here cartooned as archery targets (see key in **c**), with rings denoting abdominal segments 1–7 (A1–A7) and the anterior (A8a) and posterior (A8p) parts of abdominal segment 8. As the terminalia rotate, the ejaculatory duct revolves around the rectum. Reproductive organs (sperm pump and paragonia omitted) are gray; digestive tract is black. Genitalia are drawn as viewed from the rear in **d**. The rotation must somehow be steered by *myoID* [2146], which encodes a class I myosin [1203], but we do not yet know how myosin ratchets the turning [1681]. One model [771] postulates that MyoID alters cell shapes by binding Dachsous (an atypical cadherin), but alterations in cell shape alone cannot explain the shearing movements observed here, which operate like the tumblers of a combination lock.

a. Status before rotation. A1–A7, A8a, and A8p are labeled as clock faces (3, 6, 9, and 12 o'clock), with adjacent rings clamped to one another (arcs).

b. First stage of rotation. The inner L/R module (A8p) detaches (dashed arc) from the outer one (A8a) and turns 180° clockwise (note the upside-down "12"). Detachment requires apoptosis at the interface of the modules [1215,2215]. As a result of the rotation, the anus (a) and penis (p) become inverted because the "bull's eye" is affixed to A8p and turns with it [1972].

c. Second stage of rotation. The outer L/R module (A8a) rejoins (solid arc) the inner L/R module (A8p) but detaches (dashed arc) from A1–7, after which the A8 unit (A8a + A8p) turns 180° clockwise. As a result the original polarity of the anus and penis is restored. Externally, there is no trace, but internally the duct has been wound 360° around the rectum.

d. Dorsal view of adult male (above) and posterior view of anal plates and genitalia (below), which comprise the terminalia. Because the penis is hard to discern amid the surrounding structures, it is redrawn (inset).

Our grasp of the genetics underlying the L–R axis is at least as good for the fruit fly as for any other animal species at the present time [475]. Yet even here, our understanding of L–R evo-devo is rudimentary at best. The rotation of the male fly's genitalia (Figure 1.8) is a case in point. All of the available clues may point to a myosin-based mechanism, but we have no idea how myosin is orchestrating cell movements at the interfaces of the L/R modules so as to drive the rotation with such precision. And our ignorance of causation is as deep at the evo- level as it is at the devo- one: we may be able to explain this peculiar pirouette adaptively in terms of the fickle fads of mating postures in dipteran ancestors [1454,2215], but we still have no idea how the "180° L/R module" evolved in the first place, nor how it duplicated to achieve the full 360° twirl [928]. The challenges that lie ahead will test the mettle of even the most intrepid evo-devotees.

Nevertheless there is one major achievement worth savoring at this point. We have succeeded in seeing gross anatomy from the perspective of the genome, and it conforms nicely with our comfortable intuitions about the world in general. The ABCs of animal assembly turned out to be just the XYZs of Cartesian geometry. Evolution really did quilt the body as Leonardo drew it – along orthogonal axes. Indeed, the independence of these axes may have accelerated bilaterian evolution, because changes could be made along one dimension without altering the other two. We can console ourselves with the realization that the puzzles we face are tractable. There will be no Einsteinian corrections for the Newtonian simplicities we've found. We have attained a durable understanding of the basic architecture of the scaffold. All we need to do from here on out is to put some flesh on those bones, so to speak.

2   Nervous System

All 28 bilaterian phyla have a nervous system [2003], and all but two of them have sufficient aggregations of neurons to qualify as having a brain of some sort [562,918,1625] (but cf. [1559,1651]), so it would seem safe to conclude that urbilaterians must have had a brain also [963]. The two exceptions – Hemichordata and Echinodermata – were once thought to be basal deuterostomes [968], but the growing consensus is that their diffuse neural nets are actually derived simplifications [948,960,963,1492,1725]. Unfortunately, fossils are of little help in settling this dispute [363] because neural tissue of the right vintage (Cambrian or Precambrian) is only rarely preserved [286,1376,2489], so we must turn to other evidence of a comparative nature [79].

A common matrix of transcription factors subdivides the CNS

The human brain has nearly a million times more neurons than the fly brain (86 billion vs. 100,000 [119,1179]), but it uses the same Cartesian grid of gene expression to guide its construction as its Lilliputian counterpart [86,963]. That grid is charted in Figure 2.1. Some of those genes also govern the body axes (see Chapter 1) [335] – implying that the first central nervous system (CNS) may have merely adopted the body's pre-existing coordinate system [963]. Curiously, the opposite appears to have been true for *Hox* genes: their original role seems to have been to balkanize the CNS into modular precincts [513,746,939,1971], with their more well-known chores along the body's A–P axis as a whole evolving later.

The districts mapped in Figure 2.1 are also largely shared by annelids [505,2174], so this ground plan spans all three main branches of the bilaterian tree (Deuterostomia, Spiralia, and Ecdysozoa) [963,1662]. The baroque intricacy of the map seems sufficient to rule out convergence as an explanation for the congruence [1884,2198]. Hence, we can reasonably conclude that the nervous system of urbilaterians was similarly regionalized and centralized [81,602,2157,2257]. If so, then that primordial CNS must have later undergone simplification in certain descendant clades (see Puzzle 2.1).

Virtually all of the shared genes in the coordinate grid encode transcription factors [2158,2470], so they could theoretically offer a Cartesian (x, y) framework for "painting" neurons with distinct identities in terms of morphology, connectivity, neurotransmitter, etc. [48,1227,2082]. Is there any evidence for such a combinatorial code? Yes, there is. The most direct confirmation comes from peptidergic

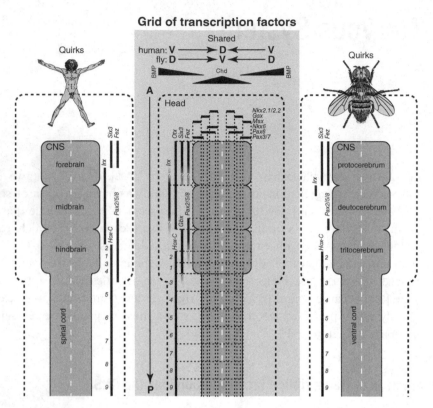

Figure 2.1 Gene expression in the central nervous system (CNS) of mammals and flies. The dark gray "totem pole" in the light gray rectangle is a consensus map of the urbilaterian CNS that is based on gene expression zones (bars) in mice, flies, and the annelid *Platynereis*, which represent all three main trunks of the bilaterian tree – Deuterostomia, Ecdysozoa, and Spiralia, respectively [963]. The CNS is drawn as if the head (dashed outline) were fileted, flattened, and pinned to a dissecting tray [84] . . . or alternatively, in the case of the mouse, as if the neural plate were seen from above before it rolls into the neural tube [345]. The ventral (V) midline of vertebrates corresponds to the dorsal (D) midline of flies due to the inversion of a basal chordate (cf. Figure 1.4). However, brains are not perfect cylinders, so the (x, y) zones of the urbilaterian were surely not as rectilinear as drawn here [2294], nor would all of the stripes necessarily have spanned the CNS. Moreover, the third ("z") axis of animal brains (not shown) emerges via a variety of time-dependent rules (cf. Figure 2.3) [336], reminiscent of how skyscrapers rise above the street level in Manhattan as they're built, one story at a time [2192,2251,2253].

Triangles denote gradients of the morphogens Bone Morphogenetic Protein (BMP) and Chordin (Chd) along the D–V axis (cf. Figure 1.4). Gradients are not drawn for the A–P (anterior–posterior) axis because the original A–P morphogens are unknown (see Chapter 1). Flanking diagrams show differences ("quirks") between mice (≈ humans) and flies that blur the termini of the corresponding bars (denoted by fading and a question mark) in the center (urbilaterian) map. The uncertainty in the *Gbx* bar is not actually due to a mouse/fly disparity, but rather to annelid data (not shown) [963]. Vertebrates have a conserved axon scaffold that is partly based on this map [2385].

Wherever the names of homologs differ (*vnd*/*Nkx2*, *ind*/*Gsx*, *msh*/*Msx*, *HGTX*/*Nkx6*, *ey*/*Pax6*, *paired*/*Pax3*, *otd*/*Otx*, *optix*/*Six3*, *earmuff*/*Fez*, *mirror*/*Irx*, *shaven*/*Pax2*, *unpg*/*Gbx*), the murine name is used. All of these genes encode transcription factors [152,448],

---

Puzzle 2.1   Did hemichordates and echinoderms "lose their minds"?

If, as seems likely from the data in Figure 2.1, urbilaterians had a CNS that was more complex than the nervous systems of hemichordates and echinoderms, then we are faced with the riddle of why those two groups forfeited most of their brains [292,1953]. They are sister phyla [701], so perhaps the CNS was modified in their common ancestor? Probably not, since their nervous systems differ starkly from one another [961,1438,1642]. In echinoderms the revamping of the CNS occurred as the body was retooled from bilateral to radial symmetry [56,1072,2526] to suit a sedentary lifestyle [2119]. Indeed, they seem to have lost their *Hox* organization for the same reason (see Figure 1.3) [1539]. Hemichordates instead adopted a burrowing lifestyle [1220], and in their case the ancient grid of transcription factors is still present, but it is tattooed into their *skin* as well as into their neural network [968,1357]! This fact has led some authors to propose that chordates imported the ectodermal platform of a hemichordate-like ancestor into their CNS [1725], rather than the other way around. However, that idea has been met with considerable skepticism [183,1642], and it has been thoroughly rebutted by Linda Holland in an incisive exegesis of the evidence [963].

---

neurosecretory cells that invariably arise at the intersection of *Six3*-expressing (A–P) and *Nkx2.1*-expressing (D–V) zones in vertebrates (hypothalamus), insects (pars intercerebralis), and annelids (medial prostomium) [531,882,1662,2174,2263]. Amazingly, 27 different families of neuropeptides have been traced back to the urbilaterian nervous system. In the quote below "pNPs" are proneuropeptide precursor

---

←

Figure 2.1 (*cont.*)

and some of them (*Nkx2*, *Gsx*, *Msx*, *Nkx6*) came from ancient *Hox*-related clusters [970,1247]. *Hox-C* stands for the *Hox* complex (cf. Figure 1.2), with individual genes numbered (e.g., *1* = *Hox1*) to mark the anterior extent of their (overlapping) expression ranges. The rule of *Hox* colinearity, which is obeyed by ectoderm along the body's A–P axis in both flies and mice (cf. Figure 1.2), is disobeyed here insofar as *Hox2* is expressed anterior to *Hox1*.

  Redrawn from [963] with *Hox* expression zones from [45,464,950,1150,1640,2157]. Similar trans-phylum comparisons have been published before [949,968,1092,1313,2318,2470]. Ref. [130] depicts these regions in three dimensions, and [1999] shows them in relation to the vertebrate neural plate and its outer rim of sensory placodes. For neural compartments and organizer boundaries (morphogen sources) see [1150,1151]. For the neural circuitry of locomotion and its corollary riddles see [522].

  Author's aside: The evo-devo community owes a huge debt of gratitude to Detlev Arendt and his research group in Heidelberg for raising the obscure annelid *Platynereis* to the status of a "workhorse" model organism [412]. Without this ability to tap so deeply into the heart of the spiralian clade, no sound conclusions like those depicted here [89] could ever be reached about our shared bilaterian ancestor [2095,2294]. Chordate–arthropod (deuterostome–ecdysozoan) comparisons alone may be expedient, but they are ultimately insufficient.

proteins from which neuropeptides are enzymatically cleaved, and "GPCRs" are G-protein-coupled receptors:

> The combined analysis of pNPs and neuropeptide GPCRs identified 27 ancestral urbilaterian pNP-receptor families pointing at a hitherto unknown sophistication of neuropeptidergic systems in the urbilaterian. These pNPs regulate several aspects of physiology, including sexual behavior and reproduction (GnRH, achatin, oxytocin, and GnIH/SIFamide), diuresis (CRF/diuretic hormone, calcitonin, and vasopressin), gut and heart activity (achatin, luqin, and orcokinin), pain perception (opioid), and food intake (NPY, kinins, neuromedin-U, galanin/allatostatin-A, and orexin/allatotropin). These pNPs may have originated concomitantly with the origin of a complex bilaterian body plan having a through gut with novel controls for food intake and digestion, excretory and circulatory systems, light-controlled reproduction, a centralized nervous system, complex reproductive behavior, and learning. The stable association of receptor–ligand families across bilateria revealed the long-term coevolution of receptor–ligand pairs. [1068]

Additional molecular fingerprints allow us to deduce further structural and functional subdivisions within the urbilaterian brain that are based on a common set of neurotransmitter molecules (Figure 2.2) [81,1492,2294]:

1. A medial serotonergic compartment that is devoted to locomotory control [86,505].
2. A paraxial cholinergic sector of efferent motor neurons [81,505,2275]. Fly motor neurons rely on the same regulatory genes (hb9 and Nkx6) as vertebrates and annelids [260,359,1227,1241,1977], but they are glutaminergic instead [1092].
3. A dopaminergic module that became the basal ganglia of vertebrates and the central complex of arthropods [643,1318,2197], both of which integrate sensory inputs into coordinated motor outputs [218,1433,2178].
4. A glutaminergic region that became part of the cortex (hippocampus?) of vertebrates [558] and the mushroom body of arthropods [2453] – both of which store memories of food smells [2290]. The memory circuitry for flies is known in great detail [107,108,2342].

Neurotransmitters

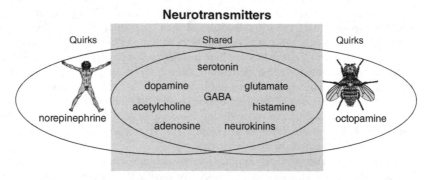

Figure 2.2 Neurotransmitters shared by humans and flies. Venn diagram of major neurotransmitters that are either jointly used by humans and flies (overlap zone; gray box) or not shared (labeled as "quirks"), based on [1618]. More than 100 neurotransmitters are known in animals [984,1009], and many of them can be traced back to our urbilaterian ancestor [1068]. GABA stands for γ-aminobutyric acid.

Those shared molecules and modules could theoretically enable neurons from distant phyla to communicate with one another if we were to artificially juxtapose them in a laboratory setting. Puzzle 2.2 offers such a fanciful thought experiment for the sole purpose of exploring this possibility and demonstrating that the truth in this case may be considerably stranger than fiction.

Despite the extensive homology that bilaterian nervous systems appear to share at the genetic level, there is a baffling diversity of developmental mechanisms at the cell and tissue level. The breadth of that diversity is amply documented in Figure 2.3 [884]. One of the most dramatic examples concerns how cells sink beneath the surface of the gastrula to become neurons. Chordates accomplish internalization via neurulation (invagination of a solid swath of ectoderm along the A–P axis) [1359], whereas flies carry out this rearrangement mainly by a process of salt-and-pepper ingression (detachment of individual neuroblasts, one at a time, from the ectoderm) [2167].

Does this panoply of morphogenetic variability undermine our conclusion that bilaterians inherited a common CNS coordinate system? Not necessarily. After all, we take for granted a comparable span of variability in *gastrulation* movements among the vertebrate classes themselves (e.g., invagination in frogs but ingression in birds) [87,2126], yet the final result of those processes is always the same: a conserved, two-ended digestive tube [85]. It is a truism that as body geometries change over evolutionary time, new mechanical constraints can force cells to pursue alternative routes to the same anatomical end [372,1137,1208,1501,1604], and the CNS may just present us with one more example of this trend [666,2327].

Moreover, the human–fly disparity in CNS formation is not actually as stark as it first appears: a large part of our nervous system comes from neural crest cells that delaminate more or less like fly neuroblasts [2268], and certain parts of the

---

Puzzle 2.2   To what extent could xenogenic neural circuits interact?

Given that flies and humans use such similar CNS area codes (Figure 2.1) and neurotransmitters (Figure 2.2), is it conceivable that parts of a fly brain could interface with parts of a human brain? Imagine, for example, that we could somehow graft the part of a male fly's brain that governs its courtship ritual onto a man's motor cortex – setting aside the obvious ethical reasons why such an experiment should never be done [622]. Would this "Frankenfly" vibrate his arms as if they were wings [1961] whenever he saw an attractive woman? As silly as this scenario sounds – though it does not depart much from the plot lines of the 1958 horror movie *The Fly* with Vincent Price and its 1986 remake with Jeff Goldblum – transplants that are comparable to it have actually been performed in birds [1120,1258]. To wit, when the rear of the quail midbrain is grafted in place of its counterpart in a chick embryo, the ensuing "quicken" looks like a chicken but sings like a quail [138].

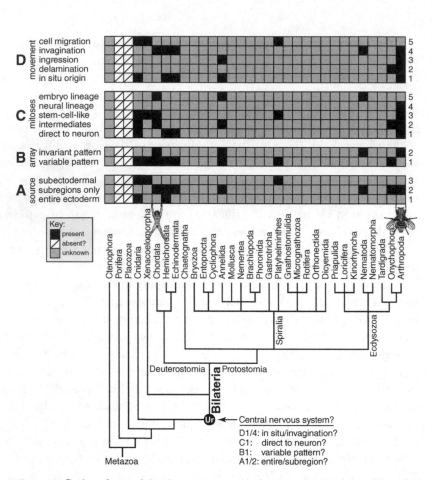

Figure 2.3 Styles of neural development among animal phyla. The four-part classification scheme presented here was devised by Volker Hartenstein and Angelika Stollewerk [884]. Styles of neurogenesis are categorized in terms of: A (whether neurons come from the entire ectoderm, from subregions thereof, or from cells beneath it), B (whether or not neural progenitors always arise in the same pattern), C (whether neural progenitors make neurons directly or indirectly via intermediates, whether the intermediates divide in a stem-cell-like way, and whether neurons follow a strict pedigree from their progenitors or within the embryo as a whole), and D (whether neurons stay superficial or move inside via delamination, ingression, invagination, or migration). Black squares signify usage of the designated style in at least some members of the indicated phylum (but not universal presence throughout the phylum), diagonal lines denote its absence in all species examined thus far (though future analyses of more species could show otherwise), and gray squares mean that available data are insufficient to draw any conclusion at present (see key).

To see how this scheme works, consider the chordate column. It was compiled as a pooled composite of cephalochordates (A2, B1, C2, D4), urochordates (A2, B2, C5, D4), and vertebrates (A2, B2, C2/3, D4), so the squares that are blackened should not be interpreted as applying to all of the species in this phylum. Indeed, it is remarkable that urochordates build basically the same CNS as vertebrates via a derived (nematode-like) fixed-lineage (C5) mechanism. Likewise, different groups of arthropods reach the same endpoint by entirely distinct origami tactics [2191,2192]. The conjectured parameters of the urbiliaterian nervous system (bottom right) are inferred from the above distributions and based on the analyses of Hartenstein and Stollewerk [884]. Letters are classes (upper left), and numbers are subclasses (upper right), with slash marks indicating ambiguity. Tree of extant phyla as in Figure 1.3.

fly nervous system (the stomatogastric and neuroendocrine moieties) invaginate somewhat like our neural tube [489,886] (cf. neural invaginations in other arthropod groups [2191,2192]). Even within our own phylum the urochordate *Ciona* somehow managed to reduce neurulation to a "digitized" neuroblast-like array without altering the ultimate anatomical outcome:

> In the urochordate *Ciona intestinalis*, neurogenesis follows a determinate lineage pattern that, similar to that found in the nematode *C. elegans* and the leech *H. robusta*, is considered highly derived. *Ciona* larvae are freely swimming organisms shaped similar to a vertebrate tadpole and possess a small number of cells that, from the beginning of cleavage onward, are all produced in a fixed lineage mechanism. The neural plate consists of 48 uniquely identifiable neural progenitors that are arranged in a regular orthogonal pattern, forming four columns and six rows. Following neurulation, these progenitors undergo another two to three rounds of mitosis before a subset of them delaminates to form the invariant, small set of larval neurons. [884]

Another iconic feature of chordate CNS development, aside from neurulation, is the induction of neurulation itself by the underlying notochord – the axial rod that gives our phylum its name [1987]. Amazingly, however, molecular profiling suggests that the notochord may not be a novelty of our phylum after all [70,916]! Our notochord is apparently homologous to the annelid axochord [1253]. Phyletic surveys confirm this notion, and further imply that urbilaterians had a medial muscle that evolved into both the notochord and the axochord in separate descendant clades [274]. No vestige of such a muscle has yet been detected in Arthropoda, so it may have been lost from that phylum, and the same is true for Nematoda. Indeed, only a few anatomical hints have been uncovered elsewhere in the ecdysozoan clade, so the antiquity of this structure and its import for CNS induction [1711] are unclear at present [274].

Neurons make excessive connections that are later pruned

In order for neurons to connect properly with one another, the tips of their growing axons must wend their way through a tangled jungle of other cells and axons to find their intended partners [373]. During this precarious journey the axons typically rely only on local cues, without the benefit of any sort of global positioning system. Not surprisingly, this haphazard process of blind searching often results in faulty wiring [393], and animals have evolved various tricks to maximize accuracy and minimize damage.

One trick that animals commonly use for editing errors is to produce an initial excess of neurons. This "Shotgun Strategy" ensures that at least some of the axons in the fusillade will hit their targets, and those axons that fail to do so can subsequently be pruned [1106,1181,2359], with or without the destruction of the cell bodies from which they emanated [436,1897,2502].

Developing neurons typically suffer colossal carnage during the pruning phase. Rates of cell death can reach 50% in some parts of the CNS [495,1690] in both mammals [2144] and flies [1926]. Excess neurons normally kill themselves by a ritualized process called apoptosis [2485] after the Greek word for dead leaves falling from trees [2444].

The steps of this suicide pathway are highly conserved (Figure 2.4) [55,77,102,696], and the same steps are used for the selective pruning of axons (without the death of the cell body) [495]. The winnowing process is inherently Darwinian because only the "fittest" (best matched) connections survive [581,1118]. The net result, ideally,

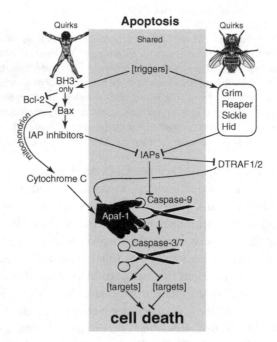

Figure 2.4 Components of neural apoptosis shared by mammals and flies. Homologous proteins are inscribed in the gray rectangle, while analogous agents are listed as "quirks" in the margins. Mammalian names are used for shared proteins; names of the fly homologs (in parentheses) are: IAPs (Diap1), Apaf-1 (Dark), Caspase-9 (Dronc), Caspase-3/7 (DrICE and Dcp-1). Abbreviations: Apaf-1 (Apoptotic peptidase activating factor-1), DTRAF (*D. melanogaster* tumor-necrosis factor receptor-associated factor) IAPs (Inhibitors of Apoptosis Proteins) [1665], and Diap1 (*Drosophila* IAP-1).

"BH3-only" denotes BH3-only proteins (upper left) [411]. Arrows indicate activation; T-bars denote inhibition. Triggers for neural apoptosis (top) tend to be intrinsic to the cells themselves [2485], but they can arise externally [106]. Caspases are depicted as scissors because they are cysteine proteases that cleave target proteins at specific sites [1217,2500]. Those targets (bottom) either stimulate apoptosis in their cleaved state (left) or block apoptosis in their uncleaved state (right) [1217]. Apaf-1 (schematized as a hand using the scissors) binds Caspase-9 to form an utterly fascinating wheel-shaped "apoptosome" [1217], whose exact structure varies among phyla [2517,2518]. Release of Cytochrome C from the mitochondrion is an integral step for mammals but not for flies [142].

This diagram is excerpted from the schematic in [2485], which compares *Drosophila* to mammals. The colorful names for the fly's IAP inhibitors (Grim, Reaper, Sickle, and Hid) [208] are typical of the whimsical (and in this case rather macabre) names that fly research-ers have coined for mutants over the years. The core pathway of programmed cell death is remarkably conserved among animal phyla [102,411,603,696,2286], but it is nevertheless unorthodox when compared to most other cellular signaling pathways because it relies on a protease cascade (cf. blood clotting [59]) instead of a kinase–phosphatase cascade [1008], although certain caspases are also regulated by phosphorylation [1217].

is a brain full of finely tuned neural circuits for sensory input, central processing, and behavioral output [495]. The overall mechanism is clearly not perfect, however, given all the congenital neurological disorders that afflict our species.

Do flies suffer from such disorders? Surprisingly, they do [1618,1679,2321,2494]. Insect behavior was widely thought to be instinctively reflexive [806] until Karl von Frisch showed that bees can teach their hivemates the locations of nectar sources through waggle dances [1483,2346], and Seymour Benzer proved that flies can associate smells with shocks via Pavlovian conditioning [107,186,1191]. As more cognitive abilities are documented with each passing decade [381,1433,2162], we have been forced to see insects less like robots and more like ourselves [805,806,1936], though it remains unclear how many of their brain functions have evolved independently [626,2454].

One brain function turns out to be nearly universal: long-term memory relies on the cAMP pathway across the animal kingdom [1103]. Flies actually resemble humans even further insofar as their long-term memories are consolidated every night, as ours are, during episodes of deep sleep [188,527,903] that we typically experience as dreams [1125,1481,2062]. Moreover, flies react to mind-altering drugs almost as we do [2452], and some of their behaviors are so much like our own that they seem to mock us (see Puzzle 2.3).

The anthropomorphic attributes of fly behavior were charmingly portrayed in Pulitzer-prize-winner Jonathan Weiner's 1999 book *Time, Love, Memory* [2404]. But the most memorable paean ever written to the similarities of humans and flies

---

Puzzle 2.3   What does memory storage have to do with alcohol sensitivity?

Ethanol affects flies as it does us [823]. Tipsy flies stumble as we do, so researchers have been able to devise clever devices that can sort flies according to their tendency to topple over at a given dose of ethanol [1540] and thereby screen for mutants with higher or lower sensitivities. The mutations that have been recovered in this way turn out to often disable genes in the cAMP signaling pathway that mediates learning [2452]. No one knows why these two aspects of CNS operation (alcohol sensitivity and memory capability) should be related [1916], nor do we even know how or why ethanol impairs balance in the first place [1172].

We are also clueless about how liquor affects libido. Pathetically, what is true for drunk men is equally true for drunk male flies: they are more prone to be aroused sexually but less able to consummate the act itself [1264]. Frustrated male flies also resemble men insofar as they tend to drown their sorrows in alcohol after being rejected by females [1317,1915,2084].

How such a simple molecule (C_2H_6O) can have such a huge impact on so much of both human and fly behavior remains a riddle [1321,2145], but ethanol is not alone is this regard. We also have yet to fully comprehend the anesthetic effects of ether ($C_4H_{10}O$), chloroform ($CHCl_3$), and nitrous oxide (N_2O) [821,1742].

was published 205 years earlier by the poet–artist William Blake (1757–1827). He included the following ode, simply called "The Fly," in his philosophical collection *Songs of Innocence and of Experience* (1794):

Little fly,
thy summer's play
my thoughtless hand
has brushed away.

Am not I
a fly like thee?
Or art not thou
a man like me?

For I dance
and drink and sing,
till some blind hand
shall brush my wing.

If thought is life
and strength & breath,
and the want
of thought is death,

then am I
a happy fly,
if I live,
or if I die.

Coordination between left and right involves commissural axons

Bilaterally symmetric animals must coordinate the two sides of their body in order to swim, walk, or fly efficiently, and the mode of locomotion is typically regulated by a central pattern generator of some sort [679,1119,2052]. A vivid demonstration of such regulation comes from a cute experiment with chick embryos. When posterior spinal tissue that controls the legs was replaced by anterior spinal tissue that controls the wings, the chicks that hatched moved their legs in unison as if they were trying to flap them in a vain attempt to fly [464,1596]. In fruit flies the circuit that flaps the wings during flight is sensibly separate from the one that alternately flicks the wings when a male courts a female [1961].

In order for a central pattern generator to choreograph muscle contractions on the two sides of the body, it must rely on hard-wired connections that span the midline [335,827]. Such lateral connectives are abundantly apparent in the flatworm CNS, where transverse commissures (≈ rungs) are spaced at regular intervals in a lovely, ladder-like array [1892].

The functional need for lateral connectives is vividly shown by mice whose V0 class of commissural interneurons has been ablated within the ventral spinal cord.

Such mice can no longer walk by alternately stepping as they normally would with their left and right legs in turn, and instead they hop like miniature rabbits [178,1845,2231].

A comparable "mirror syndrome" has been described for people who are congenitally missing their transverse connectives [2409]. Such mutant individuals tend to jointly move their arms in ways that are reminiscent of how babies flap their arms when they're excited [665,1763,2164]. Mothers tend to take this quirk for granted, but it is surely hinting at something rather profound about how our CNS installs locomotory modules as we develop.

Despite all of these various indications of the utility of trans-midline wiring, the largest bundle of transverse connectives in the human brain turns out to be not nearly as essential as we might have thought (see Puzzle 2.4) [1763].

The two halves of the bilaterian CNS are virtually identical anatomically [371] despite the Nodal marker that is used by other organs within the body to establish overt left–right asymmetries of various sorts (see Chapter 1). Thus, any commissural axon that happens to take a wrong turn when it gets to the midline would have no obvious way of telling whether the neurons in its vicinity are contralateral targets or ipsilateral imposters [648].

Puzzle 2.4   What role does the corpus callosum play in mammals?

Along the midline of the human brain is a thick bundle of axons that cross between our two hemispheres [1869]. This is the corpus callosum, and it was made famous in the 1960s by the split-brain experiments of Roger Sperry and his team [2064]. Sperry studied patients whose callosum was surgically cut in half in an attempt to cure their epilepsy [1732].

Astoundingly, these people manage quite well postoperatively without having any callosum at all [104,169]. Only when they are subjected to unilateral stimuli in a laboratory setting do they reveal subtle asymmetries in perceptive ability [615]. Hand coordination is only mildly impaired [775], and walking appears not to be affected at all [2325], presumably due to the pattern generator for bipedal walking being located in the spine rather than in the brain [335,569,2409].

How on earth can such a fat fascicle with ~250 million axons [775] be so functionally dispensable [358]? Might it have evolved merely to fine-tune the manual dexterity of tool-using proto-humans [1715,1763]? No! It arose ~100 MY ago in the stem line of placental mammals [6,1869], though its original role remains obscure [1493]. Even so, the callosum does seem to have acquired new duties as our hominin clade evolved: our ability to learn languages appears to require left–right linkages, and defects in those circuits may be partly responsible for congenital dyslexia and related reading disabilities [2323,2373].

After an axon crosses the midline it can never go back again

Urbilaterians appear to have solved the problem of telling an axon which side of the midline it is on by inventing a foolproof trick that has endured ever since in virtually all of its descendant phyla. Instead of waiting to *correct* errors after they occur, as in the "Shotgun Strategy," the CNS *prevents* them from occurring in the first place by what might be called a "Passport Strategy." First, midline cells emit signals that either attract or repel commissural axons [1107,1108]; axons are initially sensitive to attractants but immune to repellents [1333]. Then, whenever an axon crosses the midline its receptors are altered, like having your passport stamped at a customs checkpoint [358], so that it now responds in the opposite way (repulsion instead of attraction), hence causing it to avoid the midline [1605,2529]. From this point on, the axon may still lose its way, but at least it will avoid recrossing the midline accidentally and synapsing with the right cell on the wrong side [1108].

The need for this fail-safe mechanism becomes obvious from the chaotic wiring that ensues when the passport bulwark malfunctions. The gene *roundabout* (abbreviated *robo*) gets its name from the misrouting of commissural axons in loss-of-function mutant flies: the axons cross the midline but then circle back, only to cross repeatedly like cars trapped in a traffic roundabout [2043]. The protein encoded by *robo* is a cell-surface receptor that binds a ligand called Slit [264,835]. Slit is a diffusible repellent [1852,2102] that is secreted by midline cells [1149,2552]. It overrides the effects of attractants such as Netrin [200,523,2172].

The Slit–Robo contraption has been documented in vertebrates, nematodes, and planarians, as well as in flies [609], though details vary among the various taxa. In flies, for example, the Robo1 receptor is disabled in pre-crossing axons by an endosomal sorting protein (Commisureless) [755] and by heterodimer formation with a paralog (Robo2) [609,1977,2152]. In mice, on the other hand, the Robo1 and Robo2 receptors are blocked by a Robo3 paralog expressed as a Robo3.1 splicing isoform [1065], and axons acquire the ability to detect Slit after crossing causes Robo3.1 to be replaced by a permissive Robo3.2 isoform [214,365]. In neither case do we know how axons switch their states in such an all-or-nothing manner [50,625].

In the mammalian visual system, ingrowing retinal axons get segregated to either the same (ipsilateral) or the opposite (contralateral) hemisphere upon reaching the midline [600], but the decision as to whether a retinal axon crosses there is not handled by the Slit–Robo device, but rather by a separate (Zic2–EphB1) ligand–receptor pathway [717,938,1864]. Some crossing of the midline is required for higher centers to compare left and right images [1237,1807] so as to compute distances to objects in the visual field [145,2120], but even those vertebrates that totally lack stereoscopic vision send retinal axons across the midline [1250,1776]. A priori, we might have predicted the primitive state to have been a 100% *ipsilateral* projection (after all, why make axons travel further than they have to?), but it appears instead to have actually been a 100% *contralateral* projection. The adaptive benefit of this crisscrossed wiring remains uncertain (see Puzzle 2.5).

Puzzle 2.5   How did the vertebrate CNS get crisscrossed?

Chiasms are X-shaped crossings of axons named for the Greek letter *Chi* [2356]. The optic chiasms of vertebrates facilitate depth perception by allowing over-lapping visual fields to be compared by centers that get input from both eyes [145,2120]. Optic chiasms therefore make sense for animals that have binocular vision (e.g., primates) [830], but paradoxically, they also exist in walleyed ver-tebrates, which shouldn't need them (e.g., fish) [290,1890]. Indeed, basal verte-brates appear to have had a 100% *contralateral* projection [1067,1776], despite each retina being an outgrowth from the *ipsilateral* side of the brain [781]. Why?

One clue to this mystery is that the crisscross tactic is not confined to the visual system but applies to the motor control system also. For example, it is widely known that strokes in the left hemisphere impair the right side of the body and vice versa [651]. The reason for this peculiar asymmetry is that pre-motor axons cross the midline in the corticospinal tract [1276,1432] just below the medulla [1437,2409,2443]. Crossed wiring has no obvious benefit for humans [131], but it may have given our fish ancestors an escape reflex [1114,2356] wherein detection of a predator on one side causes contraction of flank muscles on the other side [334,1228,1813] so as to turn the body away from harm without any pause for conscious thought [321,1979,2092]. This scenario sounds plausible, but it raises the question of why the CNS of more basal chordates (e.g., amphioxus) is not also crossed [2216], given that they also swim like a fish [1162,2076].

Great strides have been made in the tracing of wiring diagrams for brain circuits in mice (e.g., [1074]) and flies (e.g., [107]). However, it is important to realize that genomes do not encode blueprints per se [542]. Rather, neural networks are built from the bottom-up perspective of individual cells, not from the top-down per-spective that an electrical engineer would employ [2502]. Individual axons are pro-grammed to take certain actions at certain times in certain contexts [1412], and the conditional rules take the form of "If this happens, then do that" [895,1243]. Axons execute internal commands from their own cell nucleus [297], but they also navigate via external cues from surrounding cells [1815,2528]. Once we shift our focus from the static product to the dynamic process that leads up to it, we can start to see how critical the timing of events can be for the wiring of neural circuits [2502]. It is here that evo-devo offers some useful insights.

The fly's visual system is a case in point. Flies lack the sort of *midline* chiasm that characterizes vertebrates (see Puzzle 2.5). Instead they have *lateral* chiasms that flank the midline, and the same is true for other insects and for most crustaceans [891,892,2195]. The chiasms in all of these cases occur in the following series: retina → lamina → X → medulla → X → lobula, where "X" marks a chiasm, and the lamina, medula, and lobula are ganglia in the brain's optic lobe. The two sequential chiasms arise via the kind of polarized growth that occurs in the fly's

> ## Puzzle 2.6   How essential are polarized mitoses for neuronal wiring?
>
> Conventional wisdom has long held that asymmetric mitoses are essential for CNS development in flies and mammals [2450,2487] and other animals [2351,2407], because polarized cell divisions are so pervasive in neural differentiation [516]. However, in 2006 a surprising report contradicted this paradigm, and the ensuing paradox has not yet been resolved. Indeed, the discovery was shocking at so many levels that it rocked the world of cell biology as much as it did the realm of neuroscience [1485]. The paper's title discloses the nature of its heresy: "Flies without centrioles" [157]. Really? How can any animal develop without such a key cog in its mitotic machinery? Yet mutant flies lacking centrioles develop virtually normally until their exit from the pupal case, when they collapse because they lack the cilia from the stretch-sensing proprioceptors that are needed to keep their balance (chordotonal organs; see Figure 4.5d) [124]. Nor do these flies have any obvious brain defects . . . despite that fact that ~30% of their neuroblasts fail to undergo the very asymmetric mitoses that had been thought to be so necessary for neural circuitry.

retina [1985,2194]: (1) mitoses within each ganglion occur mainly at one end, (2) newborn neurons in each ganglion connect to nascent neurons in the adjacent ganglion, and (3) chiasms occur wherever adjacent ganglia grow in opposite or orthogonal directions [589,891,1753,1792]. Hence, they are a natural outcome of organ geometry and growth polarity, despite looking rather counterintuitive to us (see Puzzle 2.6). As for where these chiasms came from, our best guess is that they evolved when a precursor ganglion underwent duplication, delamination, and reorientation of its derivative layers [2078].

Neurites emanating from the same neuron repel one another

In order for neurons to weave themselves properly into the fabric of a circuit, they must have a way of distinguishing the dendrites of their target neurons. Historically, the most influential theory for how neurons recognize one another has been the Chemoaffinity Hypothesis proposed by Roger Sperry in 1963 [2150]. Sperry imagined that (1) molecular tags decorate the surfaces of neurites (i.e., axons and dendrites [709,1268,1280]) and (2) these tags must vary from one neuron to another. When an axon encounters a dendrite, he argued, it would only form an enduring synapse if its "key" fit the "lock" of that dendrite. Ever since this clever conjecture was published, neurobiologists have been searching for evidence of the tags that Sperry predicted.

Some of us thought that this Holy Grail had been found when reports were published about a strange fly gene called *Dscam1* [1001]. *Dscam1* is expressed in neurons, and its protein product is present on the surfaces of their neurites.

So far so good, but what really astounded everyone was the potential variablity of the Dscam1 protein itself [2006,2451]. The trick that *Dscam1* uses to generate such variants is alternative splicing [1266]. Four of the gene's 24 exons have a set of alternative forms, but only one variant within each set remains in each cell's mature mRNA. The process is like dealing a poker hand by picking one card from each of four decks. The number of possible combinations of "cards" – literally protein isoforms – can be computed by multiplying the number of options among the decks [2451]: 12 (exon 4) × 48 (exon 6) × 33 (exon 9) × 2 (exon 17) = 38,016 possible versions of the overall protein, though the true number of possible ectodomains is half this number (19,008) because exon 17 encodes the transmembrane domain.

A similar exponential combinatorial logic had earlier been found for neurexin and neuroligin proteins in vertebrates [2223], but our hopes that they might be Sperry's elusive locks and keys [1511,2313] were dashed when in-vivo knockouts of their respective genes were shown to have no functional effect [1206], though there is some recent evidence that might revive the idea [1851,2015].

Our initial hopes for *Dscam1* likewise turned out to be wishful thinking [2554]. *Dscam1* does endow each neuron with a unique identity, but that identity is *not* used for selecting a mate [990]. Rather, it is used for preventing dendrites of the same neuron from getting intertwined with one another [1778]. When any two dendrites bearing the same Dscam1 "barcode" (i.e., sister dendrites from the same neuron) come into contact, homophilic binding causes them to subsequently repel one another [1445]. In this way the dendritic tree of a neuron is forced to spread more broadly so as to cover more territory [2551], and axons use the same trick to maximize the distribution of their collateral branches [2372].

Vertebrates use the same basic strategy (albeit without alternative splicing) but employ the protocadherin gene family instead of *Dscam1* [721,1267,1947]. These respective neurite "self-avoidance" devices are therefore convergent, rather than conserved [1517,2554].

Even so, vertebrates do have *Dscam* homologs that perform functions similar to those of *Dscam1* [1306]. No gene or gene family has yet been found that exactly fulfills Sperry's prophecy [2308], though Ephrins and their receptors do come close [945,1169]. Absence of evidence may not be evidence of absence, but it still seems prudent to begin to rethink our notions about neuron matching [2554]. Meanwhile, we can console ourselves with the richness of the *Dscam* story itself, which offers a cornucopia of riddles worth thinking about (see Puzzles 2.7–2.10).

Puzzle 2.7   What role does *Dscam* play in Down syndrome etiology?

"*Dscam*" stands for *Down syndrome cell adhesion molecule*. Humans have a *Dscam* homolog on our chromosome 21 [988], and its extra dosage in trisomy-21 individuals may be contributing to the mental deficiencies that are associated with this disorder [697,760]. However, we do not yet know which neural pathways are affected.

Puzzle 2.8   What was the primordial function of Dscam in Bilateria?

Dscams play various roles aside from neurite self-avoidance [91,258,899,2005], but we do not yet know which of these duties came first [797,2005]. For example, Dscams in the CNS facilitate axon pathfinding at the midline [1331] by serving as netrin receptors [65], and Dscams in the *peripheral* nervous system enable retinal neurons to undergo axon tiling [1499], cell spacing [698], and laminar selection [2481]. Immunological functions for Dscams exist in *Drosophila* [2397], mosquitoes [535], and other arthropods [1617,1716], though it is not yet clear whether the *Dscam* exon repertoire is used for antigen recognition in the same way as the mechanism that produces vertebrate antibodies [92,534,2546].

Puzzle 2.9   How did Dscam evolve so many cute little jigsaw puzzles?

When sister neurites meet, their outstretched Dscam1 proteins "shake hands" across the intervening gap [2005]. In such an antiparallel orientation, the overlapping amino acid sequences must be complementary in both shape and charge in order to mediate homophilic binding via non-covalent interactions [1109]. The fact that the fly's Dscam1 isoforms can form 93 (= 12 + 48 + 33) self-binding jigsaw-puzzle pieces (*each one unique!*) [1478,1991] is astounding [2060,2554]. How on earth did they evolve? Novel exons are known to arise via unequal crossing over [1263], but we do not yet know what selection pressures could have been high enough to refine those exons over the eons [2371] in the face of such a huge redundancy in the repertoire [2451].

Puzzle 2.10   How is only one version of each exon randomly selected?

The question of how cells make random choices is a great mystery in biology (see Puzzle 3.6) [1082]. Binary choices are relatively trivial, since they can be implemented by instabilities that resolve into 50:50 ratios – essentially letting the noise bias the signal [895] – and animals typically use *Notch–Delta* circuitry for this sort of "flipping of a coin" [654,658,792]. However, the task of randomly choosing a single item from a large menu of possibilities is another matter entirely. That problem is more like rolling dice or spinning a roulette wheel, and there are at least a dozen possible outcomes in each of *Dscam*'s three sets of ectodomain exons.

We have only a few clues so far as to how an exclusive choice is made: an essential RNA hairpin has been found near the exon 4 cluster (12 options) [1202], and a conserved RNA "docking site" has been mapped near the exon 6 cluster (48 options) [796], but we don't yet know how either of these motifs work, and we are completely ignorant about the exon 9 cluster (33 options) [935,2005]. Nevertheless, we do have proof that the choice is random for the exon 4 cluster [1517], and we are starting to grasp why individual neurons "spin the roulette wheel" more than once to produce up to 50 isoforms each [906,1517,2123,2554].

As for the enormous number of available exon combinations (19,008), research has shown that flies can survive with only a fourth as many (4752) [897]. Even this number seems gratuitously excessive, however, given that a single isoform suffices to allow the sister neurites of an individual cell to repel one another in a wild-type background [898].

A command center in the brain controls circadian rhythms

Flies sleep as we do, although they do so standing up like six-legged horses. Given all of the many parallels between the nervous systems of human and flies that have been surveyed here (Table 2.1), it should come as no surprise that researchers have learned a lot about human sleep by studying flies [533]. For example, recent experiments have helped illuminate (1) how the memories we store from each day's activities get consolidated overnight [188,527,903,1855] and (2) why sleep disturbances may be contributing to neurodegenerative disorders [2337] – including an affliction that we would normally never associate with flies: Alzheimer's disease [1091,1466,1582,2337]!

Sleep is regulated by circadian rhythms, and those rhythms are controlled by command centers in the brain [64,553,1213,1934]. In humans the circadian oscillator resides at the midline in the suprachiasmatic nucleus (above the optic chiasm) of the hypothalamus [289,1891]. In flies it is relegated to two bilateral chains containing six neuron clusters each [528,874,2029,2495], but there is a separate circuit in the mushroom body that controls sleeping and waking [64,2110].

Seymour Benzer (a founder of behavioral genetics) and his collaborators spent decades dissecting the genetics of circadian rhythms in *Drosophila* [875], and the corpus of their work stands as a lasting testament to the power of the genetic approach in general [263,849,2404]. Little did those early pioneers suspect that many of the cogs and gears that they inventoried in the fly's clockwork would later be found in its human counterpart [348,395,1717,1791,2512]. In both cases the pacemaker uses an autoregulatory feedback loop between the *period* gene and the protein it encodes (Figure 2.5) [2516]. Indeed, the more closely we examine the clocks of flies and mice, the more homologs we find (see Puzzle 2.11) [932,1063,2165].

Regardless of which species we choose to consider, the circadian clocks that govern its metabolism will continue to cycle with a period of ~24 hours even when the individuals are consigned to total darkness, and the phasing of those clocks can invariably be reset just by modulating the light–dark regimen [1556]. This susceptibility of all of life's clocks to entrainment by day–night cycling attests to how dependent all of us earthlings are upon both the sun's rays and the Earth's rotation.

Table 2.1 Features of the nervous system that are shared by vertebrates and flies[1]

Shared feature	Conserved mechanism?	Vertebrate quirks	Drosophila quirks	References
A–P partitioning	The brain is divided into three sections along its A–P axis. A nerve cord extends posteriorly.	Boundaries between the sections secrete morphogens and act as classic organizers.	Patterning is not affected by *Pax2/5/8*-LOF.	[949,1023,1092,1102,2318,2470]
Rostral identity	*otd/Otx, optix/Six3, earmuff/Fez, mirror/Irx, Pax2/5/8,* and *unpg/Gbx* specify fates along the A–P axis of the brain.	*Six3*-expressing neurons secrete a taxon-specific set of peptide hormones.	*optix*-expressing neurons secrete a taxon-specific set of peptide hormones.	[15,86,963,1076,2173,2174,2455]
Caudal identity	*Hox* genes serially specify caudal fates in the brain and nerve cord.	Metameric hindbrain *Hox* zones are shifted relative to somite *Hox* zones.	Metameric neural zones coincide with body segments.	[45,464,950,963,1092,1640,1884]
D–V allocation	Chordin (or other BMP antagonists) determine the boundaries of the neurogenic ectoderm.	Most of the neurogenic ectoderm invaginates, though neural crest cells delaminate.	Neuroblasts ingress from neurogenic ectoderm, though a few zones invaginate.	[884,963,1492,1520,1521,1624]
D–V partitioning	*vnd/Nkx2* specifies fates in medial CNS; *ind/Gsx* specifies fates in paraxial CNS; *msh/Msx* specifies fates in lateral CNS and dorsal PNS.	Same D–midline (roofplate) morphogen (Dpp/TGFβ) but different V–midline (floorplate) morphogen (Hh/Shh).	Same D–midline morphogen (Dpp/TGFβ) but different V–midline morphogen (Spitz).	[86,963,1071,1859,1950,2319, 2449]
Hemisphere splitting	Hh/Shh (from foregut and visceral mesoderm) bisects the brain into hemispheres.	Foregut is endodermal. Shh is also emitted by notochord and floorplate.	Foregut is ectodermal.	[1071,1711]
Proneural licensing	*SoxB* enables cells to become neurons, as do bHLH-encoding genes in the *achaete-scute* and *atonal* families.	Early-acting bHLH transcription factors differentiate neurons vs. glia, instead of neurons vs. epidermis.	*NeuroD* genes were lost in the lineage leading to flies, and the role of *neurogenin* was reduced.[2]	[191,1145,1817,1842,1944,1950, 1982,2097,2302]
Neuron spacing	*hairy*-related genes block neuron initiation, and the *Notch* pathway dictates neural spacing via lateral inhibition.	Delta and Serrate (Notch's main ligands) form stripes in the spinal cord but random arrays in midbrain.	Lateral inhibition in the PNS can yield salt-and-pepper arrays of bristles or single macrochaetes.	[168,1093,1229,1244,1842,1875, 2103]

Neuron pruning	Caspases cause neuronal suicide unless blocked by "Inhibitors of Apoptosis" proteins.	Prevention of apoptosis (paradoxically) does not cause much neural hyperplasia.	Prevention of apoptosis causes severe neural hyperplasia.	[1217,1665,1964,2335,2484,2485, 2500]
Neuron survival	Spätzle2/BDNF is secreted by target cells and blocks apoptosis.	GDNF (instead of BDNF) promotes survival of CNS motor neurons.	Effects of LOF mutations are greater on the CNS than on the PNS.	[2214,2264,2543]
Neuron identity	Temporal changes in amounts of the RNA-binding proteins Imp and Syp assign fates based on birth order.	Evidence for comparable usage of this clock device in mice is circumstantial and not yet proven.	This clock device has been thoroughly documented for the mushroom body of the fly brain.	[1336,2274]
Mitotic asymmetry	numb enables sister cells to adopt different fates.	Sister cell fates in numb-LOF and numb-GOF mutants are (strangely) identical.	Sister cell fates in numb-LOF and numb-GOF mutants are opposite.	[339,435,586,831,1133,2421, 2487]
Mitotic termination	prosperol/Prox1 lets stem cells differentiate.	Blocks production of astrocyte glial cells.	LOF results in neuroblast tumors.	[380,1097,1447,2179]
Motor neurons	Hb9 and various LIM-class homeoproteins specify motor neurons.	All motor neurons rely on these agents for their cell-type identity.	Only a subset of motor neurons rely on Hb9. Others use Eve instead.[3]	[341,1227,1241,1670,1977,2082, 2275]
Pioneer axons	Erection of a ladder-like scaffold of descending, commissural, and circumoral axon tracts.	Circumoral tract now encircles part of the brain (infundibulum) due to D–V inversion.	Circumoral tract still encircles stomodeum (as in common bilaterian ancestor).	[84]
Axon insulation	gcm specifies glial cells that ensheath axons of the CNS or PNS.	gcm has other roles in the placenta, etc. Glia:neuron ratio ≈ 50:50. Myelin sheath.	gcm has a separate role in the immune system. Glia:neuron ratio ≈ 10:90.	[336,670,887,1035,1682,2143, 2411,2555]
Axon pathfinding	Netrins provide guidance cues (either attractive or repulsive) for axons.	Netrins also play roles in angiogenesis, morphogenesis, and cell migration.	Netrins help guide the axons of motor neurons to their muscle targets.	[76,242,378,608,1320,1512,1753]
Axon pruning	Caspase cascade destroys excess axons.	Assisted by Schwann cells in the PNS.	Associated mainly with metamorphosis.	[1106,1181,1897]
Midline crossing	Slit is secreted by CNS midline cells. It binds Robo receptors to repel axons from the midline.	The floorplate emits signals (aside from Slit) that are not used by flies (e.g., Shh).	Commissureless allows axons to cross the midline by destroying Robo1.	[358,524,609,1151,1343,1605, 2529]

(cont.)

Table 2.1 (*cont*.)

Shared feature	Conserved mechanism?	Vertebrate quirks	Drosophila quirks	References
Chiasm formation	No. Convergent trait: axon pathways cross over one another at X-shaped junctures.	Corticospinal chiasm at midline implements contralateral motor control by the brain.	Optic chiasms on either side of midline are attributable to differentiation waves.	[892,1980,1985,2078,2195,2216]
Dendrite tiling	Mostly convergent, but vertebrate Dscams do play roles similar to their fly homologs.	Combinations of cell-surface proteins from genetically clustered protocadherin paralogs.	Combinations of Dscam1 isoforms that are produced via alternative splicing.	[1267,1306,1517,1947,2554]

[1] "Quirks" are derived traits (apomorphies) that differ in vertebrates and flies. Slash marks denote homologs (fly/vertebrate) [677]. Gene (*italic*) and protein (roman) names: BDNF (Brain-derived Neurotrophic Factor), Dpp (Decapentaplegic), Eve (Even-skipped), *Gbx* (*Gastrulation brain homeobox*), *gcm* (*glial cells missing*), GDNF (Glial Cell-derived Neurotrophic Factor), Hh (Hedgehog), *Irx* (*Iroquois homeobox*), *msh* (*muscle segment homeobox*), Msx (Msh homeobox), *otd* (*orthodenticle*), *Otx2* (*orthodenticle homeobox 2*), *Prox1* (*Prospero homeobox 1*), Shh (Sonic hedgehog), TGFβ (Transforming Growth Factor β), *unpg* (*unplugged*). Other abbreviations: A–P (anterior–posterior), bHLH (basic Helix–Loop–Helix), CNS (central nervous system), D–V (dorsal–ventral), GOF (gain-of-function), LOF (loss-of-function), PNS (peripheral nervous system). For more fly/vertebrate contrasts see [345,884,1040,1970]. For genetic circuitry of A–P and D–V axes see [464,1071,2046,2470]. Other shared aspects of CNS construction include: (1) modularity in connectivity [370,1095,1179,1413], (2) differentiation waves for colinear wiring [1985], (3) reliance on stem cells [336,586], (4) dichotomy of the somatic vs. visceral nervous system [194,1641], and (5) a blood–brain barrier [461,670]. CNS modules that appear to share deep homology include: (1) vertebrate hypothalamus ≈ fly *pars intercerebralis* [2270], (2) vertebrate basal ganglia ≈ arthropod central complex [643,1318,2197], and (3) vertebrate pallium ≈ mushroom bodies of flies and annelids [2290,2453]. CNS functions that seem to be conserved include: (1) cAMP-dependent formation of long-term memories [1103], (2) circadian rhythmicity [932,2165], (3) circadian control of sleep [64,553,1213,1934], (4) sleep-dependent energy conservation [2001], and (5) sleep-dependent memory consolidation [188,527,903,1855]. Indeed, flies may even dream as we do [1125,1481]. Behaviors that may have a common basis include aggression [1317,1915], compulsion [859,2042,2488], and stress [2494]. Flies may even feel pain as we do [1014,1105]. Unsolved mysteries include: (1) the origin of neurons [16,80,598,1417,1502,1705], (2) the origin of neurotransmitters [984,1204], (3) the origin of synapses [47,1561,1955], (4) the origin of glia cells [887], (5) the origin of the telencephalon [965], (6) the origin of brain ventricles [1360], (7) the origin of contralateral motor control [1980], (8) the origin of neural superposition in the fly eye [1243], (9) the origin of the neocortex in mammals [1112] and its regionalization [46,558,1339,1664], and (10) the expansion of the neocortex in humans [754,1853,2129,2251]. The last riddle holds the key to intelligence [591,1766], and we may be close to pinpointing the genes responsible [125,238,650,663,1885]. For cognitive abilities of flies vs. mice see [391,2044].

[2] These differences (and others) are mainly based on a study of the annelid *Platynereis dumerilii* [2097]. Because that species is a spiralian – the third great branch of bilaterians aside from deuterostomes (e.g., humans) and ecdysozoans (e.g., flies) – this research has allowed us to more accurately discern which features are conserved, lost, or derived.

[3] At one time flies were thought to differ from vertebrates insofar as they express Hb9 in interneurons as well as motor neurons [1670], but vertebrates were later found to use Hb9 in their interneurons as well [272].

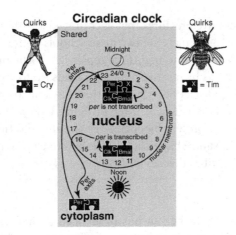

Figure 2.6 Components of the circadian clock shared by humans and flies. Homologous proteins are cartooned as jigsaw-puzzle pieces in the gray rectangle. Protein names (inscribed) are the same for humans and flies (Per = Period; Clk = Clock), except for Bmal, which is called Cyc (= Cycle) in flies. Protein "X" is not conserved. Its identity is given under "Quirks" in the margins (Cry = Cryptochrome; Tim = Timeless). This diagram is based on the schematic in [2516], which compares flies to mice, but human names are the same as those for mice [1791]. Clk and Bmal are transcription factors (bHLH/PAS class) that bind E-box (CACGTG) motifs next to *per* and other clock genes (not shown). Clk–Bmal heterodimers activate *per* transcription during the day, but are prevented from doing so at night when Per X heterodimers enter the nucleus (circle) and bind them in an inactive complex. T-bar denotes inhibition. Other components of the circuitry have been omitted for the sake of clarity; many of them are also homologous [932,2165].

Amazingly, the *per* homologs of flies and humans still share four splice sites inherited from the progenitor gene [2348]. For more on homology see [745,873,2556], for other conserved genes see [620,1998], for fly circuits see [528,1308], for mammal circuits see [289,1346], and for target genes under circadian control see [172,1063].

Puzzle 2.11   How old is the *period*-based pacemaker?

The extensive homologies of the components – including the *period* master gear – in fly and human oscillators suggest that this feedback device is at least 600 MY old [856,2294]; but it is probably no older than that, because the circadian clocks of plants and fungi appear to operate so differently [177,2513]. Moreover, the pacemaker of cyanobacteria seems so divergent from those of eukaryotes [1556] that it has been widely thought to have also evolved independently [1930].

This inference of convergence at the kingdom level has recently been called into question by the surprising discovery of an auxiliary (*period*-independent) circadian oscillator [583] that works the same way in prokaryotes as in eukaryotes. The heretical implication of these new findings is that *some* features of circadian timekeeping are 2.5 billion years old [1351]!

As for the selective pressures that encouraged a 24-hour clock "app" to evolve in the first place [1929], there are two main schools of thought: (1) daytime damage to DNA by UV rays led to nighttime repair [730,1930], or (2) daytime damage to cellular components by reactive oxygen radicals led to nighttime repair [179,1497].

Incredibly, in at least one case, evolution was able to cobble together a 24-hour oscillator clock from as few as three proteins. Every student majoring in chemical engineering or protein biochemistry should read the 2015 paper from Carl Johnson's lab at Vanderbilt entitled "Circadian clocks: unexpected biochemical cogs," which describes this device [1556].

The extent to which the sun has influenced animal evolution over the eons is most obvious, of course, in the structure of our eyes (cf. [161]). Their job, put simply, is to harvest incident photons in a coherent manner so that our brain can construct an internal representation of our environs – i.e., to "see." Thereby hangs a tale, which is told in the next chapter.

3   Vision

Keats was wrong when he accused science of ruining the beauty of a rainbow by "unweaving it" [476]. On the contrary, Newton *enhanced* that beauty by deftly explaining the prismatic essence of color [316]. Indeed, no topic in all of science is more poetic than vision. Vision spans the orders of magnitude between a star that dwarfs the Earth and a molecule which welcomes ethereal emissaries from that star after they've traveled 93 million miles and bounced off nearby objects.

Photons are detected by 11-*cis*-retinal (a vitamin-A derivative)

Virtually all animals that are capable of sensing light see the world by means of the same 20-carbon molecule, or minor variants thereof, regardless of the type of eye they possess [2161]. This molecule is a pigmented derivative of vitamin A called 11-*cis*-retinal [1165]. It looks like a kite whose tail of unsaturated carbons has a kink in the middle (Figure 3.1). The kink straightens via bond rotation whenever a photon is absorbed by the tail's π-orbital system [1602], thereby yielding an all-*trans* isomer [1591,2051]. In contrast, single-celled organisms tend to use other chromophores (light-sensing molecules) that operate quite differently [1090]. Given that other chromophores could have worked equally as well as retinal, it seems odd that animals should have chosen a molecule that they typically cannot synthesize metabolically on their own (see Puzzle 3.1).

The tip of 11-*cis*-retinal's tail is covalently attached to the seventh helix of an opsin protein [1674] that is embedded in specialized membranes of every photo-receptor cell (PRC). When the kink at carbon-11 straightens, the nearby helices of the opsin spread apart to reveal a binding site for a G-protein [1992,2028]. The G-protein then amplifies the signal and conveys it to the PRC's ion gates. From this point on, the PRC acts as an ordinary neuron [1863]. What it tells the next neuron in the visual processing chain is: "I've been hit by a photon!" [1142].

How did opsin become the photonic Prometheus of the animal kingdom? Opsin belongs to a large family of integral membrane proteins called G-protein-coupled receptors (GPCRs) [1931]. Most GPCRs act like garden-variety, cell-surface receptors insofar as they embrace their ligands (e.g., neurotransmitters) non-covalently and release them quickly after a fleeting kiss. Opsin seems to have departed from this tradition by marrying its ligand covalently and waiting for a third party (the photon)

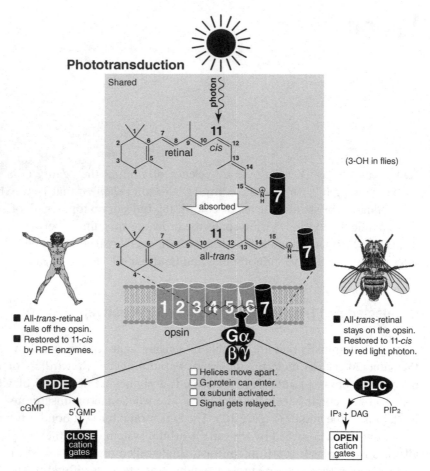

Phototransduction

Shared

photon

11
cis

retinal

(3-OH in flies)

absorbed

11
all-trans

opsin

Gα
βγ

All-*trans*-retinal falls off the opsin.
Restored to 11-*cis* by RPE enzymes.

All-*trans*-retinal stays on the opsin.
Restored to 11-*cis* by red light photon.

☐ Helices move apart.
☐ G-protein can enter.
☐ α subunit activated.
☐ Signal gets relayed.

PDE

cGMP   5'GMP

CLOSE cation gates

PLC

IP3 + DAG   PIP2

OPEN cation gates

Figure 3.1 How photoreceptors transduce light into neuronal signals. Features depicted in the shaded rectangle are common to humans and flies, whereas those outside are not. Proteins do not typically absorb light in any part of the visible spectrum [718], so animal eyes have relied upon an adduct in their visual systems [1090]. Remarkably, the same chromophore is used by all animal eyes. That molecule is 11-*cis*-retinal or a variant thereof [766]. One minor variant, which is found in flies, has a hydroxyl group on its C-3 carbon [256].

The 11-*cis*-retinal is invariably bound covalently to a lysine in the seventh transmembrane helix (black cylinder) of an "opsin" protein. Opsin is ~340–500 amino acids long [237,1760]. When a photon is absorbed, the kink at C-11 straightens to yield all-*trans* isomer [38], and opsin's sixth helix shifts to unveil a binding site for the Gα subunit of a trimeric G-protein [2028], though that site actually involves helices 2 and 3 as well (not shown) [1992].

All-*trans*-retinal leaves the opsin in vertebrates but usually stays attached in flies [1234,1537,2376]. It is converted back to the 11-*cis* isomer enzymatically by the retinal pigment epithelium (RPE) in vertebrates [40,541,1165], but the reversion in *Drosophila* is primarily achieved *non*-enzymatically by subsequent absorption of a second (~570 nm) photon [100,441,1234] – which helps explain why their eyes are red [1117,2169]. Recently, flies were found to possess an auxiliary enzymatic pathway for restoring the 11-*cis*-retinal [100,2376], and that pathway operates in pigment cells like the RPE of vertebrates.

Puzzle 3.1   Why didn't we ever evolve the ability to make our own retinal?

It seems strange that human vision should be so dependent upon a chemical that we must *eat* [1080,1772] – namely, β-carotene or related carotenoids [837,2376]. After all, food sources have changed quite drastically as our chordate ancestors shifted between herbivory, carnivory, insectivory, and omnivory in adapting to different habitats over the eons [159]. Were the enzymes that are required to synthesize 11-*cis*-retinal de novo really so difficult to evolve [2347]? Or is β-carotene so abundant that selection pressures were simply too weak to reward such metabolic tinkering? This mystery is not merely academic: vitamin-A deficiency is a major cause of childhood blindness in developing countries around the world [2130].

---

to trigger the divorce [1800]. Based on these clues, the following scenario has been proposed for opsin's origin [441,1234,2071]:

1. An ancestral GPCR used all-*trans*-retinal as its non-covalent ligand for a neural role other than vision [1765,2056] – perhaps for chemoreception [733]. The mere act of binding was enough to reshape the GPCR so that a G-protein could fit an allosteric site [1545] and relay the signal, "The ligand has arrived!"
2. Assuming that this same binding pocket was able to fit 11-*cis*-retinal as a competitive inhibitor, the GPCR would have bided its time in an OFF state until a photon turned 11-*cis*-retinal into all-*trans*-retinal, thus triggering the GPRC's ON state and allowing a G-protein to transmit the message, "A photon has arrived!"

←

Figure 3.1 (*cont.*)

In flies the Gα signal is relayed to phospholipase C (PLC), which converts phosphatidy-linositol-4,5-bisphosphate (PIP$_2$) into inositol-1,4,5-trisphosphate (IP$_3$) and diacylglycerol (DAG), resulting in the *opening* of cation gates [78,868,1117,1537]. In vertebrates, on the other hand, the Gα signal is relayed to phosphodiesterase (PDE), which converts cGMP to 5'GMP [951,1081], resulting in the *closure* of cation gates [78,1537,2071]. It seems terribly wasteful for human cPRCs to be perpetually ON (i.e., firing action potentials with ion gates opened) in the dark (when we are asleep, for heaven's sake!), but it turns out that fly rPRCs use just as much energy when they are OFF [2497].

Different subclasses of Gα protein are used by c- versus r-opsins (not shown) [441]. The helices of opsin are drawn aligned, but they actually form an irregular bundle [470,1731]. The cavity between helices 5 and 6 is the Gα binding site. Gray ovals with zigzag lines are membrane phospholipids. Omitted are opsin's N- and C-terminal tails and inter-helix loops [1846,2261]. Parts are not drawn to scale. These alternative pathways are described in Gordon Fain's *Sensory Transduction* [617], as are many unanswered questions. For more mysteries of vision, see Sönke Johnsen's splendid book *The Optics of Life* [1080].

NB: Animal opsins seem unrelated to the microbial opsins [735] that permit circadian shuttling up and down in the water column [255,275], but their structures are eerily similar [1379]. We should avoid jumping to a conclusion here, because structure can actually be conserved when amino acid sequence is not [1013].

3. Over time, some mutations altered the pocket to exclude all-*trans*-retinal but admit 11-*cis*-retinal, and other mutations led to a covalent bond. This GPCR became the first opsin, and the neuron employing it became the first PRC. As Darwin realized long ago, "any sensitive nerve may be rendered sensitive to light" [463,786].

This "Modified GPCR" argument is supported by the ability of 11-*cis*-retinal to still act as a light-sensitive ligand even after its lysine anchor has been detached (provided the Schiff base is restored) [2545], and it is made more plausible by the recent discovery of a second retinoid binding site inside bovine opsin [1390]. That site is near the pocket for 11-*cis*-retinal, but it does not involve a covalent linkage, nor is there yet any indication that the helices around it are displaced when the ligand binds [1518].

If opsins did evolve from GPCRs, then how can we explain light-activated *bacterial* proteins that also have seven transmembrane helices and a retinal (all-*trans* isomerizing to 13-*cis*) bound to a lysine in the seventh helix . . . but *lacking DNA homology* [599]? Until 2013 it was assumed that microbes and metazoans must have independently converged on the same protein scaffold because it is optimal in terms of light detection. In that year Kristine Mackin and her coworkers at Brandeis called this conjecture into question by showing that they could scramble the helices without sacrificing activity [1379]. Might these two classes of opsins be homologous after all? Perhaps. Any vestiges of sequence similarity could have been erased long ago due to functional divergence [2160], especially because microbial opsins serve mostly as ion pumps [1115]. Additional alternative origins have been proposed [641,1647,2066], so the issue of opsin's evolutionary ancestry is far from settled.

Detection is facilitated by elaborate phospholipid membranes

A photon cannot excite a photoreceptor unless it hits an 11-*cis*-retinal molecule therein, but this bull's-eye is tiny, not only relative to the PRC but even relative to the opsin. PRCs solve this "Robin Hood Dilemma" by erecting a gauntlet of opsin-studded membranes [571] – rather like assembling a stack of pepperoni pizzas thick enough that an arrow shot into the stack should intersect at least one pepperoni by chance alone. Plants use the same trick to catch photons with chlorophyll, which is embedded in comparable stacks of thylakoid membranes sequestered inside chloroplasts that endow plants with their inherent green color [112,1584,1821].

PRCs fall into two distinct categories based on how they construct their stacks of membranes. *Ciliary* PRCs (cPRCs) create their stacks from the membrane of an apical cilium [1428], whereas *rhabdomeric* PRCs (rPRCs) rotate their apical microvilli to form a lateral "rhabdomere" column (Figure 3.2) [1344]. The rhabdomeric (r-type) strategy is relatively simple, but the ciliary (c-type) process has enigmatic intricacies (see Puzzle 3.2).

Aside from the stark differences in how their membrane stacks develop, cPRCs and rPRCs also differ *biochemically* in (1) the kind of opsin that they employ [641,1812],

(2) the type of α-subunit of the G-protein that they use to bind their opsin, (3) the nature of the signaling cascade that lies downstream of the G-protein, (4) the effects of that cascade on their ion gates (opening vs. closing) [78,1537], and (5) the behavior of the retinal molecule after it isomerizes to an all-*trans* configuration.

Until 2010 it was thought that all-*trans*-retinal detaches from opsin in vertebrates and is recycled back to its 11-*cis*-retinal state enzymatically [1165], but that it stays attached in flies and is recycled in situ by absorption of a second photon [1234]. However, in that year it was reported that flies *do* have an enzymatic pathway for replenishing 11-*cis*-retinal in their rPRCs, and this pathway turns out to reside in the adjacent pigment cells – a division of labor that is histologically similar to the replenishment process in vertebrates [100,2376].

How on earth did cPRCs and rPRCs become so different in the first place? In a remarkably prescient 1965 essay entitled "Evolution of photoreceptors," Richard Eakin, who pioneered the study of light-sensing cells with electron microscopes, argued that (1) the first PRCs were *cPRCs* and that (2) *rPRCs* arose as an *offshoot* in protostomes (arthropods, mollusks, etc.) when they diverged from deuterostomes (chordates, echinoderms, etc.) [571,572]:

> At what point in evolutionary history did the rhabdomeric line arise from the more primitive ciliary one? My present conjecture would be at the branching of the ancestral bilateria into the two principal lineages: one leading to the deuterostomes (and retaining the ciliary type photoreceptor) and the other giving rise to modern acoels, pseudocoels, and protostomes (and developing the new rhabdomeric type photoreceptor). [571]

Eakin was acutely aware of glaring exceptions to this basic trend of mutual exclusivity. In fact, he and his coworker Jane Westfall had just demonstrated the coexistence of *both* cPRCs *and* rPRCs in amphioxus [574,1225] – the most basally branching extant member of our chordate phylum within the deuterostome clade [192]:

> Indeed, the first well-documented exception to my thesis was provided by work in my own laboratory, namely a study of the fine structure of photoreceptors in amphioxus (Eakin and Westfall, 1962b) in which we showed that the receptors in the eyecups of Hesse are rhabdomeric. Additionally, we found that certain cells in the wall of the cerebral vesicle called Joseph cells were likewise rhabdomeric . . . However, other cells in the roof of the cerebral vesicle, also believed to be photosensitive, are ciliary in type. [571]

Nevertheless, he dismissed such cases as either atavistic throwbacks to cPRCs in protostomes or independent offshoots to rPRCs in deuterostomes, in either case occurring *after* the divergence of protostomes and deuterostomes:

> The occasional appearance of a ciliary type photoreceptor in a protostome indicates to me the evolution of an adventitious ciliary photoreceptor in the rhabdomeric line . . . The occasional occurrence of rhabdomeres of microvilli in a deuterostome would signify to me that the group which possessed them had at some point in its evolution lost the ancestral way of forming photoreceptors and mutated to the newer pattern. [571]

Modern genomic studies confirm Eakin's contention that the c-type pathway is the more ancient of the two [1799,1800,2208]. However, the mounting instances of rPRCs

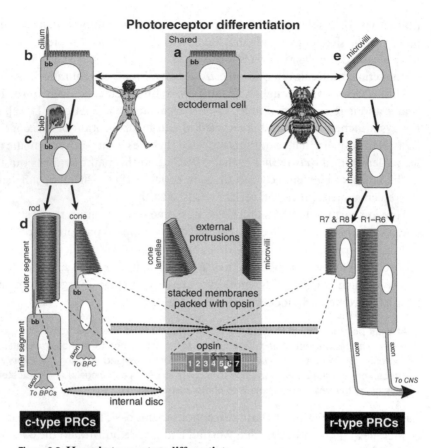

Figure 3.2 How photoreceptors differentiate.

a. Structures in the shaded rectangle are common to humans and flies, whereas those outside are not. In both humans and flies, photoreceptor cells (PRCs) begin as ordinary ectodermal cells [403,2041] with a lawn of finger-shaped microvilli (actin-filled protrusions) on the apical surface. The oval is the nucleus, and "bb" denotes a basal body [1426]. Mature PRCs contain stacks of membranes packed with opsin at a density of ~4×10^4 proteins per square micron [619]. This combination of high surface area with high opsin density maximizes the chance that a photon will be absorbed as it passes through the cell. The photosensitive membranes of rod PRCs (flat vesicles) are internal, while those of cones (flat lamellae) and fly PRCs ("toothbrush" filopodia) are external. NB: There is another type of PRC whose opsin-laden membranes are internal like a rod PRC, but this "phaosome" is a derived type of rhabdomeric PRC that evolved in the stem lineage of clitellate annelids during their "Dark Ages" when they burrowed into ocean floor [537].

b–d. Development of ciliary (c-type) PRCs in mice and, by inference, humans [427,1142, 2212,2320]. b. A cilium [741] grows out from the basal body [113]. c. The plasma membrane of the cilium inflates. This "outer segment" fills with flattened vesicles (of uncertain origin) that are roughly parallel to the cilium. Vesicles are more irregular than depicted. d. The outer segment elongates, vesicles reorient ~90°, and more are added. Rods and cones are named for the shapes of their outer segments. Rods appear to have evolved from cones [237,1234,1801,2071] in stem vertebrates [109]. Vertebrate PRCs use an enigmatic synapse to innervate bipolar cell (BPC) interneurons [617,1863]. In the central retina of primates, each cone synapses with one BPC, whereas rods innervate

---

Puzzle 3.2   How do cPRCs build such a neat stack of flat membranes?

Two alternative hypotheses have been proposed. The "Evagination Model" claims that opsin-bearing membranes come from the *plasma* membrane, while the "Fusion Model" asserts that they arise via the fusion of *internal* vesicles [389,1142]. Neither model explains how discs achieve uniform sizes in rods [344,1428,2436], nor how lamellae acquire diminishing diameters in cones [427].

Another issue is raised by the extremely narrow neck between the PRC's inner and outer segments. How on earth are raw materials shipped through this "Panama Canal" [1794,2247] and then assembled into a "skyscraper" [1142,2425] of a genetically prescribed height on the other side [1295,1427]?

Meanwhile, there is yet another complication to contemplate here. Each skyscraper has its top tenth lopped off each day and replaced from below [1583]. Our best clues to this treadmilling phenomenon come from a 2015 study which tracked the entry of opsin into the rod outer segment and the shedding of discs at the rod tip [987]. (Thylakoid organelles are equally dynamic and enigmatic [62,1033,1821].)

←

Figure 3.2 (*cont.*)

multiple BPCs, which reduces resolving power [642,2393]. The ciliary axoneme spans the outer segment of cones but not rods [1583]. Photosensitive membranes of rods ("discs") have no connection with the plasma membrane, but those of cones ("lamellae") are contiguous with it [1583]. Cones are 1000 times noisier in the dark than rods [1888,1906]. NB: PRCs are actually thinner than depicted (e.g., rod outer segments are ~2 μm wide × ~25 μm tall) [446,1142,1239], and they have more discs or lamellae (~1000) [389,1583]. Surprisingly, opsin density per disc varies axially [987]. In the absence of opsin, discs and lamellae develop abnormally in both mice [1272] and flies [350,403], so opsin may play a structural role [403,1212]. NB: All views are cross sections except the central schematics, which are three-dimensional. Some vertebrate PRCs have drastically different shapes from the ones drawn here [1234].

e–g. Development of rhabdomeric (r-type) PRCs in flies [1344,1783]. **e.** The apical fringe of microvilli pivots by 90° [1781], though not as neatly as shown here [353]. **f.** Eventually, microvilli cover one flank to form the rhabdomere. **g.** The rhabdomeres of R1–R6 (mono-chromatic) PRCs elongate to ~80 μm with ~30,000 microvilli [872,1781,2135], while most R7 and R8 (color-detecting) PRCs reach only half this size. The shorter microvilli of R7 and R8 afford greater acuity but less sensitivity [2483]. R7 and R8 PRCs in the dorsal rim area (see Figure 3.5) are exceptions: they have larger rhabdomeres for sensing polarized UV light [869,871,2304,2419], which they use as a solar compass [2405]. Unlike our PRCs, fly PRCs send axons directly to the CNS [353,2026], rather than via interneurons. One rhabdomeric lamella is enlarged to show how it is dotted with opsins, though it actually has a 10-fold lower density than vertebrate rods [619]. NB: Fly rhabdomeres twitch when exposed to light, so their transduction chain may entail a mechanical step [872], though the amplification is still mainly catalytic. The rhabdomere stack tapers from ~2 μm in diameter distally to ~1 μm proximally (R. C. Hardie, personal communication). No one knows whether opsin could work as a *cytosolic* (vs. transmembrane) protein [1965], but vision can be restored in blind mice by expressing opsin in *non*-PRC neurons [340], so this notion is plausible, and water-soluble sensors do exist in plants [1524].

in deuterostomes [78,573,719,2314] and cPRCs in protostomes [88,222,1221,1627,1741] refute his phylogenetic argument. Instead, these "crossover" cases favor the hypothesis that urbilaterians possessed both cPRCs and rPRCs *before* the protostome–deuterostome split [90,1800].

The most startling exception to Eakin's exclusivity proposal was encountered in 2002: rPRCs were found in the mouse retina [189,896] and later in the human retina as well [1580], where only cPRCs had been thought to exist. Strangest of all, these rPRCs were not even located in the same layer as the cPRCs. Rather, they constituted a small fraction (~1%) of retinal ganglion cells (RGCs). Until that time RGCs were thought to act only as interneurons to relay input from rods and cones to the brain [529], but the data now showed that a small subclass of RGCs in the mammal retina transduce light autonomously by an *r-type* G-protein cascade [579,791,1718,1827]. The history of how these clues were pieced together is as intriguing as an Agatha Christie novel [1789].

How could these intrinsically photosensitive RGCs (ipRGCs) have been overlooked by previous researchers? The reason is that they lack the conspicuous membrane specializations that we expect for PRCs [529,1827]. This absence makes sense adaptively because light must pass through the RGC layer before it reaches the rods and cones, so any extra absorption of photons at the RGC level would degrade the image-forming capability of the underlying cPRCs [1364,2497]. For the same reason, perhaps, ipRGC dendrites, which are also photosensitive, form a gossamer (photon-porous) web as they fan out across the retinal surface [896,907,1828]. (An analogous lacework of tiled dendrites exists in the skin of fly larvae [722,2474].)

Consistent with the overall r-type disposition of its transduction apparatus [791], the ipRGC uses a genuine r-type opsin like the opsins of the fly's eye [233,1580,1619]. This "melanopsin" was first noticed in frog melanophores in 1998 [1827], and a contemporaneous "News & Views" *Nature* essay captures the astonishment of the scientific community upon learning of its discovery (boldface added):

> A study by Provencio *et al.* . . . now shows that light-sensitive pigment cells in frog skin, called melanophores, express a molecule, melanopsin, which is similar to the rhodopsins used to detect light in the eye . . . All of this is hardly unexpected. Rather, **the real surprise comes in Provencio and colleagues' finding that melanopsin is more closely related to the rhodopsins in invertebrate eyes . . . than to those in vertebrate eyes, including the frog's own eyes.** [95]

Because melanopsin is configured to monitor ambient brightness rather than image contrast [271,529], it allows ipRGCs to regulate circadian rhythms [1405,1760] and pupillary reflexes [364,2477] under predominantly bright light conditions. Both of these subconscious tasks are handled by the brain's suprachiasmatic nucleus [1364,1557], but ipRGCs project axons to the brain's imaging centers as well [1789].

Based on what we have learned recently about the phylogenetic distribution of PRC types [78,619,1802,2349] (Figure 3.3), opsin types [641,1799,1812,2071,2261], and photic pathways [1800,2497], a consensus is emerging about the location of the c/r branchpoint (and its associated events) within the tree of animal life [735,1234,1632,2071,2349]:

1. G-protein cascades arose in early metazoans (including sponges), probably in association with chemoreception, rather than vision.
2. Opsin originated in a cnidarian–bilaterian ancestor (sponges lack them), but initially there were no membrane stacks or image-forming eyes.
3. Various G-proteins may have docked with this opsin, but the only ion channels affected were those gated by cyclic nucleotides as in cPRCs.
4. By the time urbilaterians appeared, the *c-opsin* gene had duplicated, and this paralog ultimately went on to found the *r-opsin* gene family.
5. The *r-opsin* paralog was expressed in a different physical location where it could shape the subsequent evolution of a new cell type: rPRCs.
6. Both types of PRCs persisted in various branches of the bilaterian tree, but cPRCs tended to be lost by protostomes and rPRCs by deuterostomes.

This proposed scheme for c/r divergence differs from Eakin's conjecture in (1) its *timing* of the c/r branchpoint *before* bilaterians arose and (2) its supposition that there were no subsequent instances of cPRCs becoming rPRCs (or vice versa). Together, these axioms imply that the cPRC was still relatively unspecialized when it begat the proto-rPRC (steps 1 and 2) [1632]. If so, then the rPRC would not have needed to climb down, so to speak, from a high adaptive peak (e.g., ciliary membrane stacks) before heading in a new direction (rhabdomeric stacks) [490,632]. Eakin's hypothesis, on the other hand, would have forced it to do so . . . via a much less parsimonious route.

Duplication and divergence (steps 3 and 4) have routinely fostered novelties at every biological level (genes, cell types, and organs) across the gamut of living things [809,1847], so it is natural to suppose that they might be involved here too [1666,1837]. The basic idea is that redundancy allows flexibility: one copy (e.g., rPRC) can vary freely because the other copy (cPRC) fulfills the original function. What remains unclear is where these cell types were deployed at the *tissue* level (see Puzzle 3.3).

Redundancy tends to erode after ~10 MY due to deleterious mutations [795,1423] unless it manages to enhance robustness [664,1769], so the mere fact that the rPRC has persisted for $\geq$600 MY implies that it adopted a divergent role soon after it arose [328,1630]. We do not yet know whether that role was (1) spatial imaging or (2) phototaxis or (3) circadian cycling or (4) some combination thereof [661]. But in any case, one or the other type of PRC (c- or r-) might have been lost later (step 5) during bottleneck episodes of ecological change [440] when its specialized light-sensing properties [619] were not needed in that particular clade at that particular time [1233,1632].

Why did protostomes forfeit their cPRCs while deuterostomes tended to lose their rPRCs? This enigma has nagged us ever since Eakin documented the c/r dichotomy circa 1960. One plausible answer to this question was proposed in 2010 by

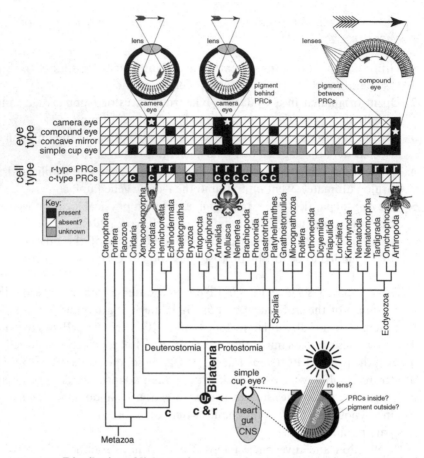

Figure 3.3 Distribution of light-sensing cells and organs across animal phyla. Black squares indicate presence of the trait in at least some members of the phylum (but not universal presence throughout the phylum), diagonal lines denote its absence in all species examined thus far (though future analyses of more species could show otherwise), and gray squares signify insufficient data to draw any conclusion at present (see key). Data on c- versus r-type photoreceptors (PRCs) come from [78,619,810,1741,2349]. The mere presence of c- or r-opsin genes does not warrant a "c" or "r" label here, which is only given if stacked membranes are also present. Thus, the "r" in the chordate box is not due to humans (our ganglion cells lack rhabdoms), but rather to our protochordate kin [69,90,1225]. Arthropods possess c-type opsins [1812], but no definitive c-type PRCs have yet been found, with the possible exception of the (freaky) male scale insect [296,557] and its parasitic wasp [2491], whose stacked membranes do not come from microvilli. (The simple eyes encircling the scale insect's head may be remnants of compound eyes [2025].)

Based on the prevalence of c- and r-type PRCs among phyla, it is likely that both types of PRC were present in Urbilateria [90,1812]. Data on eye types come from Land and Nilsson's *Animal Eyes* [1239] as well as personal communication from Dan-Eric Nilsson, except that the "a–g" categories in their Figure 1.7 are pooled here: camera eye (c, d), compound eye (b, e, f, h), simple cup eye (a, a+) – whose lens (if present) boosts brightness but fails to focus an image on the retina – and concave mirror (g). For 10 phyla we can rule out camera, compound, and mirror optics but don't yet know (until behavioral experiments are done) whether the organs are small cup eyes (capable of image formation, shadow detection, or directional light-sensing) or merely "eye spots" (with minimal PRCs,

---

Puzzle 3.3   Where were the first rPRCs located within the bilaterian body?

At their inception or soon thereafter, the rPRCs may have resided in a distinct location outside the old c-type eye (step 4 above) [1632,1837]. The reason for thinking so is that species in diverse phyla are known to have physically separate c-type and r-type eyes (e.g., annelids [88], mollusks [1627], and chordates [574,1225]).

Alternatively, the proto-rPRCs might have occupied a distinct *layer* within an otherwise c-type eye [78,1234]. This speculation is based on the following facts: (1) the mammalian retina has rPRCs (ipRGCs) in front of its cPRCs (rods and cones) [1789], (2) the retina of the scallop (phylum Mollusca) likewise has apposed layers of rPRCs and cPRCs [2497], and (3) the retinas of some fish species have photosensitive r-type horizontal cells (non-RGCs) atop their cPRCs [367].

The former (ectopic) hypothesis received support in 2013 with the discovery of r-opsin-expressing cells in the body segments of an annelid (*Platynereis*) and along the flanks of the zebrafish [123], and it got another boost in 2015 when light-sensing, r-opsin-expressing neurons were found in octopus skin [1858]. However, we don't yet know whether these extra-ocular rPRCs are conserved or derived.

$\leftarrow$

Figure 3.3 (*cont.*)

screening pigment, and pixel resolution). The prevalence of simple cup eyes (also called "pit" eyes [1197]) suggests that the first bilaterian eyes were of this kind – lined with PRCs and cloaked with shielding pigment [1087,1632,2025]. They may have resembled those in scyphozoan cnidarians [1782]. Such a cup can detect the direction of light because its pigment layer casts shadows on its PRC layer (inset) [2299], and it could even see images if the opening (pupil) were to shrink to a *Nautilus*-style pinhole [1239,1672]. Based on the anatomy of both amphioxus [1225] and onychophorans [1449], urbilaterians may have had a single median eye as depicted here [1632], rather than a pair of lateral ones, though its median eye may have been ventral instead of dorsal (cf. [123]).

Ecological needs can drive upgrades from simple cups to fancier eyes [440,1632], but truly sophisticated eyes only evolved three times: in chordates, cephalopod mollusks, and arthropods (starred boxes) [1632]. Octopus eyes look eerily like human eyes (cartooned above) – complete with cornea, iris, ciliary muscles, and eyelids (omitted for clarity) – despite having evolved separately [1671]. Spiders and scorpions (phylum Arthropoda) have camera eyes [1239,1482,2389], while insects and crustaceans rely mainly on convex compound eyes (sketched above). Even so, flying insects also use camera-like ocelli to gauge their posture relative to the horizon [1201,2389]. Camera eyes invert images (arrow), whereas compound eyes do not. Vertebrates put shielding pigment behind the PRCs, whereas insects wedge it between them.

Tree of extant phyla as in Figure 1.3. Controversy persists as to whether Ctenophora is the most basal animal phylum [1069] and whether it has bona fide PRCs [98]. Encyclopedic compendia of eye optics can be found in Land and Nilsson's *Animal Eyes* [1239] and Schwab's *Evolution's Witness* [2025]. For deeper reflections on eye evolution see reviews by Gehring [735] and Nilsson [1632]. Only a few surveys of the *Pax6* retinal gene network (within and among phyla) have been conducted thus far [466,1209,2349], so those data are omitted.

Puzzle 3.4   Were the eyes of primitive bilaterians simple or compound?

Compound eyes are so conspicuous that it is easy to forget that insects also have smaller, camera-type eyes like our own. A fly has three such "ocelli" on the top of its head (see Figure 3.5). Those rudimentary sensors are used as stabilizers to monitor the horizon in order to orient the body during flight [1201].

Velvet worms, which diverged from arthropods near the base of that phylum [936], only have eyes of the ocellar variety [1449]. Hence, this sort of eye might have been the primordial condition of arthropods, if not of ancestral bilaterians [634,2025,2559]. Consistent with this inference, cnidarians also have camera-like eyes [720,1634,2403].

Aside from the question of morphology, we would also like to know *where* on the body the eyes first resided [734,1239] (cf. Puzzle 3.3). Were they located laterally (as in velvet worms [1449]) or medially (as in amphioxus [1225])? Until we find fossils of the right age we cannot know for sure [182,285], but one thing is certain from the fossils we do have: it did not take long for compound eyes to evolve. The top predator of the Early Cambrian ~515 MY ago was a meter-long arthropod with as many as 16,000 ommatidia per eye [1265,1746]. The only deuterostomes to have (independently) evolved a compound (vs. camera) eye are starfish [1238].

Gordon Fain, Roger Hardie, and Simon Laughlin [619]. Based upon a re-evaluation of the data, they infer that the first cPRCs were more like our cones than our rods in their sensitivity to light. If so, then any animals that foraged in deep waters (protostomes?) should have relied more on their rPRCs (which work well in any light), while those that lived in sunnier habitats (deuterostomes?) should have favored their cPRCs (which only operated well in bright light before rods evolved in primitive chordates [109,1676]).

The modern world's c/r divide may thus have begun with some sort of niche partitioning among early bilaterians. Urbilaterians themselves probably had simple cup eyes (Figure 3.3), but it may not have taken long for imaging lenses to evolve [1087], leading to the emergence of simple and compound eyes in different clades at different times (see Puzzle 3.4).

Color vision relies on a subclass of dedicated photoreceptors

When 11-*cis*-retinal is free in solution (but still protonated as a Schiff base) it absorbs light at a wavelength of ~440 nm (Figure 3.4) [1602]. When it is cradled in the embrace of its opsin host, on the other hand, 11-*cis*-retinal can be tuned to various other wavelengths by means of charged or polar amino acid residues that are situated nearby [1619]. Those residues modify the fluidity of electrons in retinal's π-orbitals [1602,1965,2375,2544]. The more energy that is needed to excite those electrons, the more the absorption peak will shift toward the violet end of the spectrum [1619]. The less energy that is needed, the more it will shift toward the red end.

Humans and flies use the same "division of labor" strategy for assigning duties to our PRCs [415] in accordance with the inescapable trade-off between resolution and sensitivity [440]. In both cases a minority of PRCs is dedicated to color vision in bright light, while the remaining (monochromatic) majority function well under both dim and bright light conditions to detect contrast independently of color [29,1134,1368,1784].

> In certain respects the outer and central photoreceptor cells [of flies] bear similarities to human rods and cones, respectively. Both R1–6 cells and rods are very sensitive to light, express a single visual pigment, and make up the majority of photoreceptor cells. By contrast, R7–8 cells and cones are less sensitive to light, express multiple pigments, and comprise a high-acuity system. [1537]

This functional dichotomy explains why everything looks gray to us in the dark [1906,1976], and also why motion and color are processed separately to some extent in the fly brain [1543,2482,2483], though there is a surprising amount of inter-channel cross-talk [2384,2415].

Fly PRCs all look alike (see Figure 3.2g), but the same is not true for vertebrates. Our color-detecting cones use lamellae, but our monochromatic rods use internal discs that are detached from the plasma membrane (see Figure 3.2d). This correlation between structure and function was thought to be causal until 2015 when one of the two types of PRC in the sea lamprey was shown to have the physiological features of a rod (single-photon detection [1888]) but the lamellar form of a cone [1563] (see Puzzle 3.5). This finding also implies that functional rods, which can detect very dim light, arose before the agnathan stage of vertebrate evolution [109,2387].

Most animals (including humans [2116]) use separate PRCs for different wavelengths of light, because colors can only be perceived by comparisons between wavelengths [1910,2127,2334,2482]. Curiously, this "one-receptor-one-neuron" rule is broken by a subset of PRCs in both flies and mice (Figure 3.5). In the dorsal ~30% of the fly eye, every R7 PRC in the yellow subclass expresses two opsins (Rh3 and Rh4) [1453], and in the ventral ~80% of the mouse eye, every cone PRC expresses two opsins as well (S-type blue and M-type green) [75].

Wherever opsin co-expression is confined to one part of an eye the obvious implication is that it is serving a distinct function for that portion of the visual field [2260], but the reasons may differ for mice and flies [1030,1909]. Mice are nocturnal, so any selection pressure to maintain color vision would be negligible [1005,1632]. Hence, the genetic wiring required to enforce the "one-receptor-one-neuron" rule may be decaying in mice due to the "use it or lose it" principle [974,1423]. Flies, on the other hand, are diurnal (like us), so that argument cannot apply. Having UV-light detectors of different bandwidths – broad (Rh3 + Rh4) vs. narrow (Rh3 alone) – may confer

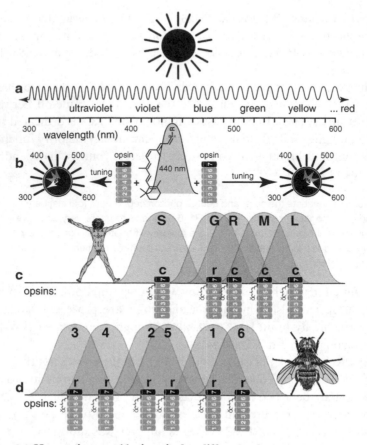

Figure 3.4 How opsin tunes 11-*cis*-retinal to different colors.

a. Spectrum of ultraviolet and visible light [765]. Across the scale are listed what we (for unknown reasons [244,1044,1326]) perceive as colors from violet to red, though red (~620–750 nm) is really beyond the right end, and not all labels are centered in their ranges [11,1045].

b. Peak wavelength (λ_{max} = 440 nm) that is absorbed by "naked" 11-*cis*-retinal (the protonated Schiff base unmoored from its opsin) [1602]. When tucked into the bowels of an opsin protein (helices drawn as cylinders; cf. Figure 3.1), the 11-*cis*-retinal can be tuned to other wavelengths by charged or polar amino acid residues in its vicinity [901,1004,2504,2544]. Opsin thus acts like the knob of an AM/FM radio dial, allowing the species (≈ listener) to receive selected frequencies of light (≈ sound) relevant to its niche [256,470,1136,1701].

c. Spectral sensitivities of our three (c-type) cone opsins, which absorb light at short (S), medium (M), and long (L) wavelengths [1045,1602], plus our (c-type) rod (R) opsin [1909] and the melanopsin pigment used by a subset of our (r-type) retinal ganglion (G) cells [129]. Our mammalian ancestors suffered a "nocturnal bottleneck" and thereby lost two opsin genes though disuse [470,744], but our primate ancestors regained trichromatic vision by duplicating one of the remnant opsin genes and shifting its λ_{max} [470,560]. Unsolved mysteries include: (1) why red is such a vivid color despite no PRC being maximally tuned to it [1239], (2) why we enjoy rainbows and other colorful things [1188], (3) why we have so few S-cones [34], (4) what *non*-visual opsins do [676], and (5) why evolution tuned the peaks of PRCs more often via opsin than via retinal, given that changing retinal's bonds could have worked just as well [618]. Melanopsin works like the r-opsins in flies [1364,1580], but its role in mammals is to measure brightness rather than to help form images [49,861]. NB: Absorbance curves are broader and less symmetric

> ## Puzzle 3.5   Why do vertebrate rods and cones differ in shape?
>
> This discovery of "hybrid" PRCs in lampreys [1563] raises the question of whether other salient differences between rods and cones might be equally inconsequential. For example, why should a cone-shaped outer segment be any better at color detection than a cylindrical one? One thing is certain: the shape of a PRC is *not* attributable to the opsins that it expresses [1833], because (1) tiger salamanders use the very same opsin for their green-appearing rods as for their blue-sensitive cones [1375] and (2) chameleons express rod-type opsins in cone-shaped PRCs [1456].
>
> To further explore this riddle, transduction components have been experimentally swapped between rods and cones to determine whether they operate just as well in differently shaped cells [360,693,1134]. Unfortunately, we don't yet have a firm answer. We still have much to learn about why cells have distinctive shapes in general [1770] and about the extent to which those shapes are maintained by selection based on the adaptive roles that they may play.

←―――

Figure 3.4 (*cont.*)

than depicted here (and in **d**) [129,1375,1700,2007], so humans can actually see light beyond 700 nm [1080]. Moreover, every opsin has a minor absorbance shoulder in the UV range (not shown) [470,1080,1424]. For that reason, people who are fitted with ordinary plastic lenses after cataract surgery magically acquire the "super power" of being able to see in the UV range [2166], unlike the rest of us, whose yellowish lenses absorb UV [236,744,1424,1736]. Most mammals lack UV-tuned PRCs [470,1004], unlike honeybees and other insects, which use them to identify flowers [480].

d. Spectral sensitivities of the six (r-type) fly opsins Rh1–Rh6 [629,1968,2482] that are expressed in ocelli (Rh2) [1806] or in compound eyes (Rh1 and Rh3–6; see Figure 3.5) [766,870]. The absence of opsins in the yellow–red range is due to screening pigments that give fly eyes their red color [1153]: flies use orange light (~570 nm) to convert all-*trans*-retinal back into 11-*cis*-retinal [1117,2169,2170]. Red-sensitive opsins are rare in nature, possibly because they are susceptible to thermal noise [710] from infrared radiation [1080,1369]. Flies actually use their Rh1 opsin as a heat sensor in the dark [2066]. Red light seems alien in the blue ocean [470,1006,2388], but dragon fish can emit red light and manage to see it by using chlorophyll molecules that activate their PRC opsins [539,1141]. Indeed, mice become more sensitive to red light when injected with chlorophyll derivatives [2391]. Mantis shrimps hold the "world record" for the most differently tuned PRCs (up to ~21) of any animal, which they achieve via pigments in their lenses [229,1424,2269]. One butterfly species can actually tune a single opsin to multiple (red) wavelengths via such (non-opsin) pigments [2519]. For a wider survey of spectral sensitivities across the animal kingdom, see the spectacular book *Visual Ecology* [440] by Thomas Cronin, Sönke Johnsen, Justin Marshall, and Eric Warrant (especially their Figure 7.2). NB: The 11-*cis*-retinal chromophore is actually buried inside the protein [237], not on its surface as drawn (for clarity) here (and in **c**). The electric-field effects that are exerted by side groups in retinal's vicinity are reminiscent of how performers play music "in thin air" on a theremin device without ever touching the instrument itself [2112]. Because there are only a few amino acids close enough to retinal's tail to have sufficient influence [237,2503] it is not surprising that the same "hot spots" have been modified in the same way separately in different species [470,1429]. Strangely, that is the case for flies and cows [1967] and apparently also for birds and butterflies [674].

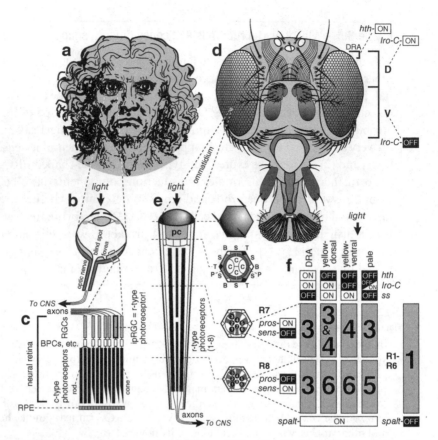

Figure 3.5 How eyes function in vertebrates and flies. Modified from [923,925]. Genetic circuitry is discussed in [798,867] for humans and [1084,2224,2305,2417,2418] for flies. See [1046,2115,2413] for fiendishly clever experiments, [353,1728] for *cis*-enhancers of *opsin* genes, [2418] for how form follows function, [1243,1753] for how R1–R8 PRC axons find their targets, [2007] for how flies can use their monochromatic R1–R6 PRCs to see color, [1702] for how flies process color beyond the PRC level, [437,1663,2328] for how *Iro-C* may have created a "love spot" in the "eye of the beholder" that is devoid of R7s, and [691,1078,2128] for the "bedtime story" of how rod nuclei (once upon a time) became focusing lenses. For more riddles see [2415].

a. Humans and other vertebrates see the world through single-chamber "camera" eyes. The resolving power of our eyes is impressively ~100 times better than a fly's [1316], in part due to the relatively gigantic scale of our body [2389]. Surprisingly, however, a human-size fly would still not see as sharply as we do, because their ommatidia suffer inherent diffraction limitations [1239]. Nor are flies able to perceive depth in the same way we do, because their left and right visual fields hardly overlap [402,2024,2343].

b. Light rays are refracted by the cornea (outer hump) and lens (gray oval) onto PRCs in the neural retina (light gray layer). Photons that escape capture are absorbed by the retinal pigment epithelium (RPE; dark gray layer). Acuity is highest in the fovea (dimple) but absent in the blind spot where the optic nerve exits the eye [798].

c. Enlargement of boxed area in **b**. The optic nerve (at top) conveys axons from retinal ganglion cells (RGCs; front layer) to the central nervous system (CNS). About 1% of mammalian RGCs are intrinsically photosensitive (ipRGCs) insofar as they possess an r-type opsin and an r-type transduction cascade [88,451,662]. However, they lack stacked

an adaptive advantage for the subset of their PRCs that point skyward to sense polarized light [161,440,871,2419].

The scofflaw genes that allow opsin co-expression in flies are members of the *Iroquois-Complex* [1453], but the homologous *Iroquois* genes in mice seem to be innocent bystanders [981,2537]. Hence, this aspect of color vision probably evolved

---

Figure 3.5 (*cont.*)

membranes (cf. pineal cPRCs that also lack stacks [1405]) and hence can only operate in bright light [529,1789,2002]. After light traverses the RGCs, bipolar cells (BPCs), and sundry other interneurons, it finally reaches the rod and cone (c-type) PRC layer [765]. (See [925] for how evolution inverted our retina.) Each eye has ~7 million cones and ~120 million rods [1142,2560] ($\approx 20{,}000\times$ more than the ~6000 PRCs per fly eye [234]). Mammals use a combinatorial code (not shown) [2089,2558] to assign identities to retinal cells [280,1421,1562,2122,2522], as do flies (**f**).

d. A fly looks alien to us because its eyes are compound, immobile, and huge relative to its face [234]. Each eye has ~750 ommatidial facets [923]. Flies also have three camera eyes (white ovals) atop the head [26] that are too unfocused to form images. These ocelli detect contrast [1519] and inform the fly about its tilt relative to the horizon [1201]. (Our reptile ancestors also had a median eye, but this third eye sank into the brain to form our pineal gland [1405].) The adult fly has two tiny remnants of its larval eyes (not shown) [295], which regulate circadian rhythms [677,955,2156]. Unlike our retina, which is an outgrowth of our brain [2294], the fly's retina develops separately from its brain [923], coming instead from an eye disc.

e. Each ommatidium is a tapered cylinder that is capped by a cuticular dome [353]. After light rays enter the dome ($\approx$ cornea) they pass through a pseudocone (pc) lens secreted by four "cone" cells (C in the cross section; no relation to human cone PRCs). Each ommatidium has eight r-type PRCs, six of which (R1–R6) span its height and form a trapezoid [1211]. The other two are shorter and are stacked one atop the other, with R7 above R8 and turned 90° so they can jointly detect polarized light [1080]. Numbered circles in the lower two cross sections are rhabdomeres. A sleeve of primary (P), secondary (S), and tertiary (T) pigment cells encases R1–R8 and absorbs stray photons (like the RPE in vertebrates) [1239,2389]. At three of the six vertices a bristle (B) replaces a T-pigment cell [1872].

f. Flies employ five opsins, numbered 1, 3, 4, 5, and 6 in gray rectangles (cf. Figure 3.4) [2007]. R1–R6 cells use Rh1 ("Rh" stands for Rhodopsin), while R7 and R8 use Rh3, Rh4, Rh5, or Rh6 based on a binary "barcode" of multiple transcription factors [1910] that works as follows:

1. *spalt* distinguishes R7–R8 (*spalt* ON) from R1–R6 (*spalt* OFF).
2. *prospero* (*pros*) and *senseless* (*sens*) distinguish R7 (*pros* ON, *sens* OFF) from R8 (*pros* OFF, *sens* ON) [2475].
3. Expression of *homothorax* (*htx*) in the dorsal rim area (DRA; upper right in **d**) forces R7 and R8 to use Rh3.
4. R7–R8 pairs outside the DRA are either "yellow" (Rh6 in R8) or "pale" (Rh5 in R8) [2414], depending on the ON/OFF state of *spineless* (*ss*) [2416].
5. The yellow and pale R7–R8 pairs are arranged randomly [176,1494] in a ~70:30 ratio of yellow:pale [2266], though the ratio does vary among species [943].
6. The *Iroquois-Complex* (*Iro-C*) [1144] forces yellow R7 cells to co-express Rh3 and Rh4 in the dorsal third of the eye (**d**) [338,1785,2104], but it has no effect on pale R7s [1453].
7. R8s are induced to adopt a certain opsin (Rh5 or Rh6) by their R7 partner [1495].
8. Expression of *ss* is reduced in yellow-dorsal R7s [2266].

independently. However, one deeply conserved gene does assign PRCs the duty of color detection in the first place [479]. The fly's *spalt* gene and its *Sall3* vertebrate homolog encode a zinc-finger transcription factor that is instrumental in specifying the PRC subtypes that are used for color vision (boldface added):

> **The strong activation of S-cone-specific transcripts by Sall3 implies a surprising functional homology with the Spalt gene complex of *Drosophila*** in the regulation of photoreceptor differentiation. In *Drosophila*, Spalt genes are necessary for specification of the inner R7 and R8 photoreceptors, which are responsible for color discrimination and in many respects form a cone-like photoreceptor class in the compound eye . . . Notably, Spalt genes are selectively expressed in blue-sensitive Rh5-positive R8 photoreceptors and are required for expression of Rh5 opsin.
>
> In mice, we observe that Sall3 is both necessary and sufficient for the expression of blue-sensitive cone opsin but not green-sensitive opsin. Such a direct conservation of gene function in photoreceptor development is unusual, and even more surprising because the blue-sensitive visual opsins of insects and vertebrates evolved independently. **Although this observation might represent evolutionary convergence, it could alternatively imply that ancestral bilateria possessed a dedicated short wavelength-sensitive photoreceptor cell type**, the differentiation of which was guided by a Spalt family gene. [481]

If *spalt* is experimentally turned OFF in a fly PRC that normally expresses it, then this cell will adopt an R1–R6 (monochromatic) fate, whereas if *spalt* is turned ON in a cell where it is normally OFF, then this cell will become either an R7 or R8 (color-sensitive) PRC. The particular opsin that is expressed by an R7 or R8 cell (Rh3, Rh4, Rh5, or Rh6; Rh stands for Rhodopsin) also depends upon the ON/OFF states of other genes that encode other transcription factors (Figure 3.5). Collectively, these genes constitute a binary code, but the fates of the R7 and R8 cells are also dictated by unknown random factors [2305] (see Puzzle 3.6).

Puzzle 3.6   How do R7 PRCs choose a subtype fate so randomly?

The process by which color-sensitive PRCs in humans express a single opsin gene via a locus control region is relatively well understood [415,2116], but we don't yet know how a similar outcome is achieved by R7 PRCs in the ventral half of the fly eye [1909]. In that region ~70% of R7s express Rh4 (with *ss* OFF), and ~30% of R7s express Rh3 (with *ss* ON) in a salt-and-pepper mosaic distribution [176,2416]. After the R7 cell selects an identity, it signals its R8 partner to adopt a compatible state [1495], so the R8 PRC does not participate directly in this lottery.

It is easy to imagine how cells could "flip a coin" to yield binary states [1494,2334], but it is harder to understand how they are able to bias a decision away from this 50:50 equivalence [413]. Similar biases also exist in the ratios of cone subtypes in humans [954,1082,2414].

Solving this riddle might shed light on how cells compute digital outcomes in general [301,1452] and why genomes leave some patterns to chance alone (e.g., fingerprints) [925]. Curiously, fish and birds assign PRC states by deterministic rules instead of probabilistic ones (see Puzzles 3.7 and 3.8).

Humans and fruit flies are both able to discern details in the center of their visual fields with high acuity, but the foveal devices that they use to do so differ so starkly from one another that they must have evolved separately. Humans use a dense array of cone cells [798], whereas flies superimpose the inputs from ommatidia that reside just above and below their mirror-image equator [869]. Thus, we have a foveal *spot* while flies have a foveal *stripe*. Other insects possess similarly specialized subregions of the eye for locating mates [978,1424] or catching prey [440,521,1219].

Outside the fovea proper, the spatial arrangement of PRCs should be of little consequence as long as the various wavelength-detectors are distributed at roughly uniform densities. A priori, therefore, we might expect a salt-and-pepper scatter of PRC subtypes, and that turns out to be true for our own eyes [954,2022] and more or less true for fly eyes as well [176,1494]. However, the regularity of spacing for each subtype varies considerably among mammals [413,2415], owing, perhaps, to odd constraints on dendritic tiling [1880], and for some unknown reason the PRC subtypes of fish and birds are arranged in astoundingly orderly geometric patterns (see Puzzles 3.7 and 3.8).

Puzzle 3.7   How (and why) are cones arranged so precisely in fish?

Unlike the disorderly cones of humans [954,2022] and the hexagonally arranged cones of macaques [483,2432], the cone PRCs of the fish retina are aligned in orthogonal rows and columns like a two-dimensional Rubik's cube [51,236,1871]. Zebrafish, for example, have four types of cones – R (red), G (green), B (blue), and U (ultraviolet) – that alternate in each row (...B-U-B-U-B-U... or ...R-G-R-G-R-G...) and form repeating palindromes in each column (...U-G-R-B-R-G-U...) [1870]. It is unclear why fish need such a fancy input screen for detecting colors in their aquamarine underworld [34,1487,1759,1784], though color filtering is only a significant problem in deeper waters [1121,1291,1760].

Puzzle 3.8   How are the five types of cones patterned separately in birds?

The bird retina poses a geometric riddle that is at least as intriguing as that of the fish retina [237,1075,1909]. It contains five different types of cone PRCs, and each of them is organized in a hexagonal lattice. However, the different lattices exhibit their own characteristic spacing intervals, and heterotypic cone neighbors seem oblivious to one another's presence, so the five lattices must be tiled independently of one another [1200]. We do not yet know whether they are established at different times, though a rigid sequence does seem likely [818]. Nor do we know which cellular behaviors are involved in creating this pattern: cell repulsion (via filopodia?), cell movement (within the epithelium?), cell division, or cell death [1340,1879]. Solving this retinal riddle might illuminate how genomes enable cells to tesselate epithelia in general [2,975,1581] and how they deploy modular subroutines in space and time to create periodic patterns of various kinds [893,922].

When did color vision arise? In the lineage leading to our own species the ability to perceive color goes back to stem vertebrates (fish) [237,404,674]. Subsequently, mammals suffered attrition of the opsin repertoire during their nocturnal "Dark Ages" [129,440,744,1200], but those losses were largely remedied by our primate ancestors, who bequeathed trichromatic vision to us [1603]. In the fly's case, color vision has been traced back to the dawn of the arthropod phylum [256,1195].

The paired-homeobox gene *Pax6* governs eye development

In 1995 Walter Gehring and his collaborators in Basel stunned the evo-devo community. They showed that the fruit fly's *eyeless* (*ey*) gene can induce extra eyes on the legs and elsewhere when it is artificially expressed in those body regions [844]. Even more amazing was their discovery (reported in the same *Science* paper) that *Pax6*, the orthologous gene in mice [1844], can elicit these same sorts of ectopic eyes when it is misexpressed in flies. Hence, it seemed as if an ancient master gene for bilaterian eye development had been unmasked [135,736,1198].

> It is really remarkable that you can take a tissue that would normally make a wing or an antenna and by turning on one [gene], make that into a complex thing like the eye . . . This is like someone finding a [gene] that would turn a kidney into a liver. – Gerald Rubin [147]

Pax6 contains two DNA-binding motifs: a paired box [1712] and a homeobox [1786,2107], though, surprisingly, the homeobox turned out to be dispensable for eye induction [1197,1835]. Other properties of the gene were also found to violate the standard *Hox* paradigm [1612,2398]. Most notably, when *Pax6* is disabled, the eye fails to develop [1205], rather than transforming into some other type of organ [136,844]. This phenotype surely did not conform with what we had come to expect for LOF alleles of *Hox* genes like *Ubx*, for example, which convert a haltere into a wing just as its GOF alleles turn a wing into a haltere.

Subsequent experiments uncovered several downstream genes (*sine oculis/ Six1&2*, *eyes absent/Eya1–4*, and *dachshund/Dach1&2*) that cooperate with *Pax6* [466,1197,1198,2093] in eye development. Indeed, this integrated circuit is widely used across the animal kingdom [660] in various tissue contexts [799,1111]. Its endurance as an intact entity may be due to an intimate meshing of the protein products with one another at target enhancers [863,1111,1873], because it is hard for evolution to tinker with any part of a solid-state module without disabling the entire device [248,1300]. The various echelons of the pathway hierarchy are thought to have evolved stepwise by the gradual insertion of "middle managers" between *Pax6* and its targets [733,2299], though the details of how this occurred have yet to be worked out.

The only other animals besides vertebrates and arthropods that ever evolved sophisticated, general-purpose, image-forming eyes are the cephalopod mollusks (octopuses, squid, cuttlefish, and nautiloids) [1238,1632]. Indeed, their camera eyes are eerily similar to our own. The eye of an octopus, for example, not only has a lens

---

Puzzle 3.9   Why do brainless jellyfish have eyes?

A priori, even the most primitive marine animals might be expected to have *some* ability to sense ambient brightness in order to reflexively move up or down in the water column on a daily cycle [255,1631,2208], regardless of whether they have a centralized nervous system with which to construct images – i.e., to see things [2025,2403]. However, the box jellyfish *far* exceeds those expectations. It has no fewer than 24 eyes, some of which seem much too elaborate for sensing luminance alone [1634,1782]. Each corner of its box-shaped bell has six eyes of various types (pit, slit, or lensed) embedded in a hanging stalk. One species is known to use those eyes for navigating mangrove swamps [720], but we have no idea how its neural ganglia manage to process this visual information. Likewise, starfish use eyes at the tips of their arms to navigate edges of reefs [719,1238].

Citing cases like these, Gehring surmised that animals must have acquired eyes before they evolved nervous systems able to form images [733] (cf. [82,1898]). Admittedly, a few animals that *do* possess brains *don't* actually use them to analyze input from PRCs that lie outside their head region. For example, octopuses have PRCs in the skin of their arms but only seem to use them locally to help pigment cells adjust their camouflage to changing lighting conditions [1858].

---

and retina, but also an eyelid, cornea, iris, and an array of ciliary (focusing) muscles [1671]. Because these parts develop so differently (e.g., their retina does not induce their lens) cephalopod and vertebrate eyes were historically seen as a classic case of convergence [1710,2025]. This conclusion was *not* overturned by the trans-phylum, eye-inducing powers of *Pax6* that Gehring *et al.* demonstrated in 1995, but it was called into question as authors struggled to redefine what we mean by "homology" [1629,1666,1787]. Not surprisingly, the term "*deep* homology" was coined only two years later (in 1997) to try to deal with such murky situations [2086].

If *Pax6* was really so instrumental in the evolution of bilaterian eyes, then it should still be playing a key role in cephalopod eyes as well as in ours [2509]. In 1997 Gehring's lab confirmed that *Pax6* is indeed expressed in the embryonic eyes of various cephalopods, and showed that the squid's *Pax6* ortholog can induce ectopic eyes in flies [2289]. These results supported Gehring's notion that *Pax6* is a master gene. Moreover, a *Pax6*-like (*PaxB*) gene was subsequently found to be expressed in the eyes of a box jellyfish [1951] (see Puzzle 3.9) – suggesting that *Pax6* (or a precursor thereof) governed "eye" development even before the cnidarian–bilaterian split [1782].

But other facts undermine *Pax6*'s reputation as a master gene

The argument for *Pax6*'s supremacy as a master gene, as presented above, may seem convincing, but it actually contains glaring weaknesses that have been pointed out repeatedly by numerous authors [879,953,1197,1629,1666]. Listed below are the major criticisms that have been leveled at Gehring's master-gene hypothesis in particular and at his monophyletic theory of eye evolution in general:

1. The ability of the *Pax6* gene from a mouse or a squid to induce extra eyes in flies merely proves that the key domains of the Pax6 protein have been conserved. It says nothing about whether *Pax6* plays similar roles in mice or squid [953]. To counter this objection a bit the Gehring camp could presumably point out the ability of heterologous *Pax6* to induce extra eyes in frogs (as well as in flies) – showing that it can launch eye formation there as well [383,2558].

2. It is not enough to show that a gene is *sufficient* to induce an organ in order to certify its role as a master gene. The gene must also be shown to be *necessary*, and in that regard the evidence is equivocal. The optic vesicle fails to form when *Pax6* is inactivated in chickens [315], but it proceeds to the optic cup stage without *Pax6* in mice [815]. Moreover, *Pax6* is not needed for regenerating eyes of planarians, developing adult eyes of polychaetes, or for the Hesse eyes of the chordate amphioxus [1198].

3. Neither *Pax6* nor the other members of its network are used exclusively by eyes [660,799,1111], so *Pax6* is obviously not *dedicated* to eye development [1629]. For example, *Pax6* is needed for nasal development [815] and for pancreatic islet formation [1975] in mice, as well as for hindbrain neurogenesis in rats [2227], but some of these extra-ocular roles may be due to *cis*-enhancers that arose in the recent past [202,304,811,1101].

The authors who have raised these objections endorse a rival school of thought, which asserts that urbilaterians lacked eyes but expressed *Pax6* in the head region where eyes later evolved in certain lineages [632,879]. Facts that support this proposal include: (1) flies lose their entire heads (not just their eyes) when either of their *Pax6* homologs – *ey* or *toy* – is disabled [1205], and (2) similar mutations in nematodes severely affect the entire head [379]. In clades that possess eyes *Pax6* is presumed to have been co-opted independently for a role in eye development because it happened to be in the right place at the right time [953,1628,1999,2349].

> *ey/Pax6*, or the ancestral gene from which they evolved, might have specified some activity only in a particular location in the body (the top front of the head). Then the use of the same gene in mice and fruitflies would reflect only the fact that the two animals grow eyes in a similar body region . . . At some level, homology must exist between mice and fruitfly eyes; the question is whether the homology is at the level of eyes, or head regions. [1903]

One advantage of this co-option scenario is that it avoids the difficulty of trying to explain how an organ as useful as an eye could have been lost by so many phyla (see Figure 3.3). One flaw of the co-option theory, though, is that it fails to explain why *Pax6* should be associated with eyes that develop outside the head region [1765]. Two striking examples are found in echinoderms: *Pax6* is expressed in eyes at the tips of starfish arms [719,1238] and in the photosensitive tube feet of sea urchins [1282,2314]. What about the few animals whose cephalic eyes lack Pax6 [1198]? Conceivably, *Pax6*'s role might have been usurped by other genes in their ancestors [2262,2349].

Originally, the Pax6 protein may have activated *opsin* genes directly [1197,1787] by binding cognate motifs in their promoters [353,1728], but other homeodomain proteins now serve this function in both flies (Pph13 [1508,1911]) and vertebrates (Otx and Crx [1861,2299,2349]). Nevertheless, *Pax6* still participates in the cocktail of transcription factors that assign cell fates in the vertebrate retina [1337,1421,2522,2558], and it is expressed in the retinal ganglion cells that function as rPRCs [78].

Moreover, a direct DNA-binding role appears to persist in lens development, where *Pax6* (or a paralog) has evidently been tasked with recruiting proteins to serve as crystallins [449,1087]. Indeed, one *Pax* enhancer for a chicken crystallin can elicit expression of a reporter gene in the fly lens [217], despite the fact that fly and vertebrate lenses look nothing alike and develop entirely differently [166,2305]. The explanation may reside more at the level of cell type than organ type [2363], because cell types – and the attendant batteries of genes that run them [1671] – are more stable than developmental pathways over geologic time scales [80,2434].

This same logic may apply to other "flertebrate" genes – a term coined by Joshua Sanes and Larry Zipursky for homologous genes that are used in similar ways by flies and vertebrates [1976]. There are many such genes, and their roles in fly versus vertebrate eyes are inventoried in Table 3.1.

Table 3.1 Features of the visual system that are shared by vertebrates and flies[1]

Shared feature	Conserved mechanism?	Vertebrate quirks	*Drosophila* quirks	References
Phototransduction	Chromophore (11-*cis*-retinal) and host protein (opsin).[2]	PRCs hyperpolarize via phosphodiesterase pathway.	PRCs depolarize via phosphoinositide pathway and signalplex.	[868,1537,1676]
Cis-retinal recycling	Enzymatic pathway acting in pigment cells adjacent to PRCs.	Rod PRCs convert retinal to retinol in their outer segment.	Esterification of the retinoid does not appear to be involved.	[100,2376]
Color vision	*spalt*/*Sall3* designates the PRCs (and opsins) used for color vision.	*H. sapiens* uses three opsins in our cone PRCs to detect color.	*D. melanogaster* uses four opsins in its adult eye to detect color.	[415,481,1537]
Pupil	No. Convergent trait: aperture mechanism.	Mediated by muscular contractions of the iris.	Mediated by pigment granules pulled by Myosin V.	[1117,1986]

(cont.)

Table 3.1 (*cont.*)

Shared feature	Conserved mechanism?	Vertebrate quirks	*Drosophila* quirks	References
Fovea	No. Convergent trait: high-resolution region for image analysis.	High-acuity spot with high density of cone PRCs.	High-acuity stripe where PRC inputs are superimposed.	[25,869,1420, 1699]
Eye identity	*ey/Pax6*-dependent kernel of regulators.[3] GOF: extra eyes.[4]	Subcircuits of kernel govern development of kidney (and other organs).	A paralog of *ey* (*twin of eyeless*) outranks *ey* in the hierarchy.	[536,733,762,1197, 1421,2299, 2476]
Lens formation	*ey/Pax6* enhancer for lens protein activation.	Lens does not come from retinal epithelium.	Lens-making cells arise alongside PRCs.	[217,353,1087]
Retinal growth	*hth/Meis1&2* spurs retinal mitoses; area code selector gene.	Also specifies area code in brain (tectum).	Associated with spreading wave.	[596,914,1347, 2414]
Founder neurons	*ato/Atoh7* specifies the earliest cell type in the retina (RGC or R8).	RGCs do *not* induce later cell types.	R8 cells initiate a cascade of inductions.	[269,762,1060, 1130,1819, 2211]
Neurogenic wave	*hh/Shh* drives a wave of neural induction across the retina.	Wave expands radially to fill the retinal field; *FGF*-regulated centrifugal wavefront.	Wave moves linearly from posterior to anterior; *dpp*-regulated linear wavefront.	[1210,1393,1455, 1615,1870, 1985,2368, 2390]
Hexagonal lattice	*Notch* establishes spacing intervals via lateral inhibition.	RGCs get spaced uniformly.	R8 cells get spaced uniformly.	[708,1455,1567, 1611,1787]
Retinal cell identity	*pros/Prox1* specifies retinal cell fates.	Elicits horizontal cells while suppressing rod PRCs and Müller glia.	Overrides R8 default state to allow an inner PRC to become R7.	[353,414,570]
PRC identity	*otd/Otx2* specifies PRC fates.[5]	*otd/Otx2*-null causes PRC degeneration.	*otd/Otx2*-null switches PRC identities.	[1861,2054,2417]

Table 3.1 (*cont.*)

Shared feature	Conserved mechanism?	Vertebrate quirks	*Drosophila* quirks	References
PRC shaping	*crumbs*/*CRB1* mediates cellular reshaping.	Additional role in cone outer segment.	LOF causes shorter rhabdomeres.	[1036,1193,1762]
Membrane layering	*prominin* and *eyes shut* assemble membrane stacks.	Ciliary (vs. microvillar) membranes.	Creates spaces between adjacent rhabdomeres.	[1622]
Interneuron identity	*dVsx1*/*Chx10*, *ato*/*Atoh7*, and *acj6*/*Brn3b* specify interneurons.	Interneurons reside in the retina.	Interneurons reside in the brain.	[594,595,1633]
Non-neural border	*Mitf* inhibits retina. LOF: non-neural cells become retinal cells.	*Mitf* elicits melanin pigment in the surrounding RPE.[6]	Peripodial epithelium (where *Mitf* is active) lacks pigment.	[199,852]

[1] The term "quirks" signifies derived traits (apomorphies) that differ in vertebrates versus flies. Slash marks denote homologs (fly/vertebrate) [677]. Rescue of fly LOF phenotypes by vertebrate homologs has been shown for *ey*/*Pax6* [844], *eyes shut*/*EYS* [1622], *otd*/*Otx2* [2262], and *prominin* [1622]. Abbreviations: GOF (gain-of-function), LOF (loss-of-function), PRC (photoreceptor cell), R8 (ommatidial PRC 8), RGC (retinal ganglion cell), and RPE (retinal pigment epithelium). Gene symbols: *acj6* (*abnormal chemosensory jump 6*), *ato* (*atonal*), *Atoh7* (*atonal homolog 7*, alias *Ath5*), *Brn3b* (*Brain-specific homeobox*/*POU-domain 3b*), *CRB1* (*crumbs homolog 1*), *dVsx1* (*Drosophila visual system homeobox 1*), *ey* (*eyeless*), *FGF* (*Fibroblast Growth Factor*), *hth* (*homothorax*), *Meis* (*Myeloid ecotropic viral integration site*), *Mitf* (*Microphthalmia-associated transcription factor*), *otd* (*orthodenticle*), *Otx2* (*orthodenticle homeobox 2*), *Pax6* (*Paired box 6*), *pros* (*prospero*), *Prox1* (*Prospero homeobox 1*), *Sall3* (*spalt-like transcription factor 3*). Features omitted due to uncertainty about commonality [1976] include: (1) dorsal–ventral compartmentalization [2104], (2) induction of cell types by one another [762,1234,1567,1787,1879], (3) cell movements to refine RGC or PRC spacing [1879,1881], and (4) cell death to tighten the lattice [241,826,1879]. For similar tables of further traits see [353,536,1210,2104,2398]. For reviews see [353,762,1976]. For theories concerning eye evo-devo see [595,1629,1666]. For "eye candy" galleries of diverse animal eyes see [2025,2506,2548]. For unsolved mysteries see [34,1879].

[2] The variant of 11-*cis*-retinal that is used by flies has a hydroxyl on carbon-3 [256,868]. The term "opsin" is used here (and elsewhere in this book) to denote the host protein (with or without its chromophore) instead of "rhodopsin," which can cause confusion because its meaning has narrowed recently to apply only to rods [1619].

[3] A "kernel" is a conserved gene circuit [469]. Differences between this term and Günter Wagner's related concept of a "character identity network" [2362] are explained in [2363]. The kernel in this case contains the genes *ey*/*Pax6*, *sine oculis*/*Six3*, *eyes absent*/*Eya1*, and *dachshund*/*Dach1* [677,735,1629,2398]. These genes are not only used in the eye but in other organs as well [202,1111,1553,1836].

[4] The GOF phenotype that is observed when *ey*/*Pax6* is artificially expressed at abnormal sites is extra (ectopic) eyes. Such eyes have been induced not only in flies [844] but also in frogs [383,1689,2558].

[5] PRC degeneration is also observed in *otx*-null flies, but only when a redundant gene (*Pph13*) is simultaneously disabled [1508]. For details of PRC gene circuitry, see [798,867] for humans and [1084,2224,2417] for flies. For combinatorial regulation of opsin promotors, see [353,1911].

[6] In the chordate lineage the association between *Mitf* and melanin goes all the way back to amphioxus: the pigment cells in its frontal eye express *Mitf* and contain melanin [1234,2350].

This chapter has only skimmed the surface of what we've learned about human and fly eyes and how they are encoded genetically, built developmentally, and related evolutionarily. Even so, it should be abundantly clear by now that science has succeeded in dissecting optics without sacrificing aesthetics [735,2025] – notwithstanding Keats's rude remark. Be they simple or compound, eyes are more than just devices. They are works of art.

4 Touch and Hearing

In his poem "Auguries of Innocence," William Blake extolled the virtue of trying "to see a world in a grain of sand" or, to put it more prosaically, to see the majesty of the macroscopic in the minutiae of the microscopic. Science has allowed us to do just that for animal eyes. As shown in the previous chapter, we have been able to fathom how we see by teasing apart the complex gadgetry of the photoreceptor cell. In a similar way, the present chapter will show how we have been able to discern how we hear by teasing apart the intricate machinery of the mechanoreceptor cell.

Indeed, the convoluted contraptions in the sensory hair cells of our inner ear are so marvelous that they have occasionally inspired even the most sober scholars to gush with superlatives [150,2427]. Here is just one of their "odes" to the cochlear hair cell (boldface added), whose features will be explored shortly:

> What has been under-appreciated is that the cochlea, more specifically the hair cell, is perhaps **the most extraordinary example of precision engineering seen anywhere in the vertebrate organism**. This precision is easily seen by a careful examination of the bundle of stereocilia that extends from a hair cell at a prescribed location in the cochlea. If a specific location on cochleae of different individuals of the same species is examined carefully, we find that the length, width, number, and arrangement of all the stereocilia (which vary along the cochlea) appear identical with variations of less than 5%. The biological purpose of this precision must lie in frequency discrimination. [2281]

The hair cells of the human cochlea develop like fly bristles

Insects differ from vertebrates in how they innervate their sense organs. Generally speaking, insects use an "outside-in" strategy whereby sensory neurons arise in the periphery and send their axons *inwards* to "plug into" the CNS [1477]. Vertebrates, on the other hand, tend to use an "inside-out" strategy, whereby neurons from the CNS (or neural crest) send their axons *outwards* to connect with cognate targets on the body surface [2010]. This tactical dichotomy is especially obvious in the visual system (Chapter 3): our retina develops as a lateral outgrowth from our brain that assumes the shape of an optic vesicle and then of an optic cup, whereas the fly retina comes from an imaginal eye disc that develops separately from the CNS, albeit nearby [301].

One exception to this rule concerns the hair cells that we use for hearing [1051]. Those cells line the inside of our cochlea, and they acquire sensory innervation *directly* from the tissue of the lining itself, rather than importing their neural wiring *indirectly* via cell migration or axonal outgrowth from the CNS [18]. Indeed, the entire process that our otic epithelium uses to construct its sound detectors is remarkably similar to the recipe that the insect epidermis uses to construct its touch detectors [18]. Moreover, the corresponding cell types turn out to differentiate via the same cytoskeletal devices [2277] and to express the same molecular markers [580]. The parallels at all of these levels imply a deep homology between the hair cell module of a vertebrate and the bristle cell module of an insect (Figure 4.1) [580,681,747].

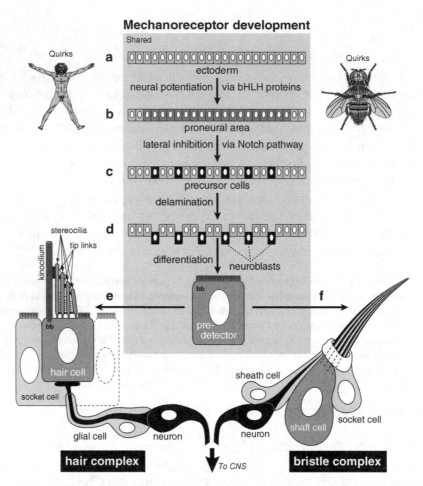

Figure 4.1 Similarities between the development of cochlear hair cells in vertebrates and bristle cells in insects.

a. The cochlear epithelium comes from the ectoderm of the otic placode in humans [52], and fly skin comes from the ectoderm of imaginal discs [923].

b. Proneural regions (darker gray) are demarcated by bHLH (basic Helix-Loop-Helix) proteins [4,191,1849]. Vertebrates use Atoh1 (a homolog of the fly's Atonal) [110,1051,1062] and Neurogenin1 (an Atonal relative) to delineate proneural fields in the cochlea [52,546,685]. Flies use Atonal for photoreceptor cells in the eye [1051] and for chordotonal organs in the ear [577], but they use Achaete and Scute for bristles in the skin [923]. Achaete and Scute form heterodimers with Daughterless [2302], and equivalent dimers are employed in the mouse ear [1051].

c. The solid proneural area resolves into punctate sensory organ precursor (SOP) cells via lateral inhibition and lateral induction (not shown) [1152,1777,1849]. Both inhibition and induction are mediated by the Notch pathway [546,580,1139,2248]. NB: The spacing of

Figure 4.1 (*cont.*)

SOPs is typically not as uniform as depicted here, nor is the ratio of SOPs to ordinary cells necessarily 1:2 [18,1330]. That ratio actually depends on the species being considered and on the location of the sensory epithelium within the body.

d. The SOPs and their progeny undergo further mitoses and rearrangements [18,546,923,2370], culminating in the assembly of a sensory module [2364] consisting of four cell types [18,685,1230]. One of those cell types is a neuroblast, which delaminates from the epithelium. Another cell type, labeled below as the "pre-detector" cell, remains on the surface and elongates its apical microvilli to form either one (**f**) or more (**e**) protrusions [2279,2540]. Those protrusions become stiff levers that pivot at their base when deflected by external force [1143]. Arrows pointing to e and f indicate that the pre-detector cell becomes the hair cell (**e**) or shaft cell (**f**).

e. The hair cell (dark gray) is a strain gauge [992]. It transduces tactile inputs into electrical outputs, which are conveyed via a ribbon-type synapse to the dendrite of an underlying neuron (black) [576]. The neuron's axon merges with the auditory nerve that leads to the central nervous system (CNS). The glial (Schwann) cell ensheathes the dendrite. This cell is the only member of the complex that does not arise in situ: it comes from the neural crest [52]. Socket cells alternate with hair cells in a tiled mosaic [1051]. They support the hair cell, and in non-mammalian vertebrates they mediate hair cell regeneration [2370,2523]. One of the flanking socket cells is shown here in dotted outline to indicate affiliation with the adjacent cluster. The hair cell converts some of its apical microvilli into a staircase of "stereocilia" [2279,2540]. That name is misleading, however, because they are not true cilia like the actual kinocilium that grows out from the basal body (bb) as a bundle of microtubules (vertical lines) [1363]. The latter (true) cilium is joined to the tallest stereocilia by a web of fibers [1132], but it vanishes before birth in mammals [174,1273,1606] and may not function in sensation even in taxa that retain it [498]. Adjacent stereocilia are lashed together by (1) a mesh of "ankle links" at their base, (2) a ladder of "shaft connectors" along their length, (3) a zone of "horizontal top connectors" near the top, and (4) diagonal "tip links" at their apices [230,270,1363,1606,2140]. When stereocilia are pushed toward the taller end of the tuft, the tip links pull ion channels open on the shorter partner [198,695,992,1594] and thus transduce motion into electrical signals [584,640]. A simplified staircase is drawn here in side view. Mice actually have dozens of stereocilia per tuft arranged in the shape of a U, V, or W, with the tallest one at the vertex [1273,1606], and chickens have up to ~450 stereocilia in a tightly packed rectangular array [2278,2281]. Our own cochlea has ~20–300 stereocilia per cell [1273,1363] and a total of ~16,000 hair cells [230,992,2027]. Our hair cells are aligned into four rows [1363]: one row of *inner* hair cells transduces the sound, while three rows of *outer* hair cells amplify the sound [454,991,1127,1349]. For genetic circuitry see [156,428,1385,1849,2370]. For cellular cartography see [156,1140]. For surprising interactions among hair cells see [1834], and for evolutionary origins see [561,683,689].

f. Unlike the hair cell complex, which is thought to arise via intercellular signaling [1152,1385,1849], the cells in the bristle complex arise via a strict pedigree that yields two pairs of sisters (shaft-socket and neuron-sheath) [1143,1230]. Transduction is also handled differently. Sensation in vertebrates occurs in the hair cell, but in flies there is a division of labor [682]: the neuron transduces downward deflections of the shaft into electrical signals via compression of its dendrite [1849]. Despite these different tactics (cf. Figure 2.3 for divergent ways that animals build a conserved CNS), the fly's shaft cell may be homologous to the cochlear hair cell [747,1062]. Both types of cells elongate their protrusions by the stacking of actin bundles (dotted lines in the shaft cell) [908] like successive floors of a skyscraper [230,2277,2281]. Those bundles disintegrate in the shaft as it matures (leaving only a legacy of tapered fluting) [1143,2282], but they persist in stereocilia (not shown) [1273].

Adapted from [18,52,923,1606].

The sensory neuron of the bristle corresponds to the sensory neuron of the ear; the shaft cell, presumably, to the hair cell of the ear (see [ref. [2282]] for the structural similarities in cytoskeletal organisation); and the socket cell to the supporting cell. The neural sheath cell has no such obvious counterpart: the glial cells in the cochleovestibular ganglion derive from the neural crest, not the otic epithelium. [18]

Homologies between insects and vertebrates are commonplace at the molecular level . . . But the homologies we have pointed out for the mechanosensory structures go deeper, including function, multicellular anatomy, development, and molecular controls. Indeed, there are few, if any, other multicellular structures where correspondences between insect and vertebrate seem so clear, detailed, and extensive. [580]

One of the molecular markers that is expressed by both the hair and bristle modules is intriguing with regard to the *Pax6* story that was recounted in Chapter 3. Just as *Pax6* offered a hint of common descent for the different eyes of various phyla, the related *Pax2* gene offers a similar clue for the mechanosensors discussed here [235,684,773]. The bristle's shaft cell needs *dPax2* for proper differentiation [1124], and the cochlear hair cell likewise requires a *Pax2* homolog [1907,2293]. Indeed, *Pax2* and its paralogs (*Pax5* and *Pax8*) regulate the differentiation of mechanoreceptors in a variety of taxa across the animal spectrum [123,773]. Evidently, this gene family has served that developmental role for at least 600 MY.

Fly bristles transduce tactile stimuli by using their shaft as a rigid lever [1425]. When the shaft gets deflected toward the surface, it compresses the dendrite attached to its base [170,757]. That compression, in turn, triggers the sensory neuron to send an electrical signal down its axon to the CNS [1061,1366].

Cochlear hair cells also rely on stiff levers [1273], but the transduction in their case occurs at the lever's tip, rather than at its fulcrum [1110]. Those levers, which are known as "stereocilia," develop from microvilli [1363] and are bundled into staircases [1606,2279,2280,2540]. Intriguingly, the height of the staircase varies linearly along the length of the cochlea (see Puzzles 4.1 and 4.2).

Stereocilia and bristles only react to forces coming from certain directions [1116], so it makes sense that their vector angles within the plane of the epidermis would have been fine-tuned by natural selection as a function of their bodily location to suit the sources of predictable stimuli. For example, flies use parallel rows of bristles on their forelegs to collect dust, and all of the bristles in those "brushes" point distally so that they can do their job effectively when the legs groom the eyes [920,2042]. The bristle polarities that we see in flies today evidently became entrenched long ago [1446], because drosophilids that were preserved in amber ~50 MY ago look remarkably like modern fruit flies in this regard [2205].

In contrast to the polarities of most fly bristles, which can be rationalized based on their function, the hair cells of our inner ear display region-specific orientations that defy any simple explanation. The polarities of sensory hairs within our utricle and saccule (described in the next section) are especially problematic from the standpoint of how a human engineer would have designed this system.

Puzzle 4.1 What causes stereocilia to vary with hair cell location?

The heights of stereociliary bundles vary in a graded manner along the cochlea: hair cells near the low-frequency (distal) end have taller stereocilia than those near the high-frequency (proximal) end [1606,2280,2281], analogous to string lengths in a Steinway piano [156,983,1271,1349].

The height gradients along this "tonotopic" axis [1363] are a function of growth *duration* in the chicken cochlea (a.k.a. basilar papilla) [2278,2280] but of growth *rate* in the hamster cochlea [1096]. In neither case do we know what dictates the temporal parameters. Nor do we know why the shapes of bundles in the semicircular canals differ from those in the cochlea. The only clue so far is that Wnt and Hedgehog pathways have both been implicated as upstream regulators [1576,1606].

Finally, we have yet to decipher how genes force the number of stereocilia per cell to vary over a ~3-fold range (~60 to 220 in the chicken) from one end of the cochlea to the other [2278]. Presumably, all of these cochlear features evolved to suit the physical constraints that are imposed by relevant sound frequency ranges, fluid dynamics, and sound amplification [454,992,1349].

Puzzle 4.2 What causes stereocilia to stop growing at a certain height?

Stereocilia could be using the kinocilium as a yardstick [2280,2281], since it resides at the highest point of the hair bundle [1606]. If so, then we need to determine (1) how the kinocilium "knows" to stop growing and (2) how it "tells" adjacent stereocilia to stop growing [1363].

Growth is controlled at the tip of the stereocilium [1597,2527], and this regulation has been shown to depend upon Myosin 15a for delivery of capping factors [230,1406,2540]. A timing device of some sort may also be involved, because taller stereocilia start growing earlier and stop growing later than short ones [1606,2281].

We also need to ascertain how growth is coordinated among stereocilia so as to build and maintain the staircases [156,575,2281], but at present we don't even know whether the mechanism is biochemical or biophysical [982,1041,1411,1695]. Nor do we know how stereocilia get aligned into rows [1363].

Some of the staircases look like tiny pipe organs [150,640,991,992], and in a way, they *are* musical instruments, albeit ones that let us *hear* music rather than ones that *play* music. Indeed, "their structural beauty has fascinated scientists since the dawn of modern science" [1901].

For more riddles see [156], as well as Barr-Gillespie's provocative essay "Assembly of hair bundles, an amazing problem for cell biology" [150]. For a survey of molecular size regulators in general see [982,1295,1427] (cf. Puzzle 3.2).

Macular hair cells display a plane of symmetry like the fly eye

Our inner ear starts out as an ectodermal placode [1386,1999], which invaginates to form a fluid-filled vesicle [1385,2010]. This spherical cyst then undergoes a grotesque transformation [156,1448,2426]: (1) a spiraling cone grows out from one point to form the cochlea, (2) three tongue-like protrusions emerge from the opposite side at right angles to one another to form the semicircular canals, and (3) two sections between these poles expand like balloons to form the saccule and utricle. Hair cells then begin to sprout in all six of these alcoves, but only at sites (stripes or spots) that will later be used for detecting motion due to (1) sound [1849], (2) head rotation [576], or (3) gravity, respectively [1367,2448].

The hair cells of our cochlea are aligned in parallel stripes, while those of the other chambers are confined to ovoid spots [1385]. The hairy spots of the saccule and utricle are termed maculae. The maculae are distinct from the spots in the semicircular canals insofar as the lawns of their hairs exhibit a line of mirror symmetry (see below) [1301,2154] instead of all pointing in the same direction as they do in the cochlea [1138,2493]. What causes such region-specific polarities?

Epithelial cell polarities across the animal kingdom turn out to be governed by the same six proteins (Figure 4.2) [621,845,2311,2493]. Those proteins comprise what has come to be called the core planar cell polarity (PCP) pathway [774,2049].

In humans this core PCP cassette orients the hair follicles throughout our skin [515,833,1166,1866] as well as the sensory hairs within our inner ear [621,1138,1533]. In flies it orients (1) the bristles throughout the skin [54,930], (2) the hairs on the upper and lower surfaces of the wings [705], and (3) the ommatidia in the eyes (cf. Chapter 3) [621].

The fly *eye* might actually help solve the polarity riddle of our inner *ear*, because it exhibits a comparable line of symmetry (Figure 4.3) [2104,2285]. In the case of the fly eye, that line is its equator. All clusters of photoreceptor cells (prospective ommatidia) initially point in the same direction as they arise from the morphogenetic furrow, but they subsequently pivot 90° either one way or the other [923], depending on which side of the equator they are on [621]. Clusters in the dorsal half rotate clockwise, while those in the ventral half rotate counterclockwise [923]. The chirality of the turn appears to be controlled by opposing gradients of the proteins Dachsous (Ds) and Four-jointed (Fj) [705,774].

Those two proteins (Ds and Fj) belong to a separate, *non*-core PCP pathway that also includes the protein Fat [705]. Interestingly, gradients of the mouse's Ds and Fat homologs – Dchs1 and Fat4 – have been observed in its CNS [2524], though they have not yet been demonstrated in the saccule or utricle [2090]. Hence, we don't yet know for sure whether the maculae are using the same PCP circuitry as the fly eye (see Puzzles 4.3 and 4.4). We do at least have *suggestive* evidence for the involvement of the non-core PCP pathway in *part* of the inner ear, because null phenotypes of *Fat4* are consistent with an effect of this gene on the polarity of hairs in the *cochlea* [1960].

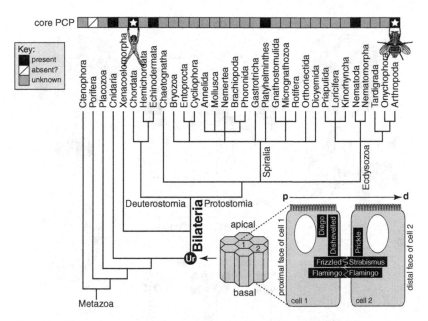

Figure 4.2 Common usage of the core planar cell polarity (PCP) pathway among animal phyla. Compiled here are the results of a 2015 survey by Rosalind Hale and David Strutt of all available data on the usage of the core PCP pathway by animal phyla. Their review was aptly titled "Conservation of planar polarity pathway function across the animal kingdom" [845]. The core PCP cassette was first studied in flies and later found to act likewise in mammals [20,705]. Although the pathway has been documented in only a few phyla aside from arthropods and chordates [845], there is at least one phylum from each trunk of the bilaterian tree (Deuterostomia, Spiralia, and Ecdysozoa), so the module likely dates back to stem bilaterians [2493]. On the other hand, the distribution of the *non*-core Fat–Ds–Fj pathway (Figure 4.3) is too poorly sampled to warrant plotting here [845,2493], but with regard to both pathways we can confidently conclude that "the similarities between fly and vertebrate PCP establishment are overwhelming" [2493].

The six proteins of the core PCP pathway are cartooned (lower right) as black rectangles, with intracellular (heteromeric) binding denoted by adjacency [1446,2203,2466,2493] and intercellular (ligand–receptor) binding denoted by jigsaw-puzzle complementarity [2493]. In the fly wing, where PCP circuitry has been thoroughly analyzed, each protein localizes to the proximal (p) or distal (d) side of each cell [2493]. Three of the proteins span the membrane and bind proteins on adjacent cells: Frizzled binds Strabismus, while Flamingo binds another Flamingo [2466]. The other three are internal: Dishevelled and Diego localize to the distal face, while Prickle resides on the proximal face. (For clarity, only one set of proteins at the interface is drawn.) These protein names are from flies, where Flamingo is also called Starry Night, and Strabismus is also called Van Gogh (Vang) [2493]. Vertebrate homologs of Frizzled and Prickle bear the same names, but vertebrate names differ for Diego (Diversin), Dishevelled (Dvl), Flamingo (Celsr), and Vang (Vangl) [1446,2466,2493]. Overall this cassette allows epithelial cells to align along an axis in response to a global cue [514], which is often a Wnt signal, depending on the tissue [514,1448,2493]. For protein circuitry see [2493]. Adapted from [845,2493]. Tree of extant phyla as in Figure 1.3.

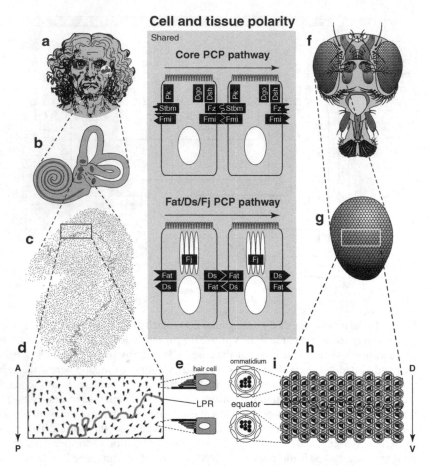

Figure 4.3 Are lines of symmetry created by planar cell polarity (PCP) mechanisms? The two PCP pathways that are shared by humans and flies are cartooned in the gray rectangle. *Above:* Conserved proteins in the *core* PCP pathway [342,1088] are redrawn from Figure 4.2, except that (1) abbreviations are used instead of full names and (2) proteins are depicted on both sides of each cell instead of just at the interface [2493]. *Below:* The same format is used for components of the *non*-core PCP pathway [845,1448,1960]. Wide ovals represent nuclei, while stacks of thin ovals denote the Golgi apparatus, where the kinase Four-jointed (Fj) phosphorylates Fat and Dachsous (Ds) to modulate their binding affinities [845]. Fat and Ds do not segregate to opposite poles of the cell as strongly as proteins of the core pathway, though they do exhibit asymmetric binding (not shown) in response to concentration gradients (not shown) [408,621,705,2466]. The non-core pathway acts quasi-independently of the core pathway [1441,1488,1960,2466] – i.e., in some tissues they act in series [216] but elsewhere in parallel [408,491]. Redrawn from [514,774].

The rest of this figure compares humans and flies with regard to cases where a line of symmetry separates antiparallel polarities. Available clues suggest that both species use the core and non-core pathways alike to align cells with respect to a reference line [1446,2137,2524], but much remains to be learned about the particulars of how they do so (see text) [1363]. For insightful reviews see [117,774,2493]. For the roles of cilia and micro-tubules in PCP see [1441,2108,2369].

a. Although our inner ear arises from superficial ectoderm, it eventually sinks to a depth visualized here through a hole in the cheek [2010].

←

Figure 4.3 (*cont.*)

b. Our inner ear may remind art lovers of a reclining Henry Moore sculpture, but to snack lovers it probably looks more like a snail trying to eat a pretzel. The snail is the cochlea, which enables us to hear, while the "pretzel" is the vestibule that enables us to balance on two legs [817,2113]. Between them are the utricle and saccule, which enable us to sense gravity [2448] via otoconia that act like a plumb line [576,1367]. Hair cells (see Figure 4.1) are found in all corridors of the labyrinth [52], but they are confined to five (dark gray) spots and one stripe [156,1138]. The cochlear (spiral) stripe has four rows of hair cells [1138]. Clockwise from lower left, the five spots are: saccular macula, utricular macula, and the three cristae of the semicircular canals [491,1849].

c. Utricular macula of a mouse, with each of the ~3600 hair cells drawn as a triangle to indicate staircase polarity (**d**). The wiggly line is a line of polarity reversal (LPR). This drawing, which was painstakingly traced from confocal images, is from [1301] and is reproduced by kind permission of the senior author, Ellengene Peterson. (A mouse proxy had to be used here because no such herculean feat has yet been performed for the human ear.)

d. Enlargement of the boxed area in **c** to show a section of the LPR (gray line). A→P (anterior–posterior axis). Each arrowhead denotes a hair cell with its kinocilium at the triangle's tip. Most of the hair cells point toward the LPR, which gives the overall array a line of mirror symmetry, albeit a wobbly one. If this were a diagram of the *saccular* macula, the hair cells would point *away* from the LPR [491,614,1138]. The utricle and saccule function primarily as *gravity* sensors (for head tilt) in vertebrates, but they were co-opted to serve as *vibration* detectors in burrowing snakes [2501]. For a perplexing reversal of hair follicle polarity see [349]. For more PCP riddles at the tissue level see [858,1167,1256,2091,2377]; for riddles at the cellular level see [758,1764,1942,1959,2428]; and for riddles at the molecular level see [682,684,686]. For models as to why the mammalian cochlea is a spiral see [1407,1408]; for guesses as to why stereocilia occur in bundles see [42,1196]; and for ideas as to why some bundles are shaped like a "V" see [2246]. For clues as to how the mammalian vestibular apparatus evolved in the first place see [173].

e. Schematics of hair cells (cf. Figure 4.1) drawn to clarify the triangle symbols. In each case the staircase of stereocilia (slanted toward the kinocilium) points toward the LPR, thereby giving the overall array its line of mirror symmetry.

f. Head of a fly.

g. Right eye, as seen from the side with anterior to the right. Each eye has a line of symmetry at its equator, though it can only be detected by looking beneath the cuticle [1211,1872].

h. Zigzag line of symmetry as seen in one specimen, redrawn from Figure 20 of [1872], except that their (left eye) diagram has been reversed to simulate a right eye. D→V (dorsal–ventral axis). Filled circles are rhabdomeres of the seven photoreceptor cells that are visible in any given cross section. The seven rhabdomeres form a trapezoid that points away from the equator, with its longest side facing anteriorly [621]. This mirror symmetry allows flies to see sharper images at the center of their visual field by using neural superposition across the equator [25] – effectively giving them a "fovea" analogous to the one you are currently using to read these words [869,1420,1699]. No interfacial line in a hexagonal array can be exactly straight, of course, but this line jogs more than it has to, revealing a slight error rate [25].

i. Single ommatidia on opposite sides of the equator, enlarged to stress the mirror symmetry. Unfilled circles, half circles, and ovals denote other cell types (e.g., pigment cells) aside from photoreceptors (cf. Figure 3.5) [302]. For the role of PCP genes in *bristle* polarity see [196].

Puzzle 4.3   How did hair cells evolve their current polarities?

Hair cells only notify the CNS when force is applied to the *front* of the stereo-cilia staircase [1116]. They do not do so when pressure is exerted on its *sides* or its *back* [576,1273,2281]. It therefore makes sense, as discussed in the text, that the regional polarities of hairs would be consistent from one individual to the next so as to suit the most common "approach vectors" of incoming stimuli.

What does *not* make sense are the regional idiosyncrasies in different parts of the inner ear [525]. Why, for example, should all cochlear hair cells point the same way (perpendicular to the longitudinal axis) [614,1140,1448], while the lawns of hairs in the utricle and saccule are bisected by a curved line of symmetry in each case (Figure 4.3)? And why should utricle cells point *toward* this line of polarity reversal (LPR), while saccule cells face *away* from it [491,614,1138]? Finally, how on earth, if the LPR itself is so important, are hair cells of the chicken utricle able to regenerate with correct polarity after that boundary is excised [2382]?

Theoretically, the utricle and saccule must be equipped to detect all possible angles of linear acceleration that the head can possibly undergo [491,1391], and there should be at least three orthogonal canals in order to completely cover all head rotation angles [491]. Given these considerations, authors have historically wondered how agnathan fish are able to manage with only two semicircular canals [132,576,687,858]. However, it turns out we have been wrong to pity the lamprey for its reputed vestibular deficiencies. In 2014 a third sensory axis was found in the lamprey's labyrinth apparatus [1392]! Curiously, its anatomy deviates so much from that of jawed vertebrates that it must surely have evolved independently.

The hair cells of the mammalian inner ear bear a striking resemblance to the neuromast cells of a fish's lateral line [457], and they may have arisen from a common ancestral cell type [134,1748]. Flies have nothing like these devices. Instead, they (convergently) evolved a different kind of "gyroscope" for monitoring motion: they transformed an old pair of hindwings into a new pair of stabilizers (halteres) [520,928].

Puzzle 4.4   Why do macular hair cells violate core PCP rules?

In a 2013 paper entitled "A balance of form and function: planar polarity and development of the vestibular maculae", Michael Deans described a paradox posed by hair cells of the mouse utricle [491]. Given that those cells orient their kinocilium-bearing side toward the LPR (Figure 4.3d), we would expect the core PCP proteins to obey this same line of symmetry. However, they do not. The mouse's Fz and Pk homologs – Fz6 and Pk2 – are expressed on opposite faces of each cell, but the Fz6→Pk2 vectors do *not* reverse polarity across the LPR [1901]. In other words, hair cells on one side of the LPR express Fz6 on their kinocilium face, while cells on the other side of the LPR express Pk2 on their kinocilium face.

How are we to explain this heretical behavior of core PCP proteins? One idea that was floated by Deans is that cells on flanking sides of the LPR might have different compartmental identities (cf. [630]) that affect their polarity. Cells with identity A, say, would interpret Fz6 to mean "Make a kinocilium here," while cells with identity B would use Pk2 as the cue for that outcome instead.

This conjecture is not as farfetched as it might seem. In the fly wing, for example, the cells on either side of the A/P border (separating the anterior and posterior lineage compartments) have different states (ON or OFF) of the gene *invected*, and mutationally forcing all cells to be ON evokes a perfect line of symmetry (in terms of vein pattern and, presumably, cell polarity) at the A/P boundary [923,2098].

Curiously, neurons on opposite sides of the LPR (in both the utricle and saccule) send their axons to different CNS targets (cerebellum vs. brainstem) [1391]. This divergent routing also suggests a difference in cell identity based on compartmental affinity.

Is there a gene complex for human hair like the one for fly bristles?

How does our genome assign hair cells to six sites in our inner ear: one cochlear stripe and five vestibular spots (Figure 4.3b)? We do not yet know. Nor do we know how our genome demarcates all of the sundry hairy areas on our skin. Why, for example, do we grow hair follicles on our scalp and then, after puberty, add follicles to our pubic area and armpits? Why do men extend this trend further than women by growing beard hair, chest hair, etc.? And why are there sacrosanct zones where no hairs are permitted to sprout at all, such as our lips, palms, and soles? Although we have no answer to this riddle for *H. sapiens*, a similar riddle was solved long ago for *D. melanogaster*, and that solution might offer us some useful clues.

The bristles of fruit flies come in two main varieties: big macrochaetes and small microchaetes [921]. Both versions contain the same four cell types: a shaft cell, a socket cell, a neuron, and a sheath cell (Figure 4.1f). Hence, the difference in bristle size is not attributable to cell number. Rather, macrochaete shaft and socket cells undergo extra rounds of DNA endoreplication, causing larger cell size [111,1128,1969].

Decades of research on bristle patterning culminated in 1995 with the detailed dissection of a gene cluster at the tip of the X chromosome that is responsible for bristle initiation throughout the fly [769]. This "Achaete–Scute Complex" (AS-C) contains four paralogous genes, two of which – *achaete* (*ac*) and *scute* (*sc*) – establish virtually all of the proneural territories throughout the fly's skin (cf. Figure 4.1b).

Expression of *ac* and *sc* is targeted to the locations of future macrochaetes by eight *cis*-regulatory enhancers (Figure 4.4). They provide the kinds of "area codes" that were discussed in Chapter 1 for the *Hox* complexes, but with one major difference. The AS-C has no colinearity along any body axis: its enhancers are scattered willy-nilly along the locus relative to the surface layout of the macrochaetes they encode. In this respect the AS-C is more typical of how genes encode anatomy [1292,2083], and *Hox* complexes are the exception. For this reason alone, the AS-C offers a better exemplar than the *Hox* complexes for how our genome might be patterning our hairy territories [1079].

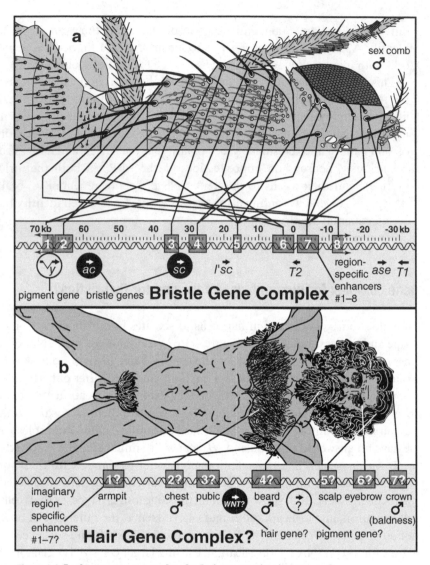

Figure 4.4 Is there a gene complex for hair patterning in humans?

a. Achaete–Scute Complex (AS-C), here termed the Bristle Gene Complex, of *D. mela-nogaster*, charted along a section (double helix in the gray rectangle) of the X chromosome, as mapped by Juan Modolell's lab in 1995 [769]. The four genes that comprise the complex are paralogs of one another: *achaete* (*ac*), *scute* (*sc*), *lethal-of-scute* (*l'sc*), and *asense* (*ase*). Within this 100 kilobase (kb) span are three other genes that don't belong to the complex proper: *yellow* (*y*), *T2*, and *T1*. Genes are denoted by thick arrows, and introns by kinks. None of the four AS-C genes has any introns. The two AS-C genes that regulate proneural fields for bristles are *ac* and *sc* (black circles), while *l'sc* regulates neuroblasts in the embryonic CNS, and *ase* acts during bristle differentiation. Numbered boxes (dark gray) symbolize independently acting *cis*-enhancer elements (cf. [97]) that turn *ac* and *sc* ON at the macrochaete sites to which they are linked in the drawing above. Macrochaetes appear to play the same remote-sensing role as cat whiskers [946,1116,1568], despite the fact that fly bristles and mammal hairs differ in anatomy

If our genome does subdivide our skin into hairy and naked areas by means of an AS-C-like "Hair Headquarters" (HHQ), then where is it? Our *ac-sc* homolog turns out to be a red herring because its function is in the nervous system and not in the skin [191,616,829]. Rather, the genes that are most likely to control hair patterning based on mutant phenotypes are those of the Wnt pathway [228,694,1922,2004,2536]. Gain-of-function alleles for genes in this pathway cause extra hair growth [1338,1599,2535], while loss-of-function alleles erase hair [63,994]. Moreover, the Wnt pathway is intimately involved in mammalian hair spacing [2088], differentiation [1498,2074], and regeneration [1031]. Humans have 19 *Wnt* genes [2322] – the same as mice [2536] – so those loci might be good places to pursue our quest for the "hairy grail" HHQ [926].

In doing so, however, we must keep in mind the route that evolution took to get to *Homo sapiens*. Proto-hominids were surely as hairy as modern chimpanzees, so the *cis*-enhancers in our HHQ are probably *inhibitory*, not *stimulatory* as they are in the AS-C. In all likelihood we lost our hair gradually over thousands, if not millions, of years as we got better at running to escape predators and chase prey [1038]. If that scenario is correct, then hair-suppressing *cis*-enhancers might have been inserted into our HHQ, one at a time, at random sites until we acquired enough to trim our fur coat down to its present dimensions. Extinct branches of the anthropoid family may have had different hair patterns, though we may never know what they looked like from their fossil remains alone.

Figure 4.4 (*cont.*)

[1922], ontogeny [568,2009], phylogeny [2468], innervation [1143,1303], and transduction [499,1116,1302,1366,2550] – thus indicating (or *screaming*) convergence. The bristle row labeled "sex comb" is a male-specific row of bristles used during courtship, and it has proven to be a veritable gold mine of evo-devo insights [1186].

b. Imaginary "Hair Gene Complex" based on the assumption that our genome designates hairy territories via AS-C-like "area codes" [1079]. Given that the Wnt pathway seems to license mammal hairs [694,1922,2004,2536] in a similar way to how the AS-C licenses fly bristles [63,994,1338,2535], one or more of our 19 *Wnt* loci [2322] might operate like the AS-C [926]. That is, (1) such a "Hair Headquarters" (HHQ) might harbor a handful of *cis*-enhancers that target *Wnt* expression to certain parts of the body, as exemplified by the connecting lines, and (2) those area codes would probably be scrambled as they are in the fly's AS-C, instead of being colinear as they are in our *Hox* complexes (cf. Figure 1.2). This scheme is probably flawed, however, insofar as it implies that evolution *added* hair to certain areas (scalp, groin, etc.), while it is more likely that evolution *deleted* hair from various areas (arms, legs, etc.) as our chimp-like ancestor (1) became bipedal, (2) began running, (3) overheated due to its fur coat, and (4) gradually lost most of its fur [1038]. Thus, the *cis*-enhancers, if they do exist, may have been installed to stop *Wnt* from being expressed in certain areas [926,1644] (cf. avian apteria [517,1079,1685]). If we *do* have an HHQ in our genome, then it is likely to dwarf the ~100 kb span of the AS-C, given the lower density of control elements in humans versus flies [2338]. For more clues to how genomes "encode" skin regions via *cis*-enhancers, see [2083,2085,2339]. For the mysterious (but related) "Case of the Feathery Feet" see [532]. And for how our brain "knows" where our skin has been touched, which is a puzzle in its own right, see [251,909,2309,2490].

From [926]. (The author apologies to Lenny, in absentia, for monkeying with his Vitruvian Man.)

Our time-honored way of thinking about the AS-C was challenged in 2015. Using an elegant blend of classical genetics and modern technology, Thomas Klein and his associates in Düsseldorf showed that flies can make bristles without either *ac* or *sc*, as long as the gene *extra macrochaetae* (*emc*) is also disabled [2302]! Their interpretation of this heretical finding is that Daughterless (Da) – the ubiquitous binding partner for Ac and Sc – can substitute for Ac and Sc by forming homodimers instead of its usual heterodimers, as long as the Emc proteins that could disable Da are absent. All four of these proteins (Ac, Sc, Da, and Emc) have a Helix-Loop-Helix (HLH) domain for binding one another, but Emc lacks the basic domain needed for binding DNA. Hence, no heterodimers that have Emc (Emc/Ac, Emc/Sc, or Emc/Da) can bind E-box targets that permit bristle initiation. Reassuringly, the bristle patterns of triply null (*ac-*, *sc-*, and *emc*-null) flies are highly abnormal – affirming the dogma that AS-C enhancers *do* control spatial patterning.

Flies use their antenna as an ear but still rely on conserved genes

Flies detect sounds via their antennae, and the sensors that they use to do so are "chordotonal" organs rather than bristles (Figure 4.5) [1126]. Chordotonal organs are stretch receptors [1425,2309] that are related to bristles evolutionarily [1061,1230].

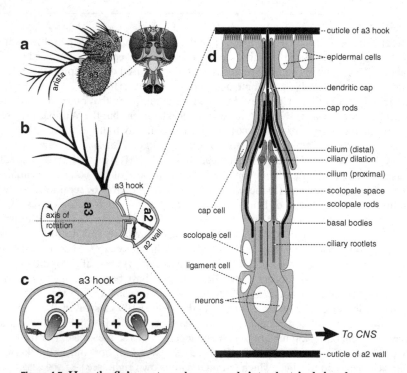

Figure 4.5 How the fly's ear transduces sounds into electrical signals.

a. The fly's ear, also known as Johnston's organ (JO), is located in its antenna [1491]. The antenna has three segments (a1, a2, and a3), plus a feathery arista that acts like a

Figure 4.5 (*cont.*)

weather vane to detect not only wind currents but also pressure fluctuations from sound vibrations [1098,1444]. Unlike the tympanal ears of other insects [1913,2478] the antennal ears of flies, bees, and mosquitoes are only sensitive to nearby sounds [1491]. Aside from its roles in hearing and wind sensing, the fly antenna also detects gravity using a dedicated subset of chordotonal organs (CHOs) [1098,1099,1444], though its chief chore is to act as a smell-sensing "nose" (see Figure 5.1) [2187]. For parallels between fly and vertebrate ears vis-à-vis transduction channels, gating springs, and adaptation motors see [171,776,1587]. For similarities in how fly and vertebrate brains process inputs from auditory versus gravity sensors see [1098], and for auditory neural circuitry see [1099,1100].

b. Cut-away view of the a2 segment of a right antenna, revealing the joint connecting a2 with a3. This joint looks remarkably like a tarsal joint [930] – not surprisingly, since antennae probably evolved from legs . . . or vice versa [445,778,1804,2086] – and it operates like the ball-and-socket joint of our shoulder. The "ball" in this case is a hook-shaped protrusion. It pivots within a "socket" that is thicker on its medial side. When the arista vibrates, the a3 segment rotates about an axis (dashed line) and causes the hook to swivel back and forth [777,1100]. Only two of the ~200 CHOs of the JO are drawn, one of which (left) is attached to the front of the hook and the other (right) to its back. Lateral is up, medial down.

c. Imaginary cross section of the a2 segment at the level of the a3 hook (cf. b) and its socket. Oscillations of the hook (left and right) alternately stretch or compress the CHOs attached to its sides [1143], thus activating (+) or hyperpolarizing (–) their neurons, respectively [43]. Only two of the ~200 CHOs are shown in this simplified schematic, and the extent of CHO stretching or compression is greatly exaggerated for clarity. Lateral is up, medial down, anterior left, and posterior right. For more detailed diagrams see [43,1100].

d. Enlarged view of a single CHO that spans the gap from the a2 wall to the a3 hook. Each CHO typically contains five cells that descend from a single mother cell via strict cell lineage, as in the bristle complex (see Figure 4.1) [1143,1230,1638]. The names of those cells (left column) are connected (dashed lines) to their nuclei (white ovals): cap cell, scolopale cell, ligament cell, and two neurons, though 10–15% of CHOs in the JO actually have *three* neurons [577]. The dendritic cap is a tubular structure made of extracellular matrix. It links the CHO to the a3 hook, while the ligament cell anchors the CHO to the a2 wall. The cap and scolopale rods are cytoskeletal girders with a core of microtubules and a thick sheath of actin bundles [227]. The scolopale space contains an endolymph whose ion concentrations mediate depolarization. Different proteins are expressed at different axial levels: NompA (extracellular matrix) and NompC (ion channel [2534]) in the dendritic cap and distal cilium, respectively [585], RempA in the ciliary dilation, Lav/Nan multimer in the proximal cilium, and Eyes shut (extracellular matrix) equatorially and basally in the scolopale space [227,1491]. Prestin is also expressed in CHO neurons of the JO [1127], but its subcellular location has not been pinpointed. For *divergent* roles of Prestin in fly and mammal hearing see text. For *convergent* roles of Prestin in bat and whale echolocation see [1335]. For convergence of a katydid's ears with our own ears see [983,1534]. For believe-it-or-not "freak shows" of (1) a one-eared praying mantis and its ability to evade bats by using a midline tympanum as an ultrasonic detector see [2479,2480], (2) a 12-eared grasshopper whose chants can be heard up to 2 km away see [2330], and (3) a tiny fly that eats crickets . . . and hunts them down by homing in on their love songs via tympanic ears that are as adept at binaural discrimination as our own ears see [1439,1600]. For the ancient role of cilia (neural or otherwise) in animal sensation in general see [2108], but see also [498,1650] for the debunking of a long-held myth. For weird ways that flies use cilia see [784].

Redrawn from [227,1100].

They differ from bristles insofar as (1) they lie beneath the cuticle rather than projecting from it [1135], (2) they usually have multiple neurons [1143], and (3) they arise via the HLH protein Atonal instead of either Ac or Sc [577]. Regarding this last point, chordotonal organs actually resemble cochlear hair cells more than bristles do [681] because the otic epidermis uses an Atonal homolog (Atoh1) for its proneural fields (Figure 4.1) [685,1414]. There must exist a substantial overlap between genetic circuits for bristles and those for chordotonal organs because mutations that are recovered in genetic screens for numbness also tend to cause deafness [578] (cf. [675]).

Given the different location of the fly's ear (in its "nose") relative to our own, it came as a shock in 2003 when Prestin, a membrane protein known to amplify sounds in mammal ears [1312,2539], was found to be expressed in the fly's ear [2401]. Prestin is a motor protein like myosin or kinesin, but it is powered by voltage instead of by ATP [454]. It enables outer hair cells of the cochlea to boost sounds [455,2402] by changing their height and stiffness in response to local vibration [453,905,991]. Initially the fly's Prestin was assumed to play a similar mechanical role, but this idea was disproven in 2015 [1127]. Flies turn out to use Prestin as an anion transporter [779], in keeping with the rest of the *prestin* gene family [454,2234], whereas mammals and birds evidently recruited Prestin for piezoelectric duty [1127,1334,1678].

If flies are not using Prestin in the same way as mammals, then how are they able to boost sounds [1491,1896]? To do so flies were found to use a transduction-based mechanism whose properties match those of fish, amphibians, and reptiles [1587].

Another surprise came in 2005 when a homolog of the *Myosin 7a* gene that is responsible for Usher type 1B syndrome in humans [2396] was shown to also cause deafness in flies [2287]. Myosin 7a is a motor protein with the actin-binding head of ordinary myosin [2492] but without its canonical tail [842,888]. It forms part of the tip-link complex of our cochlear hair cells [794], and it is expressed in the corresponding transducers of the fly ear (i.e., chordotonal organs) [2287]. In both cases Myosin 7a helps build and maintain the machinery of mechanotransduction [1826], while not actually participating directly in the transduction process itself [2288].

The ultimate proof of an extensive overlap between fly and human "hearing genes" [226,230] came in 2012 with a knockout screen for genes that require *atonal* for their expression in the fly antenna [1299]. A significant proportion (20%) of the 274 genes that were recovered in this way were found to have a homolog suspected of causing deafness in humans [2056]. The strangest finding to come out of this screen was the discovery that opsin proteins (the agents of *visual* transduction throughout the animal kingdom; see Chapter 3) are not only *present* in the fly ear but are also *required* for the transduction of sounds into nerve impulses (see Puzzle 4.5).

The acoustic properties of human and fly ears are quite similar

The best explanation we currently have concerning the functioning of the fly ear was proposed in 2008 by Björn Nadrowski, Jörg Albert, and Martin Göpfert [1587]. Their model accounts for virtually all of the measurable properties of auditory

---

Puzzle 4.5   What are opsins doing in the fly ear?

Four of the fly's seven opsin genes (*Rh3*, *Rh4*, *Rh5*, and *Rh6*) were identified in the screen for *atonal*-dependent, antenna-expressed genes (cf. Figure 3.5f) [2056]. Expression of region-specific genes in odd places occurs all the time via the frivolous activation of enhancers [1921], so this result by itself was not note-worthy. However, the researchers who performed this screen were sufficiently intrigued by this anomaly that they decided to investigate it further.

First, they showed that single mutations in either *Rh5* or *Rh6* cause partial deafness, and that the loss-of-function double mutant *Rh5-LOF Rh6-LOF* degrades auditory nerve output by 50%. Surprisingly, they then found that Rh5 and Rh6 require 11-*cis*-retinal in order to fulfill their hearing-related duties. Nevertheless, the 11-*cis*-retinal does *not* need to absorb photons to do its job, because light has no impact on auditory activity.

What role might these "dark" opsins be playing in the fly ear? One possibility is that they are acting as *thermo*sensors (cf. [2045]), since opsins are known to be capable of sensing temperature in other contexts [2066].

These ectopic opsins are thought to be remnants of a protosensory cell type that served multiple sensory modalities (tactile, chemical, and visual) before begetting the unimodal cell types that characterize modern taxa by a division of labor [688,1638,2298]. This notion may not be as farfetched as it seems, because neuromast cells of the fish's lateral line – a cell type distantly related to our cochlear hair cells – also express opsins [123]!

processing, including sound amplification. The parameters of fly and human hearing turn out to be strikingly similar (boldface added):

> Linking transducer dynamics and auditory-system performance, our analysis has mechanistic and evolutionary implications. First, **fly and vertebrate transducers share physical properties and seem to work in equivalent ways.** Molecular parameters that can be deduced from our model and those reported for vertebrate hair cells are of the same order of magnitude, with millisecond time constants of motor adaptation . . . and mechanical energies required to open single transduction channels . . . hardly exceeding the energy of thermal noise. Second, by extending transducer-based power amplification from hair cells to fly ears, our analysis supports transducer-based force generation as a widespread mechanism for amplification in hearing. [1587]

The many parallels between fly and human hearing offer hope that continued research on hearing in flies may one day lead to a cure for deafness in humans [547,737,1362,2056].

Fly and human ears do differ significantly in one key respect: the range of sounds that are audible. Flies can hear frequencies of ~100–1000 Hz [227,777,1099] with a maximal sensitivity for *D. melanogaster* at ~160–250 Hz [1143], which, appropriately, is the pitch of the male's courtship song [43,777] (cf. unique tuning peaks for other fly species [1896]). We humans, on the other hand, can hear sounds over a 20-fold broader range: ~20–20,000 Hz [992].

The accuracy of human ears actually rivals the acuity of our eyes, which, as mentioned in the previous chapter, are capable of even detecting single photons [1563]. One expert was so impressed by the auditory talents of our species that he penned the following lyrical, if somewhat technical, homage:

> Our auditory sensitivity is so great that we can detect signals that vibrate the eardrum by about a picometre. Indeed, we can hear sounds down to the level at which they are drowned in the din of thermally agitated water molecules in the cochlea. Although our hearing encompasses frequencies from 20 Hz to 20 kHz, we can distinguish the sources of sounds and the nuances of speech because our frequency resolution extends to one-thirtieth of the interval between successive keys on a piano. From the faintest audible sounds to the roar of jet engines, our hearing displays a dynamic range of an astonishing trillion-fold in acoustic power. [992]

5 Smell and Taste

Shakespeare famously observed (in *Romeo and Juliet*) that "a rose by any other name would smell as sweet," but he never addressed the question of why a rose should smell so delicious to a primate that doesn't normally eat flowers. Despite all the research conducted by the perfume industry over the decades, we still do not know why certain scents affect us so strongly [748,751]. Nor do we do know why a fragrance as complex as that of a rose should be perceived as such a gestalt [1269,2557], despite the fact that its components smell quite differently when presented one at a time [1404,2117].

Chemosensation differs from vision (Chapter 3) and hearing (Chapter 4) in one key respect: the heterogeneous stimuli within its purview cannot be reduced to a single (wavelength) dimension [2206]. To do its job, an olfactory or gustatory organ must be capable of sensing a menagerie of molecules as zany as the inventory of compounds in a college chemistry textbook [851,1017,1440]. This chore resembles the challenge faced by an animal's immune system [1529], which must somehow assess a witch's brew of epitopes that decorate the world of infectious bacteria and viruses catalogued in a clinical pathology textbook [2442].

Odors and flavors are identified via combinatorial compilation

The immunity analogy is particularly apt here because the smell and taste organs of humans and flies turn out to use the same strategy as our immune system [1530]. The solution in each case is to employ a set of lock-and-key feature detectors in combination with one another [952,976,1192,1397,1945]. Just as individual antibodies only fit *pieces* of invaders, olfactory (OR) and gustatory (GR) receptor proteins typically only recognize *parts* of molecules [851,1328,1529,2300] – like the proverbial blind men groping different parts of an elephant [1440,2354]. The brain then stitches these swatches into a composite "image" of the odor molecule [103] that can be either sharp or fuzzy [851,952] depending on the nature of the molecule.

Chemosensory organs actually have an easier task than our immune system [2442], because they can justifiably confine their scope to only those telltale molecules that signify food, mates, predators, or parasites, without having to respond to every whiff or quaff that comes their way [1473,1480]. Indeed, the menu of chemicals that can be detected by a given species generally does conform rather neatly with that animal's diet, habitat, ecology, and lifestyle [146,508,864,1856].

Smell and taste are widely thought to be separate modalities [478], but wine connoisseurs know that wine glasses are shaped the way they are to let our nose augment our tongue in savoring the richness of each sip [1387,2149,2367]. Fruit flies, it so happens, are even greater wine aficionados than we are, because fermenting fruits are the staple of their diet [1032]. It should come as no surprise, therefore, that they also sniff while sipping. The scheme that they use is clearly convergent to ours, rather than being related by common ancestry. Flies use a special pair of "noses" that seem dedicated to this chore [2354]. These "maxillary palps" can perceive nearby odors at higher intensities than the fainter doses to which the antennae are attuned [2081,2206], and the palps are aptly located on either side of the labellum (Figure 5.1c), which is the insect version of our tongue.

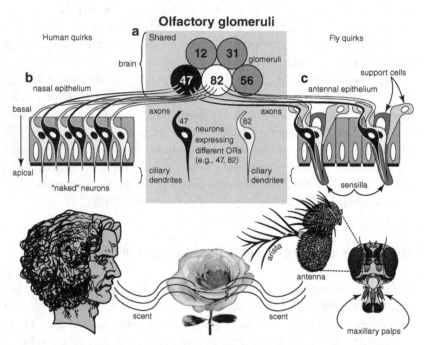

Figure 5.1 Similarities in the processing of olfactory input by mammals and flies.

a. Olfactory neurons in both mammals [1530] and flies [1304] express a single olfactory receptor (OR) [1304,1530] and send their axons to a single cognate glomerulus [711,2117]. Glomeruli are spherical conglomerates (depicted as circles) of first-order synapses [2206,2399]. Aside from their OR, fly neurons also express the generic cofactor Orco (a member of the same OR family) [864], which heterodimerizes with all OR types [148]. Two actual OR identities from flies are used here as examples [851]: neurons expressing OR 47 (black cells with white nuclei) innervate glomerulus 47 in the antennal lobe of the brain, whereas neurons expressing OR 82 (white cells with black nuclei) project to glomerulus 82 [148,850]. The other depicted glomeruli (12, 31, and 56) are imaginary. Each glomerulus actually receives input from ~50 neurons in flies and ~5000 neurons in mice [2206], and the total number of glomeruli reaches ~50 in flies [1017] and ~1000 in mice [103]. Mice use a different gene family to encode their OR proteins [184,864,2295],

Gustatory neurons are broadly tuned to only a few taste qualia

Taste is usually defined as the detection of waterborne chemicals, whereas smell entails the detection of airborne chemicals [617]. Since all phyla arose in the ocean, while only a few – including chordates and arthropods – ever ventured onto land [377,2003], we might expect humans and flies to share more in common with regard to the older of the two modalities [1053]. However, the opposite seems to be true. The taste organs of humans and flies rely on mechanisms that differ more strongly at both the cellular [346,617,704] and molecular [1325,2354] levels than do their olfactory organs.

Nevertheless, there are some common themes [60,1105]. In both humans and flies the GRs of taste receptors are tuned more broadly than the ORs of olfactory receptors (except at high concentrations) [1328]. In each case their role is to sort stimuli

---

Figure 5.1 (*cont.*)

and the glomerular taxonomy differs for their olfactory bulb [1016,1555] – the equivalent of the insect's antennal lobe [1564,2206]. OR proteins decorate the ciliary dendrites [644,2206,2255] at the tips of the sensory neurons [590]. Both mammals and flies can learn to associate particular odors with food or harm [107,2342] and respond accordingly [195,047,1176]. Higher-order neurons are omitted for clarity [1016,2206,2354,2442], as are higher-level brain centers [750,1016,1440,1555,2442].

b. We detect odors by inhaling air through our nasal passage [2206,2300]. The roof of that chamber (drawn above the head) is peppered with isolated ("naked") neurons (black or white cells) that collectively express ~400 OR proteins [1442] (vs. ~1000 in mice) [103]. This nasal epithelium is pseudostratified (not shown) [1542], and the protruding dendrites on its apical surface are bathed by mucus (not shown) [644], which coats the interior of the nasal cavity [2255,2300]. The dendrite of each neuron actually branches to form a bush of ~15 ciliary tips (not shown) [343,1123], which increases its surface area and boosts its sensitivity [617].

c. Flies detect odors primarily via the distal (a3) segment of their antennae [864,1089], though they also use their maxillary palps for intense aromas [311,2081,2206]. Unlike mammals, where the sensory neurons are distributed throughout the epithelium, flies cluster their olfactory neurons into discrete sensilla along with a cohort of non-sensory support cells [1017]. Two such (basiconic) sensilla from the antenna are drawn [2187], each of which contains two different (black or white) neurons [850]. The dendrites of those neurons are immersed in lymph that is secreted by the supporting cells [2354]. Because the different OR subtypes share the same pool of lymph they inhibit one another in a seesaw (either/or) Boolean logic circuit [1963]. Access of airborne chemicals to the lymph is mediated by pores (not shown) in the surface cuticle [1089,1259], and transport of those chemicals to dendrites is facilitated by soluble odorant-binding proteins [864,1259,2206]. Thus, flies, like humans, dissolve odor molecules before analyzing them [944] – a strategy that makes sense given the aquatic origins of animal phyla [2196]. For opinions on conserved versus convergent origins of chemosensory mechanisms see [944,1610,2054,2196,2206], and for the conserved role of homeodomain proteins in olfactory evolution in particular see [1724,1803,1804,2446]. For riddles about how ORs evolve see [311,864,1510,1856,1973]. For the mystery of why dendrites detach from sensilla that lack cilia, see [157]. For additional puzzles see [1538,2441].

Adapted from [148,1089,1269,1910,2300].

into about five basic categories [254,2496]: sweet, sour, bitter, salty, and umami in humans [443,1325] and sweet, bitter, salty, water, and CO_2 in flies [501,592,1328].

This simple palette of discrete qualia is evidently enough for the sensory palate to assess the edibility and desirability of the morsel at its doorstep, which is, of course, its main assigned task [699,2496]. When axons conveying these gustatory inputs arrive at the first relay in the brain, they do not encounter anything resembling the glomeruli of the olfactory system (see below) [2034]. Nevertheless, the circuitry at this level may still be functionally comparable to some extent [151,2354,2379].

Olfactory neurons use one receptor and project to one glomerulus

Another aspect of chemosensory logic (aside from combinatorial circuitry) that is shared by humans and flies concerns how olfactory information is parsed by the CNS (Figure 5.1a). In both cases, sensory neurons that express the same OR send axons to the same glomerulus [2206] – a globular nexus within the olfactory bulb (humans) or antennal lobe (flies) – where they synapse with second-order neurons (see Puzzle 5.1) [711,1017,1555,1564].

Each glomerulus thus amplifies the signals that presumably come from one facet of the molecule being smelled [1192], with the gain thereby achieved (in neurons per glomerulus) reaching ~50:1 in flies and ~5000:1 in mice [1440,2206]. These boosted signals are the elementary percepts that get quilted into an abstract representation [103,1269,2067] at higher levels within the brain [707,750,850].

Puzzle 5.1   How do sensory axons find their target glomeruli?

The sensory neurons that express a particular OR protein are dispersed within the olfactory epithelium of both mammals and flies, yet they somehow manage to send their axons to single glomeruli in the olfactory bulb [1016] or antennal lobe [1963]. How do they do so?

One possible strategy was discussed in Chapter 2. According to Sperry's Chemoaffinity Hypothesis [2150], axons might carry tags of some kind on their surfaces that would only match cognate labels on their corresponding glomeruli [559,2502]. The most obvious candidates for such tags would be the OR proteins themselves, and investigators initially did favor this explanation [1878], but they later abandoned it when experiments pointed to other factors [1017,2059].

Instead of *qualitative* tags, afferent axons appear to be using *quantitative* (graded) cues [1015,1778] from (non-OR) cell-surface proteins [976,1017], including semaphorins, teneurins, ephrins, Eph receptors, and transmembrane adenylate cyclases [1169,1621,2502].

Some sorting actually occurs in the axon bundle before it even reaches the olfactory bulb in mice [240,1017,1778], and electrical activity within the axons is known to play a targeting role in both mice and flies (see text) [1017,1823,2308]. Even if ORs don't directly steer axons, *OR* genes must be involved at some level for one simple reason: evolution of a new *OR* gene consistently elicits a new glomerulus [181,1015].

Ultimately, the assembled sensation of the stimulus being smelled triggers appropriate behavioral responses (appetite, arousal, or revulsion) [195,647,1176,1305] that are based on either instinct [1928,2053] or learning [108,2342], or both. One of the most dazzling displays of hard-wired instincts in flies entails "remote-control" optogenetics: fly larvae can be steered toward or away from light by activating attractive or aversive olfactory neurons that are genetically modified to respond to light [2193]. An equally elegant proof of the dependence of taste preference on GR identity is to insert human GR genes into fruit flies [17]. Such transgenic "Frankenflies" don't exactly hurry to the nearest McDonald's for a cheeseburger, but they do alter their taste preferences to a measurable extent.

The "one-receptor-one-glomerulus" modularity is such an obvious solution to the mapping problem that the vertebrate and arthropod clades could easily have stumbled upon this trick quite separately from one another [442,944,2196], and the same is true for the "one-receptor-one-neuron" rule [434,503]. This conjecture of convergence is buttressed by the fact that the genes encoding mammal and fly ORs are unrelated to each other [184,864,2295]: the ORs of mammals are G-protein-coupled receptors (GPCRs) [644,2255] like those of nematodes [1589, 1818,2429], whereas the ORs of flies are ligand-gated ion channels [369,1984, 2430].

There are ~1000 functional OR genes in mice and ~400 in humans [103,1442], but only ~60 in *D. melanogaster* [1017,1610]. The smaller cohort in fruit flies is presumably due, at least in part, to the lesser demands of their more limited diet [195,477]. The antecedents of the fly's OR genes have been traced to basal insects [1510], but, contrary to expectation, this gene family arose (as an offshoot of the GR family [508]) *after* the earliest insects adopted a terrestrial lifestyle. In contrast, the GPCR family was involved in the chemosensation business long before metazoans – or even eukaryotes – ever evolved on the planet [2055].

The decision to use a certain OR in a certain cell is more preordained by the genome in flies than in mice, where it is relatively random [1017,1304], but in both cases the olfactory epithelium is balkanized into discrete provinces by the interplay of morphogen gradients and transcription factors [456,1017,1555]. Moreover, the consequences are comparable within each of the resulting subdivisions:

1. The menu of available OR options is limited within each province.
2. Neurons that choose the same OR are scattered in their province.
3. Interspersed identities get unscrambled by OR-specific glomeruli.

The mechanism by which sensory neurons select a single OR from the multitude of options is not fully understood (see Puzzles 5.2 and 5.3).

Nor do we yet know why the "one-receptor-one-neuron" axiom must be taken with a grain of salt, so to speak, in both flies and mice. In both cases, it so happens that there is a ubiquitously expressed OR (mouse) or OR-*like* (fly) protein that accompanies each of the variable ORs to their final destination (see Puzzle 5.4).

Puzzle 5.2   How do sensory neurons choose a single receptor in flies?

The proximo-distal "area codes" for the antenna resemble those of the leg [148,1304], which is not surprising since antennae probably evolved from legs . . . or vice versa [445,778,1804,2086].

For the same reason, perhaps, the sensory neurons of the antennae and palps are housed in cuticular sensilla that look remarkably like miniature leg bristles [148,1089,2187]. Instead of one neuron per bristle, however, there are up to four neurons per sensillum [478] – all of which arise clonally from a founder cell [1910]. The ORs expressed by sister neurons appear to be dictated by transcription factors [1304,2132] that are parceled out during mitoses of the founder cells [1867]. Those factors comprise a region-specific combinatorial code [1050,1304,2132], but we do not yet know how they interact at *OR* gene promoters in such a way as to activate the chosen gene *if and only if* all of them are present [456,1049]. (In general, we need to figure out how transcription factors execute Boolean logic [1292,1293,2461].)

Rarely, more than one functional OR is expressed per neuron [2354], and in at least one case the co-expression has been maintained for >45 MY [764], but we don't yet know what function, if any, it serves (cf. GRs [456,501,1304,1910]). Despite the strict spatial regulation of sensillar subtypes, a surprising amount of intermixing occurs at zonal borders. This scatter is due, at least in part, to cell migrations within the antennal epithelium [165,2132], rather than to the "roulette wheel" mechanism that has been invoked to explain the salt-and-pepper mosaicism of OR distributions in mice (see Puzzle 5.3).

Puzzle 5.3   How do sensory neurons choose a single receptor in mice?

The olfactory epithelium in mammals is populated by naked neurons, rather than by enclosed sensilla [1017,2300], and the selection of an OR is not merely monogenic but also monoallelic (i.e., it uses the paternal or maternal allele but not both) [1530]. Most important, the choice of OR is largely random (cf. Puzzles 2.10 and 3.6) [1910].

For all of these reasons, the algorithm that makes the decision must differ significantly from the one in flies [703,1123,1910]. There are several competing models [456], but all of them invoke a negative feedback loop [2058] of a functional OR protein back onto the selection device (via an unfolded protein checkpoint?) to prevent an *OR* pseudogene from being recruited based on *cis*-enhancers alone [636,1374]. The available evidence indicates that a slow activator randomly picks one allele, whereupon a fast inhibitor prevents other alleles from being chosen [3,1530] based on motifs inside *OR* coding regions [1489].

Activation is probably mediated by locus control elements (LCEs) that only recruit one allele per gene cluster at a time [503], but if so, then how on earth is coordination achieved among separate LCEs at loci (gene clusters) on multiple chromosomes [1529]? The answer to the latter riddle may have to do with the congregating of *OR* clusters at certain foci within the cell nucleus [397] (cf. *Hox* genes, discussed in Chapter 1 [1598]).

A related issue arose in 2015 with the revelation that sensory neurons express *OR* genes from up to seven different chromosomes before settling upon one *OR* gene [860]. The researchers who made this discovery have proposed several intriguing explanations, but for now this puzzle remains unsolved.

Puzzle 5.4   Why do flies and mice need an "escort" for their ORs?

All fly ORs form heteromeric (*di*meric?) complexes with an OR-like cofactor called Orco [864,1984,2430]. This ubiquity means that Orco can't be mediating odor discrimination in the same way as canonical ORs despite its physical resemblance to them [2355]. One clue is that, unlike ORs – which are highly variable among insects – Orco is quite conserved [1249].

This riddle is reminiscent of the role played by Daughterless as the universal binding partner for basic Helix-Loop-Helix proteins (e.g., Achaete and Scute) in the proneural pathway (see Chapter 4) [2302]. Ubiquitous partners seem superfluous in general, so it is worth pondering why they evolved. Orco is traceable in the insect clade as far back as ~320 MY ago when (1) the OR family evolved [1510], (2) insects acquired wings [2459], and (3) vascular plants began emitting a diverse array of volatile chemicals [864].

In the absence of Orco, flies lose their normal response to odors, and ORs fail to localize properly to sensory dendrites [1249]. Hence, Orco appears to be an essential escort for ORs [2175]. A remarkably similar story was reported for mice in 2013. Despite the fact that the microvillar vomeronasal receptors (VRs) of mice are unrelated to fly ORs, they appear to use a similar partnering strategy [502]. Vmn2r1 (Orco's VR counterpart) is a bona fide receptor capable of sensing hydrophobic amino acids. Disabling its ortholog (OlfCc1) in zebrafish eliminates the ability to detect these (and other) amino acids.

Moreover, heterologous cells fail to insert receptors properly into their plasma membrane without OlfCc1 – implying a physical (dimeric?) association of OlfCc1 with ORs. The affinities of Vmn2r1 and its fish homolog suggest that the escort role began at the dawn of jawed vertebrates ~423 MY ago [249]. We are left with the problem of explaining how such similar escort devices (Orco and Vmn2r1/OlfCc1) evolved convergently.

Successful glomerular wiring requires sensory neuron activity

Until 2012 the glomeruli of insects and vertebrates were thought to develop differently [1017], but in that year new data revealed similarities that had been overlooked [2399]. Earlier investigators had contrasted how glomeruli develop in fly *adults* (during metamorphosis) with how they arise in vertebrate *embryos*, but that sort of analysis was inherently as flawed as comparing apples and oranges. The new study focused instead on how the *original* wiring of the antennal lobes is regulated in fly *embryos*. For the first time it was now possible to see how the fly creates its olfactory wiring *de novo* – without any residual platform of larval connectives that is left over from the embryo. When the new findings were compared apple-to-apple, so to speak, with what happens in vertebrate embryos, the following aspects of development were found to be shared [1823]:

1. Olfactory sensory neurons (OSNs) fire spontaneous action potentials. This autonomous behavior begins before mature synapses form.
2. Electrical activity of OSNs is needed in order to confine OSN axons to their cognate glomeruli. (Silencing causes axons to spread out diffusely.)
3. Electrical activity of OSNs is critical for second-order ("projection") neurons to be able to connect properly with third-order brain targets.

Puzzle 5.5   How do flies revamp their olfactory system during metamorphosis?

One question about olfactory systems that can be addressed in flies but not in mice is: How does an existing system get reconfigured during metamorphosis?

Consider, for example, the obvious fact that larvae have no need to detect sex pheromones, because mating is the furthest thing from their little minds, but adults are inclined to care about all things sexual [209,1750,2547]. Hence, pheromone genes must get roused from their slumber when adulthood dawns. There is only a small overlap between the repertoires of *OR* genes expressed during larval versus adult stages of the life cycle [1254], despite the fact that adults partake of the same food as larvae [2464].

We do not yet know how larva-specific genes get turned OFF and adult-specific genes get turned ON during the pupal period. As mentioned above, we have made some inroads into understanding how the wiring diagram of the larval olfactory system gets transformed into the adult version, but many mysteries remain unsolved [73,1255,1854,2267].

Pheromones are processed separately from other kinds of odors

Insects use pheromones (in addition to visual and auditory signals) to convey sexual or social messages to other members of their own species [1773,2295,2353], and these chemical cues are handled by dedicated neural pathways in the CNS of the recipient individuals [1469,1963]. That is, the pheromone-processing circuits tend to be separate from those that analyze other types of scents [707,1066], though exceptions do exist where food and sex intermingle [311,1893,2207]. The same logic applies to mice [250,253,1310], which go even further in segregating such inputs: mice and other vertebrates [819,820] use a separate sensory device called the vomeronasal organ for detecting pheromones even before neural processing commences [692].

We humans begin to make a vomeronasal organ when we are embryos, but nearly all traces of this rudiment are lost by the time we become adults [2303,2533]. Nevertheless, men's odors can affect women's cortisol levels [2471], and odorless compounds from women's armpits are apparently able to alter other women's menstrual cycles [2185]. Hence, *Homo sapiens* may have pathways for processing pheromonal signals [1919] (but cf. critiques [2472]). If so, then the neural circuits undoubtedly differ from those for ordinary scents [1311,1749,2378], because the physiological effects are largely subconscious [1002,2295].

In 2009 a new family of chemosensory receptor was reported for *Drosophila*, aside from the well-known OR and GR classes [185,2153]. These ionotropic receptors (IRs) were first noticed in a specific (coeloconic) type of antennal sensillum [1123], and they were conjectured to constitute an auxiliary odor-sensing modality of some kind [14,2148].

In 2014 one subgroup of IRs was found in gustatory neurons, and several of its members were implicated in pheromone detection by dint of their expression in sensilla that males use to contact the female during courtship [1178]. Unlike the fly's *OR* gene family, which is confined to insects, the fly's *IR* gene family has been documented in all protostome phyla thus far examined [442], indicating that it is at least 550 MY old [1957]. Indeed, the ionotropic glutamate receptor superfamily (from which *IR* genes arose) must be *billions* of years old, because it is found not only in animals but also in plants and bacteria [442].

Pheromone concentrations tend to be much lower (picomolar range) than those of garden-variety odorants, and accordingly, the pheromone receptors of both insects and vertebrates manifest a greater sensitivity than ordinary odor receptors [838], although they do so via divergent strategies [1123].

The epitome of sensitivity is the notorious ability of male moths to detect even the faintest scents of conspecific females at enormous distances [1434,1820,1966]. Just as researchers have marveled at the exquisite precision of photoreceptors (Chapter 3) and mechanoreceptors (Chapter 4), the literature is riddled with tributes to chemoreceptors.

Here is one such "ode" from an article entitled "Extreme sensitivity in an olfactory system" [68], where a miniature electrocardiography device was attached to the chest of a restrained male moth. As this excerpt makes clear, the poignancy of the suitor's ardor, which is measured as the rate of his heartbeat, is acutely aroused by the pungency of his beloved's perfume, as if this scent were driving him mad to intercept his sweetheart in midair (boldface added):

> We recorded olfactory-induced cardiac responses to evaluate olfactory response thresholds to behaviourally relevant odours in a moth. Specific antennal receptor neurons enable insects to detect biologically meaningful odours such as sex pheromones and host-plant volatiles. The response threshold values demonstrated here are well below anything earlier reported in any organism. **A heart response was triggered by less than six molecules of the most efficient odours hitting the antennae of the insect.** [68]

Or, to paraphrase the final two lines of Shakespeare's play about the star-crossed lovers:

> For never was a story of more yearning for intercept

> Than this of a lepidopteran Romeo for his Juliet.

6   Limbs

The superficial differences between humans and flies are nowhere more evident than in our appendages. We have two arms and two legs, whereas they have six legs plus a pair of wings. Our hands and feet have five digits each, whereas their legs have two claws each. We have a bony endoskeleton, whereas they have a chitinous exoskeleton. Despite these obvious differences in gross anatomy, it turns out that homologous genes assign regional identities to cells along our respective limb axes (Figure 6.1). Indeed, when Neil Shubin, Cliff Tabin, and Sean Carroll coined the term "deep homology" in 1997, they used animal appendages as the centerpiece for their argument (boldface added):

> We propose that the regulatory systems that pattern extant arthropod and vertebrate appendages patterned an ancestral outgrowth and that these circuits were later modified during the evolution of different types of animal appendages. Animal limbs would be, in a sense, developmental "paralogues" of one another; **modification and redeployment of this ancient genetic system in different contexts produced the variety of appendages seen in Recent and fossil animals.** [2086]

Appendages use a common set of genes to delineate their axes

The signaling molecule Hedgehog (Hh) organizes the anterior–posterior (A–P) axis of each fly leg, and its homolog Sonic hedgehog (Shh) performs a comparable function in vertebrates. Within the fly's imaginal leg discs, Hh is synthesized in the posterior compartment, whence it diffuses a few cell diameters into anterior territory. There it activates secondary morphogens that "relay" its signal throughout the rudiment [923]. In the vertebrate limb bud, Shh *itself* diffuses across virtually the *entire* A–P axis [2276] to *directly* assign digit identities for each hand or foot [925,2217].

Polydactylous anomalies in humans similar to those in Ernest Hemingway's famous six-toed cats [1242] have been traced to mutations at our *Shh* locus [61,506,1394] – yet another indication of its patterning role. Indeed, by thoroughly dissecting this locus genetically we are now on the verge of finally grasping why evolution has so rigidly constrained the number of digits that vertebrate hands and feet are allowed to manufacture during development [496,1283].

The proximal–distal axis of both vertebrate legs and insect legs is subdivided into three cardinal domains by the proteins Hth/Meis, Dac/Dach, and Dll/Dlx

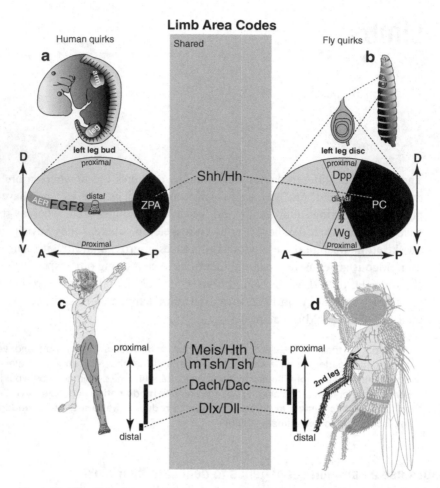

Figure 6.1 Shared features of leg development in vertebrates and flies. Proteins with similar domains of expression are listed as mouse/fly homologs in the gray box (top to bottom): Shh (Sonic hedgehog)/Hh (hedgehog); Meis (Myeloid ecotropic viral integration site)/Hth (Homothorax); mTsh (murine Teashirt)/Tsh (Teashirt); Dach (Dachshund homolog)/Dac (Dachshund); and Dlx (Distal-less homeobox)/Dll (Distal-less). Directions: D (dorsal), V (ventral), A (anterior), and P (posterior). Differing features ("quirks") are diagrammed in the margins. Black sectors of ovals (above) and black bars (below) denote zones of expression by the indicated proteins. See text for discussion of protein functions. The tactic of gated nuclear import that is used by the Hth/Meis protein (see text) [13,21,190,319,1486] is unusual but certainly not unique. For example, *Drosophila* uses this same trick along the D–V axis of the embryonic blastoderm [1939,1949,2186].

a. Left side of a human embryo with arm and leg buds labeled and somites demarcated along the future spine (after [2010]). The left leg bud is cartooned below as an oval. The bud's proximal–distal axis extends from the perimeter of the oval to its center (foot icon), with the other axes (A–P and D–V) indicated by arrows. Expression zones of signaling proteins (FGF8 and Shh) are inferred from data in mice [1348]. Abbreviations: AER (apical ectodermal ridge), FGF8 (Fibroblast Growth Factor 8), and ZPA (zone of polarizing activity).

b. Left side of a fly larva with the second-leg disc enlarged and cartooned below as an oval. The disc's proximal–distal axis extends from the outer edge of the oval to its center (see the legend of Figure 6.1 for abbreviations) [12,67,318,1486,1832]. All of these proteins are transcription factors. Hth/Meis and Dll/Dlx have homeodomains, whereas Dac/Dach belongs to a separate (winged-helix) family.

Hth/Meis is especially intriguing in its mode of action [1018,1648]. It must cooperate with the related protein Exd/Pbx as a heterodimer [1064,1956] in order to activate its target genes [1484]. However, Exd/Pbx cannot enter the nucleus without first binding Hth/Meis [319,1401,1904]. This peculiar strategy of cofactor-gated nuclear entry [13,21,190] operates the same way in vertebrates as in flies [1486].

Proof that Hth/Meis specifies proximal identities has been adduced for both vertebrates and flies by driving expression of this heterodimer in distal cells. Under these conditions Hth/Meis forces the distal tissue to make proximal structures [1486]. In both cases Hth/Meis also uses the same downstream (zinc-finger) transcription factor Tsh/mTsh to carry out its instructions [337,2465].

Dll/Dlx is much more famous in the world of evo-devo than either Hth/Meis or Dac/Dach, thanks to phyletic surveys that revealed its usage along the proximal distal axis of a wide variety of animal appendages [1723]. Examples include not only the arms and legs of vertebrates (phylum Chordata) and the legs of insects (phylum Arthropoda) but also the lobopodia of velvet worms (phylum Onychophora), the tube feet of sea urchins (phylum Echinodermata) [1626], and the parapodia of polychaetes (phylum Annelida; but cf. [1457,2447]).

The dorsal–ventral (D–V) axis of the vertebrate limb bud is bisected transversely by a prominent "apical ectodermal ridge" (AER), which serves as a signaling zone for outgrowth and patterning [1585,2292]. There is no corresponding stripe of any kind in the fly's *leg* disc (see Puzzle 6.1), but an analogous signaling zone does exist along the prospective wing margin (WM) in the fly's *wing* disc (Figure 6.2).

Figure 6.1 (*cont.*)

(claws), with the other axes denoted by arrows. Abbreviations: Dpp (Decapentaplegic), PC (posterior compartment), and Wg (Wingless). Dpp and Wg domains are quite different in onychophorans [1059].

c. Proximal–distal axis of human leg, showing inferred zones of expression of "area code" transcription factors based on studies in mouse embryos [1486]. All six Dlx paralogs are expressed in the AER at the growing tip [1724,1917]. For gene circuitry and further details see [1832].

d. Proximal–distal axis of the fly's second leg, where expression zones of "area code" (or "gap gene" [1865]) transcription factors are charted for the pupal period. This tripartite pattern is also seen in the unsegmented legs of velvet worms [1057], but the smaller-scale zonation of the fly tarsus (not shown [12]) is not [1684]. The two large dorsally pointing bristles at the leg base actually belong to the sternopleura, not to the leg proper, though the sternopleura does come from the leg disc [923]. For gene circuitry and further details see [12,923,1182,1832,1865]. For an intriguing theory of how these proximal–distal selector genes might have been redeployed from an earlier usage in the bilaterian head see [1277], and for the most well-understood case of gene redeployment (*Hox* genes in tetrapod limbs) see [1270].

This figure is adapted from [1832].

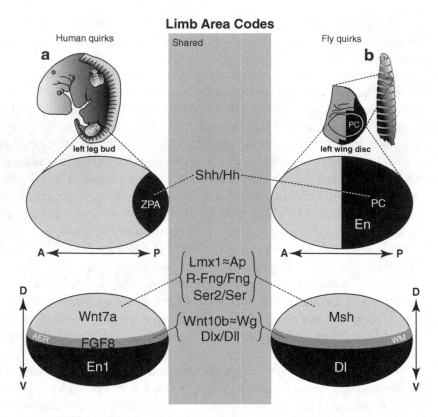

Limb Area Codes

Figure 6.2 Shared features of development in vertebrate legs and fly wings. Proteins with similar domains of expression are listed in the gray box as mouse/fly homologs or as mouse≈fly paralogs in the same protein family (top to bottom): Shh/Hh (Sonic hedgehog/Hedgehog); Lmx1≈Ap (LIM homeobox 1≈Apterous), R-Fng/Fng (Radical-Fringe/Fringe), Ser2/Ser (Serrate2/Serrate), Wnt10b≈Wg (Wingless-integration-site-1 10b≈Wingless), and Dlx/Dll (Distal-less homeobox/Distal-less). Each of the four ovals represents a flattened cone (outgrowth), whose proximal–distal axis extends from the perimeter (proximal) to the center (distal) as in Figure 6.1. See text for discussion of protein functions. For more genetic similarities of vertebrate limbs and fly wings see [1361]. Major differences between vertebrate limbs and fly wings include (1) the orthogonal roles of En/En1 along A–P versus D–V axes (see Puzzle 6.2), (2) the exclusively vertebrate reliance upon *Hox* genes along both the A–P [496,1348] and proximal–distal [1850,2086] appendage axes, and (3) the greater upregulation (at the D/V border) of Ser2 in response to R-Fng in vertebrates [1252,1920] than of Ser in response to Fng in flies [121,122,2301]. Even greater differences exist in annelid parapodia – implying that those appendages do not rely on the entire cassette sketched here . . . and questioning the antiquity of this circuitry [2446].

a. Left side of a human embryo, whose leg bud is schematized below as two ovals, with the upper one showing expression of proteins (inferred from studies in mice and chicks) along the A–P axis [1348] and the lower one doing the same for the D–V axis [1252,1734,1920,2467]. Wnt7a acts upstream of Lmx1 to confer D identity upon limb bud cells [1635]. Wnt10b is the mouse counterpart of chick Wnt3a, which foreshadows and activates FGF in the AER [2221]. FGF8 drives limb outgrowth [2063]. Wnt5a (not shown) is expressed in V mesoderm [1735]. All six Dlx paralogs are expressed

---

Puzzle 6.1   Why does the vertebrate limb bud resemble the fly's leg disc in certain ways but the fly's wing disc in others?

Based upon the parallels inventoried in Figures 6.1 and 6.2, the vertebrate limb bud seems to be a jumble of features from the fly's leg and wing disc [1832,2221]. It is therefore of some interest that the fly's leg and wing disc arise in the early embryo as a composite disc, which later splits into wing and leg fragments [399,782,783,923].

While the wing and leg may *develop* together, the insect wing did not *evolve* from the insect leg. Rather, it evolved from a separate paranotal lobe located *above* the leg [556,928,1637]. Because insect wings are a derived trait disconnected from the leg, it is unclear why their D–V circuitry should resemble that of the vertebrate limb at all (Figure 6.2). Some clues have come from branched legs in other arthropods [919,1077,2438], but the mystery remains.

Whereas Dll is required for fly *leg* outgrowth [401], it is not required for fly *wing* outgrowth [308,780] – a fact that violates the causal link between Dll/Dlx and outgrowth that was inferred from trans-phylum correlations [1626,1723], and further violations exist [1673,2218,2447]. Indeed, Dll's role in the fly wing is limited to promoting bristle differentiation along the margin [308,706,780].

Another riddle concerning Dll's mode of action still lingers from some decades-old experiments that have largely been forgotten by now. Namely, why should an ectopic leg be induced when Dll expression is artificially targeted to the central area of the wing, far away from the margin where one might expect an outgrowth to occur [780]?

←

Figure 6.2 (*cont.*)

in the AER [1724], and at least two of them – Dlx5 and Dlx6 – are necessary for outgrowth [1917]. En1 represses *Wnt7a* and *R-Fng* [1079,1252,1920]. For further details of the regulatory circuitry see [2344]. Abbreviations: AER (apical ectodermal ridge), En1 (Engrailed1), FGF8 (Fibroblast Growth Factor 8), and ZPA (zone of polarizing activity).

b. Left side of a fly larva with its wing disc enlarged alongside. The wing pouch (prospective wing) portion of the disc is cartooned below as two ovals, with proteins expressed along the A–P and D–V axes diagrammed separately (as in **a**) [706,2467]. Abbreviations: Dl (Delta), En (Engrailed), Msh (Muscle segment homeobox), PC (posterior compartment), WM (wing margin). Msh acts downstream of Ap to confer D identity upon wing cells [1496], analogous to Lmx1 in the limb bud [2341]. The vertebrate homolog of Msh (Msx1) is also involved in limb development (in the progress zone), albeit not as a D-state enforcing agent [1361]. Wg is the paramount morphogen for the wing's D–V axis [923]. Dll is expressed most strongly in the WM but weakly throughout the wing pouch [308,780,923]. Dl (Delta) is only expressed (at high levels) in the V half of the wing disc during the second instar stage, after which time it is confined to the WM [1170,2301].

Redrawn and updated from [2086], except that only the wing pouch (not the whole disc) is represented.

The D and V surfaces of the fly wing are virtually identical (albeit inverted) versions of one another [923], and the same mirror symmetry is also obvious in fish fins [925]. However, the D and V surfaces of the tetrapod limb deviate considerably from perfect symmetry – e.g., the fingernail versus palm sides of our own hands [925].

Dorsal identity is specified in both vertebrate limbs and fly wings by LIM-homeodomain selector genes: *apterous* (*ap*) in the wing disc [518] and *Lmx1* in the vertebrate limb bud [1902]. The vertebrate ortholog of *ap* is not actually *Lmx1* but *Lhx2*, which is expressed on *both* the D and V sides of the bud rather than on the D side alone [1908]. This fact does not necessarily preclude deep homology, because one paralog can sometimes usurp another's role over geologically long periods of time [2221]. *Lmx1* is activated by *Wnt7a* [1635] – a morphogen that is likewise expressed on the D side of the bud, albeit in the ectoderm (vs. *Lmx1*'s expression in the mesoderm) [1902].

Dramatic confirmation that *Wnt7a* and *Lmx1* dictate D identity comes from mice, where loss of *Wnt7a* function yields a double-V limb phenotype [1734], and ectopic (ventral) expression of *Lmx1* yields a double-D phenotype [361,1902], and equivalent effects have been documented for the chicken leg [2341]. Rarely, people are born with comparably symmetric deformities of their hands and feet [36,37,925], and one such syndrome has been traced to mutations in the human *Lmx1* ortholog [361,2219]. (These weird phenotypes must really be seen to be believed!)

The fly WM and vertebrate AER are both induced by the Notch signaling pathway at the boundary between the region where the Fringe/Radical-Fringe (Fng/R-Fng) enzyme is expressed (D half) and the region where it is absent (V half) [1028,1252,1920]. In each case Notch activity is restricted to this border because Fng/R-Fng glycosylates the Notch receptor so as to make it insensitive to its D-region ligand (Serrate) without affecting its sensitivity to its V-region ligand (Delta) [2301].

As its name implies, Fringe plays a conserved role in creating boundaries for a wide variety of different organs [726,1261], not just limbs [1026,2467]. Strangely (considering the A–P role of its homolog in flies), Engrailed1 represses *Wnt7a* and *R-Fng* in the ventral region of the limb bud [1079] (see Puzzle 6.2).

But the case for appendages in urbilaterians is not compelling

Despite all of the evidence that fly and vertebrate appendages are built upon a similar genetic scaffold (Figures 6.1 and 6.2) [1830], the consensus is that they are probably *not* homologous [601,2221]. This conclusion rests mainly on the fact that extant protochordates have neither legs nor fins: it is hard to imagine that their ancestors bore limbs, only to lose them and then regain them later [1092,2086,2087]. Moreover, if urbilaterians *did* have limbs, then the D–V inversion of the body that occurred at the base of the chordate lineage (cf. Figure 1.4) should have flipped the D–V order of our gene expression zones upside down relative to those of arthropods, which is clearly not the case (Figure 6.2) [2086,2221].

Puzzle 6.2   How did vertebrate and flies come to use Engrailed along different limb axes?

One odd difference between vertebrate and fly limbs (Figure 6.2) is the role of the homeodomain protein Engrailed/Engrailed1 (En/En1). En specifies P-state identities of cells along the *A–P* axis of leg and wing discs by activating Hh [923], but its vertebrate homolog En1 specifies V-state identities of cells along the *D–V* axis of the limb bud (by repressing Wnt7a and R-Fng) [1079,1252,1920].

   The P-state function is certainly more widespread in the animal kingdom [1744], but En/En1's ancestral task, like that of Dll/Dlx [1804], appears to have been in the nervous system [1415]. Hence, En/En1 could have been co-opted for a limb chore from its CNS role independently in the chordate and arthropod clades without any awkward rewiring of the genetic circuitry for the limb rudiment. Even so, the question remains: Why would the genome redeploy *CNS-specific* genes (*Dll*/*Dlx* and *en*/*En1*) for *limb-specific* roles in the first place [2221]?

We conclude that there has been no continuity of any structure from which the insect and vertebrate appendages could be derived, i.e., they are not homologous structures. However, there is abundant evidence for continuity in the genetic information for building body wall outgrowths and/or appendages in several phyla which must date at least to the common, potential appendage-bearing pre-Cambrian ancestor of most protostomes and deuterostomes. [2221]

The researchers who favor this view nevertheless argue that urbilaterians probably did use a common Dll/Dlx-based cassette for outgrowths of some kind [601,1277], but that those first outgrowths were likely to have been *sensory* (e.g., antennae) [1077,1516,1804], rather than *locomotory* in nature [1723,1724,2086].

The expression of the same transcription factor (out of the thousands available to be exploited in evolution) specifically in distal cells of appendages in six distinct phyla would represent a truly remarkable convergence. The more parsimonious explanation is that a *Dll/Dlx* gene was already involved in regulating body wall outgrowth in a common ancestor of these taxa . . . A scenario we favor is that an ancient ancestor of tetrapods and arthropods had primitive appendages, perhaps antennae, whose formation was under the control of a network of genes that included *Dll/Dlx, Shh/hh, BMP/dpp,* and *R-fng/fng.* [2221]

Consistent with this scenario, Dll/Dlx is widely expressed in mechano- and chemoreceptors that could have decorated the postulated protrusions before Dll/Dlx assumed full responsibility for their distal outgrowth [1043,1516,1626,2446,2447]. Moreover, the *cis*-enhancers for *Dll*'s expression in the sensory bristles, wing margin, and distal leg of *Drosophila* are separate from one another [706] – suggesting that *Dll* may have been successively recruited (for CNS, then sensilla, then outgrowths?) by adding modular enhancers to its locus, one by one [1077].

Odd-skipped transcription factors are critical for joint formation

The phylum Arthropoda is named for its jointed legs, but humans also have jointed limbs, as do nearly all of the other tetrapod species in our phylum Chordata. (Snakes are a glaring exception [928].) Even if we grant that arthropod and tetrapod legs evolved from an outgrowth in a common ancestor, there is no reason to think that such an outgrowth was segmented. Indeed, the first arthropod legs

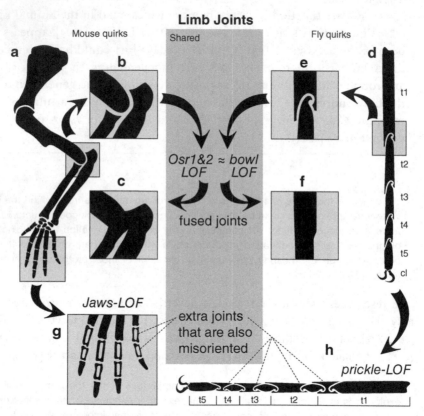

Figure 6.3 Similarities of limb joint development in mice and flies. Mutant genes with similar phenotypes are listed as mouse/fly homologs in the gray box. The wavy equals sign indicates that *bowl* (*brother of odd with entrails limited*) is not the ortholog of *Osr1* and *Osr2* (the ortholog is actually *odd*; see text), but *bowl* does belong to the same gene family as *Osr1* and *Osr2* (*Odd skipped related-1* and *-2*) [865].

a–c. Left foreleg (≈ arm) skeleton (dorsal view) of a normal mouse (**a**), with elbow joint enlarged (**b**) and compared to the corresponding joint of a mutant mouse homozygous for loss-of-function (*LOF*) alleles in both of its *Osr* paralogs – *Osr1* and *Osr2* (**c**). The mutant fails to form any "interzones" [493] (gaps between skeletal elements; **b**) that prefigure the synovial joints [1345]. As a result of this failure the humerus fuses with the radius and ulna (**c**) [712]. Schematic based on [712] (their Figures 3J and 3L). For unsolved puzzles of synovial joint development see [1709].

are thought to have resembled the *unsegmented* "lobopod" legs of velvet worms [1696,1830] in their sister phylum Onychophora [562]. Onychophoran legs have *exactly* the same sequence of Hth/Meis, Dac/Dach, and Dll/Dlx along their proximal–distal axis as fly legs (cf. Figure 6.1) despite having no joints whatsoever! This odd fact implies that tripartite zonation preceded leg segmentation in the panarthropod lineage [1057].

Given these inferences, it may come as a shock that both arthropods and tetrapods use the same zinc-finger transcription factor (or a close relative thereof) for joint initiation [1830]. The *odd-skipped* (*odd*) gene is most famous for its role in *body* segmentation of the fly embryo, where it functions in the pair-rule echelon of the overall hierarchy [429]. However, in 2003 *odd* was shown to also mediate fly *leg* segmentation [865]. This correlation supports Alessandro Minelli's "Axis Paramorphism" theory that legs evolved by co-opting genetic circuits from the body axis [1503].

Three other genes in *D. melanogaster* have zinc-finger DNA-binding motifs with strong homology to the corresponding sequence of *odd* (not to mention funny names): *brother of odd with entrails limited* (*bowl*), *sister of odd and bowl* (*sob*), and *drumstick* (*drm*). All four of these genes play a role in leg joint development: ectopic expression of any one of *odd*, *sob*, or *drm* induces joint-like invaginations, and loss-of-function mutations in *bowl* cause fusions of tarsal segments by eliminating all tarsal joints (Figure 6.3d–f) [802,865].

Figure 6.3 (*cont.*)

d–f. Tarsus of the left second leg (anterior face) of a normal fly (**d**), with the joint between the first (t1) and second (t2) segments enlarged (**e**) and compared to the corresponding joint of a mutant fly homozygous for a loss-of-function (*LOF*) allele in the *bowl* gene (**f**). Ventral is to the left and dorsal to the right. Surface bristles are omitted, and joints are drawn as if from a longitudinal section of the leg along its dorsal–ventral plane, with each white curved line being the gap between a ball and its socket [2225]. In such a rendition, the tarsus looks solid, but in fact it is fluid-filled, with an apodeme inside (not shown) that moves the claws (cl) [2125]. The evo-devo history of joints is full of puzzles in its own right [802,822,1194,2080,2226].

g. Digits (minus thumb) of the left hand from a mutant mouse homozygous for a LOF allele of the *Joints abnormal with splitting* (*Jaws*) gene. Note the extra synovial joints in two of the three phalanges per digit. Unlike normal joints, which are oriented transversely, these extra joints are oriented longitudinally, though in this diagram they may look deceptively like hollow centers alone. See text for further details. After [2124].

h. Tarsus of the left second leg (anterior face) of a fly homozygous for a LOF allele of the *prickle* (*pk*) gene called *spiny legs* (*pk^{sple-LOF}*). Ventral is up, and dorsal is down. Tick marks denote normal joint locations. Note the extra, inverted joints in segments t1, t2, t3, and t4, but not in t5. The extra joint in t1 is not always present nor as complete as depicted here, but the other joints are consistently well formed [930]. Redrawn from [923,929,930].

In mice there are two homologs of *odd* called *Osr1* and *Osr2* [713]. Because they act redundantly, it is necessary to disable them both in order to decipher their function. When such a double-knockout was accomplished in 2011, they were found to be essential for joint formation [712]. For example, when *Osr1* and *Osr2* are both prevented from functioning in the elbow, the humerus fuses with the radius and ulna (Figure 6.3a–c).

This common usage of the same family of Odd-related transcription factors for joint development in flies and mice is surely peculiar, but what makes it even stranger is that the respective joints develop so differently. The mouse elbow, for instance, starts as a solid Y-shaped condensation of mesenchymal mesodermal cells that is then carved by a dense "interzone" [493] at the future interfaces between the humerus, radius, and ulna [712]. Eventually a synovial joint emerges, made of articular cartilage, a fibrous capsule, and bounding ligaments [1345]. In contrast, the fly's leg joints arise from a cylinder of ectodermal epithelial cells that invaginate to form cuticular articulations ensheathed by a thin intersegmental membrane [1507,1809]. These ontogenies are as different (not to say disjointed) from one another as the recipe for coq-au-vin is from that for pecan pie.

All four members of the *odd* gene family in *Drosophila* act downstream of the Notch pathway, but they implement only a subset of tasks in the overall joint-forming process – viz., invagination and apodeme initiation [865]. Additional target genes of Notch are responsible for subsequent, more elaborate duties, such as sculpting the ball and socket of each tarsal joint [2225,2226]. Indeed, the Notch pathway appears to regulate limb segmentation throughout the entire phylum Arthropoda, regardless of the nature of the individual limbs or joints [1829,1830].

Notch is also involved, to some extent, in *vertebrate* joint development [979,1983], but the promiscuous reliance upon Notch for drawing metamere (and other) boundaries across the bilaterian spectrum [126,1027,2467] (e.g., somites [485,1754,1816,1962,2249]) suggests that its usage here is probably nothing more than a humdrum case of parallelism or convergence – unlike the more interesting (albeit perplexing) situation for *odd/Osr*.

Extra-joint syndromes are oddly correlated with polarity shifts

Part of the fun in doing genetics is finding weird mutants that prick your curiosity [1281]. Case in point: Ed Lewis's quest to understand the freaky fly with the four wings. His 1978 dissection of the Bithorax-Complex [1296] arguably begat the Modern Age of evo-devo [327].

To step out of my third-person narrator role briefly . . . there was an episode that gave me the same thrill as Ed surely felt in following his four-winged "white rabbit" down into that *Hox* Wonderland. I was entranced by the mutant *spiny legs*, whose wings had been analyzed in a 1982 paper by David Gubb and Antonio

---

Puzzle 6.3   How do extra, inverted joints arise in *spiny legs* flies?

More than 30 years have elapsed since extra, inverted joints were reported for *spiny legs-LOF* (*sple-LOF*) mutations of the *prickle* (*pk*) gene in fruit flies [930]. Sadly, their etiology remains enigmatic [825,923,924].

As mentioned in Chapter 4, the *pk* gene functions in the planar cell polarity (PCP) pathway (cf. Figure 4.3), so it is not surprising that $pk^{sple-LOF}$ would disorient tarsal joints as well as leg bristles (whence the name "spiny" legs) [930], but why should it be causing *extra* joints? This riddle was nearly solved by Amalia Capilla *et al.* in 2012 [320], who showed that each tarsal segment is girded by a ring of the Notch ligand Serrate. Cells along the ring's distal edge can "hear" the Serrate signal and make a joint, while cells along its proximal edge cannot do so because Pk makes them "deaf." If Pk is disabled by $pk^{sple-LOF}$, then cells on *both* edges of each ring make joints, and the proximal joints wind up being upside-down.

So far, so good, but we still need to figure out (1) why joints get flipped 180° so *precisely* while nearby bristles are oriented so *chaotically* [930], (2) why extra joints can be *perfect* despite the *intricacy* of ball-and-socket morphogenesis [803,1400,2080,2225,2226], and (3) why there are no extra joints proximal to the tarsus despite the Notch pathway being employed between segments there too [211,390,425,1865].

Aside from these remaining conundrums at the *tissue* level, there are many paradoxes at the *molecular* level. The chief one concerns how Prickle isoforms interact to produce various mutant phenotypes [1323] when their relative amounts are imbalanced [825].

García-Bellido [824]. I asked Dave to send me the stock, which he did, and two of my students proceeded to assist me in examining the mutant's legs.

What we saw was a series of upside-down tarsal joints (Figure 6.3h) that Dave and Antonio had failed to notice – a fact that Dave freely admitted and sorely regretted in later correspondence. We published our findings in 1986 in a paper brimming with puzzles we hoped would soon be solved by other workers in the field [930]. To my dismay, only a few aspects of this phenotype have ever been explained [320,924,2080] (see Puzzle 6.3).

Contortionists are commonly described as being "double-jointed," but their flexibility is not actually due to extra joints. Indeed, no anomalies have ever been reported in any vertebrate that mimic the phenotype of *spiny legs* mutant legs. Nevertheless, a mutation called *Jaws* has been found in mice that causes extra joints in phalanges [2124]. These extra joints manifest abnormal polarity, but instead of being inverted by 180° as in *spiny legs*, they are misoriented at a 90° angle relative to normal interphalangeal joints (Figure 6.3g; see Puzzle 6.4).

> ### Puzzle 6.4   How do extra, perpendicular joints arise in *Jaws* mice?
>
> In 2008 a mutant mouse strain was analyzed that exhibits extra synovial joints in the digits of the fore- and hindpaws [2124]. Strangely, these extra joints are oriented at right angles to the normal joints, as if the phalanges were split down the middle (Figure 6.3g). Based on this phenotype, the mutation was named *Joints abnormal with splitting*, or *Jaws* for short. This mutation was isolated in an insertional mutagenesis screen and was found to disable a protein phosphatase that regulates chondrogenesis. The homozygous LOF *Jaws* syndrome includes not only digit defects, but also cleft palate, reduced ribcage, and stunted limbs.
>
> We do not yet know (1) what causes the extra joints in *Jaws* paws, (2) how the joints manage to arise in the middle of otherwise normal-appearing phalanges, nor (3) why nearby metacarpals and metatarsals fail to show comparable anomalies, despite belonging to the same autopod "module" of the limb skeleton [2220,2238].
>
> This mutant is clearly trying to tell us something about how the genome assigns area codes to the skeleton, but no one yet knows what. Autopod researchers were recently stunned by evidence for a Turing-like device dictating digit number [2068,2069], so the answer might be buried somewhere inside that fascinating mechanism.

Some vertebrate and arthropod taxa can regenerate their limbs

In 1976 an ingenious article appeared in *Science* with the inscrutable title "Pattern regulation in epimorphic fields" [673]. In that paper Vernon French, Peter Bryant, and Susan Bryant proposed a new theory which, at the time, was rather shocking. They showed that certain vertebrates (salamanders) and insects (flies and cockroaches) regrow their limbs (or parts thereof) according to a common set of rules, which they used as the basis for a novel "Polar Coordinate Model." This model, with its clever explanations and testable predictions, electrified the then-dormant field of evo-devo [276], and the amperage was boosted the following year by Stephen Jay Gould's book *Ontogeny and Phylogeny* [785]. In the ensuing four decades some aspects of the model have been refuted [923], revised [276], or reinterpreted [307], but it remains a convenient framework for thinking about how limbs develop and regenerate [277,928]. So far none of this theorizing has led to any clinical cures, but that may soon change (see Puzzle 6.5).

Clearly, the next generation of evo-devo scientists has their work cut out for them if they hope to help amputees regrow limbs. The secret might lie in an offbeat phylum. More than half of all bilaterian phyla reside in the spiralian branch of the protostome clade, yet our grasp of appendage evo-devo in those taxa is pathetically meager [637,2446]. We would love to know, for example, whether octopus arms develop like human arms (or fly wings) in terms of their axial regulatory mechanisms [1639], but all we know so far is that squid tentacles do not use their

---

Puzzle 6.5   Why don't human limbs regenerate?

Despite decades of research by legions of investigators into the mechanisms of regeneration, people who lose arms or legs through war, injury, or disease are still being fitted with prosthetic devices, instead of being given a yet-to-be-devised elixir that could enable them to regrow their own limbs [1585,2188,2235,2292].

The main obstacle to progress in this area seems to be that unlike salamander tissue, which forms a blastema in response to amputation [2100], human tissue undergoes scarring that precludes distal outgrowth [259,1575,1578,2236].

Efforts to overcome this roadblock are under way on various fronts [761,1184], including pluripotent stem cells [635,804,1319], growth-promoting morphogens [1010], and the manipulation of *cis*-enhancer elements [1104].

Some mammals manage to bypass the scarring pathway, and they are being studied intensively to see how they do so [94,2047,2048]. Given the similarities between vertebrate and fly limb development (Figures 6.1 and 6.2) and wound healing [53,540,1566,1616,2374], the next breakthrough is arguably as likely to come from a fly lab as from a hospital clinic [1019,1814,2189,2237].

apterous homolog in any way akin to either fly or vertebrate limbs to specify dorsal regional identity [623].

Thus, there are some grounds for hoping that evo-devo research will lead eventually to limb regeneration, but the solution is likely to be a long way off. There is an even more critical problem crying out to be solved, however – and in this case the solution may not be so far away. As explained in the next chapter, congenital heart disease is the most common human birth defect [2326], and recent advances in the evo-devo analysis of heart gene networks offer hope that we may soon be able to repair broken hearts with interventions that are based on our growing grasp of gene circuitry [317,1011,2008,2134].

7 Heart

Geoffroy's 1822 hypothesis that arthropods are upside-down vertebrates (quoted in the Introduction and discussed in Chapter 1) was based partly on the fact that the heart is dorsal in arthropods but ventral in vertebrates [486,787,788]. He presumed that arthropod and vertebrate hearts are homologous, but he did not concern himself with one troublesome nicety: the apparent disparity in the complexity between the two types of hearts.

The fly heart is a simple tube [438,2531], whereas our heart has intricate chambers and valves [197,1768]. It is hard to believe that such different structures could be homologous, but our heart, like that of all vertebrates, begins development as nothing more than a plain (fly-heart-like) tube at the midline [890,1403], and the fish heart barely departs from this simple state even in its adult form [645,656,814]. Moreover, the fly heart actually harbors some subtle complexities of its own [1352]. Thus, Geoffroy's notion of cardiac homology may not seem so preposterous after all. A more rigorous test, however, had to wait until the modern era of molecular genetic analysis [225,889].

The critical gene for specifying heart identity seems to be *tinman*

The fly's *tinman* (*tin*) gene was so named because its null alleles cause a missing-heart phenotype reminiscent of the Tin Woodman in *The Wizard of Oz* [224,2521]. Its vertebrate ortholog was named *Nkx2.5*, based on the *NK2* type of homeobox that it contains [153]. In both flies and vertebrates *tin/Nkx2.5* governs the specification and differentiation of myocardial cells (Table 7.1) [607,2239,2521].

The *tin/Nkx2.5* gene functions as part of a cassette whose other members encode transcription factors in the GATA [1353,1774,1883,2138], T-box [265,1509,1840,1887,2176], and LIM-homeodomain families [1721,1988]. The intimate cooperation among these partners helps to explain why ectopic expression of *tin/Nkx2.5* alone fails to induce extra hearts in the same way as ectopic expression of *Pax6* induces extra eyes [607] (see Chapter 3). Thus, *tin/Nkx2.5* fails to satisfy the criteria for a "master" gene [1402,1612,2473] despite its rank at the top of the regulatory hierarchy [588,2244].

The interactions of *tin/Nkx2.5* with its cofactor (and target) genes (e.g., [1958]) were nicely summarized in a 2002 review by Richard Cripps and Eric Olson entitled

Table 7.1 Features of heart, vessels, and blood that are shared by vertebrates and flies[1]

Shared feature	Conserved mechanism?	Vertebrate quirks	*Drosophila* quirks	References
Heart induction	Dpp/BMP4 induces cardiac progenitor cells near the body midline.	Occurs at ventral midline. Second heart field regulated by Wnt and retinoic acid.	Occurs at dorsal midline.	[438,723,1168,1475, 1541,2176,2520]
Heart identity	*tin/Nkx2.5* specifies cardiac muscle identity.[2]	Expression is restricted to heart tissue per se vs. broader mesoderm.	N-terminal domain (lacking in Nkx2.5) needed for fly heart.[3]	[58,438,607]
Organogenesis	Cell clusters arise bilaterally and migrate to the body midline to form a plain straight tube.	Heart tube undergoes asymmetric looping, septation, and vascular branching to create a closed system.	Heart tube remains linear, symmetric, and segmentally modular within an open circulatory system.	[127,245,438,890,1474, 1886,2176,2244]
Myogenesis	A *tin/Nkx2.5*-dependent kernel of regulators[4] governs heart myogenesis: *pnr/GATA4*, *Doc1–3≈Tbx5*, and *Islet1*.	*Islet1* also helps to establish the secondary heart field.	*Hox* genes are also involved in dictating regional cell fates.	[265,438,1353,1721, 1774,1886,2176]
Differentiation	*Mef2* acts downstream of *tin/Nkx2.5* to regulate actin and myosin gene expression.	Four different *Mef2* paralogs are expressed in overlapping muscle cell lineages.	Single *Mef2* gene functions as a master regulator for muscles throughout the body.	[213,438,1532,1686]
Stratification	Mononucleate myocytes contain T-tubule-like structures.[5]	Outer myocardial and inner (non-contractile) endocardial layers.	Single layer of contractile myocytes around lumen.	[438,1474]
Blood vessels	Pvr/VEGF promotes formation of blood vessels (vertebrates) and heart valves (vertebrates and fly).	Smooth muscle cells surround non-muscle cells. Lumen comes from apical surface.	Single contractile (mesothelial) cell type. Lumen comes from basal surface.	[347,354,530,883, 1315,1474,1745, 2201,2531]

Table 7.1 (*cont.*)

Shared feature	Conserved mechanism?	Vertebrate quirks	*Drosophila* quirks	References
Blood cells	Blood cells and vessels come from a common "hemangioblast" cell,[6] some of whose descendant cell types are specified by Lz≈AML1.	Blood cells transport oxygen and do not enter the coelomic cavity.	Blood cells circulate throughout the coelom but do not transport oxygen.	[605,606,812,883,1245, 1260,1531,1694, 2136]
Oxygen transport	No. Convergent trait: metal atoms capture and transport oxygen.	Iron is held in a heme ring inside hemoglobin protein contained in blood cells.	Copper is bonded to a hemocyanin protein dissolved in hemolymph.	[27,293,2329]

[1] "Quirks" are derived traits (apomorphies) that differ in vertebrates and flies. Slash marks denote homologs (fly/vertebrate); wavy equals signs denote related, but not orthologous, genes or proteins. Gene (*italic*) and protein (roman) names: AML1 (Acute Myeloid Leukemia 1), BMP4 (Bone Morphogenetic Protein 4), *Doc1–3* (*Dorsocross 1, 2,* and *3* genes), Dpp (Decapentaplegic), *GATA4* (*GATA-binding protein 4*), Lz (Lozenge), *Mef2* (*Myocyte enhancer factor 2*), *Nkx2.5* (*NK2 homeobox 5*), *pnr* (*pannier*), Pvr (PDGF- and VEGF-related), *Tbx5* (*T-box 5*), *tin* (*tinman*), and VEGF (Vascular Endothelial Growth Factor). For details of genetic circuitry see [438,1353,1774,1886,2521], and for a recent review see [1352]. For involvement of microRNAs see [1436,1475,1841], and for Wg/Wnt signaling see [1475,2469]. For a survey of hearts across phyla see [197,1721,1768,2473], and for a survey of vertebrates alone see [1070]. For chamber evolution see [400,1070,1189,2099], for valve development see [93,2017,2241], and for lumen formation mediated by Slit and Robo (cf. Table 2.1) see [931,1474]. For possible deep homologies of the heart's pacemaker see [387]. For clinical applications see [1720].

[2] *Nkx2.5* belongs to a homeobox cluster called NK that is even more ancient than the *Hox* complexes discussed in Chapter 1 [298,1247,2460]. NK genes mainly function in patterning the mesoderm [971], with *Nkx2.5* having assumed the duty of specifying *cardiac* mesoderm in particular [438]. The odyssey of NK cluster genes in bilaterian evolution is at least as interesting as that of the *Hox* cluster [1365].

[3] The lack of this N-terminal domain in Nkx2.5, along with the resulting inability of *Nkx2.5* to rescue heart defects in *tin*-null flies, has been interpreted as evidence that the original function of *tin*/*Nkx2.5* was in visceral mesoderm more generally, with a specific role in *cardiac* mesoderm having evolved separately and convergently in arthropods and vertebrates [953,1862]. However, the recent discovery of a cardiac role for *tin*/*Nkx2.5* in mollusks [588] and annelids [1989] lessens the appeal of this scenario and suggests an earlier (conserved) cardiac function.

[4] A "kernel" is a conserved gene regulatory network [469]. This term is sometimes used interchangeably with "character identity network," but they are not equivalent [2363].

[5] In contrast to cardiac muscle cells, which are mononucleolate, skeletal muscle cells are syncytial [409]. A deeply conserved genetic pathway orchestrates the fusion of myoblasts during the differentiation process [5,1154,1918]. For the origin and evolution of muscle cell types see [16].

[6] In both vertebrates and flies the decision on the part of a hemangioblast cell to become a blood cell instead of part of a blood vessel is enforced by high Notch activity, while the opposite (vessel) destiny is dictated by low Notch activity [1398] (V. Hartenstein, personal communication).

"Control of cardiac development by an evolutionarily conserved transcriptional network" [438,1352]. The wiring diagram that they drew was later verified (at least in part) by converting fibroblasts into cardiomyocytes via artificial expression of the key core genes [1011]. This tour de force has tangible implications for medical approaches to heart repair [317,2008,2134].

Congenital heart syndromes are the most common birth defects in our species [2326]. Not surprisingly (given the extent of evolutionary conservation), they have been traced to dysfunctions in the same genes that govern the fly heart [392,1436,1513,1613,2163]. The prospect of being able to study this ancient cassette in a model system as tractable as *Drosophila* holds promise for clinical cures in the years to come [58,206,1352,1474,2244].

Further evidence for the notion of a conserved network of heart genes came to light in 2006 when cephalopod mollusks (squid, octopus, and cuttlefish) were shown to express a *tinman* homolog in their two-chambered hearts [588,1721]. This finding was significant because mollusks belong to the superphylum Spiralia – the third great branch of bilaterians aside from Ecdysozoa and Deuterostomia, which had already been sampled through genetic investigations of flies [279,1886] and vertebrates [438,1686], respectively.

Annelids are also spiralians, and in 2008 they joined the chorus of support. The polychaete annelid *Platynereis dumerilii* was reported to express a *tin* homolog in its pulsatile dorsal vessel – a simple heart [1531] – just as insects do [1989]. The authors stressed that this homolog is not transcribed in any of the other contractile vessels of the body. This fact makes it unlikely that the *tin* homolog is playing a wider role in myoepithelial tissue beyond just that of the heart. That is, it appears to be a heart identity gene sensu stricto.

NK4 is not expressed in contractile lateral vessels, showing that *NK4* is not just a cell differentiation gene linked to the presence of autonomously contractile myo-epithelial cells but that it has indeed a true patterning function, as in the fly and vertebrates. [1989]

The striking parallels between the transcriptional networks involved in heart development across vast phylogenetic distances support the idea that the evolutionary emergence of hearts with increasing complexity occurred through modification and expansion of an ancestral network of regulatory genes encoding cardiac transcription factors. [1686]

But nematodes use *tinman* for a non-cardiac (pharyngeal) pump

Nematoda is the only phylum studied thus far to sound a discordant note. Nematodes have no heart. Nevertheless, they express most of the core components of the cardiac control network (including *tin/Nkx2.5*). What seems strange (at first) is that these genes are expressed in a pulsatile pharynx [900,1677,1686] that is derived from ectoderm rather than mesoderm [1626].

Is it conceivable that the *tin/Nkx2.5* network originally served an all-purpose *pumping* function in *heartless* urbilaterians (or earlier metazoans [2073]) and later acquired *circulatory* duties (via co-option) in arthropods, chordates, mollusks, and annelids? Some authors have explicitly endorsed this idea [889,953,1768].

To other authors, however, it seems more parsimonious to imagine (1) that urbilaterians did have a heart [212,1531], (2) that it was later lost in certain lineages, and (3) that nematodes independently recruited the cardiac gene circuit to pump food through their gullet [2473]. This issue is far from settled.

VEGF plays a regulatory role in circulatory system development

The vascular endothelial growth factor (VEGF) signaling pathway has long been known to promote the proliferation of blood vessels in vertebrates [530,1745] (cf. amphioxus [1739]). In 2004 it was also shown to be involved in the formation of vertebrate heart valves [347,1235]. The first hint of deep homology with *Drosophila* came in 2007 when the pathway was found to be necessary and sufficient for making the fly's heart valves as well [1352,2531], despite that fact that they look nothing like their vertebrate counterparts [366].

The vascular role of the VEGF pathway extends to Spiralia also. In 2003 the medicinal leech – an annelid – was shown to make new hemocoelomic channels ($\approx$ blood vessels) in response to injection of human VEGF, owing to the presence of VEGF receptors on vascular cells [2265], and in 2010 a VEGF receptor that is expressed specifically in blood vessels was identified in the squid *Idiosepius paradoxus* [1577,2508] – a cephalopod mollusk. Based on this sample of species from all three trunks of the bilaterian tree, we would be justified in concluding, at least tentatively, that the VEGF pathway was originally used by urbilaterians (or earlier metazoans [2050]) to construct some sort of circulatory system [606].

However, as in the *tin* case, the phylum Nematoda bucks this trend and challenges this conclusion [1163]. Nematodes lack a vascular system and instead express VEGF receptors on their neurons [1810] – a type of usage that is also seen in crustaceans [702]. This usage has prompted some authors to guess that the original role of the VEGF pathway was *neural* rather than *vascular* [702,1163,1808,1810]. Their reasoning is buttressed by other facts as well [1933]: (1) nervous systems preceded circulatory systems in evolution [2003], and (2) blood vessels grow along the routes that are pioneered by peripheral nerves during development [325,587,1571].

A third proposal, which draws on its own body of circumstantial evidence, is that VEGF was originally used as a guidance mechanism for cell migration [1163], perhaps associated with a rudimentary *respiratory* system [1577]. This too is plausible.

. . . And so we arrive at another cloudy conclusion. Readers who had hoped for a happy ending are undoubtedly disheartened, but that is the state of this field. The notion of deep homology is fraught with uncertainty and roiled by controversy. Those who are too faint of heart to stay the course should give up and go home. The rest of us will batten down the hatches, lean into the wind, and brave the tempest.

Epilogue

In 1965 Arno Penzias and Robert Wilson reported their seminal discovery of the cosmic background of microwave radiation [1767] left over from the Big Bang ~14 billion years ago [519]. In a similar, albeit more prosaic, way *Deep Homology?* has probed the genetic background of the "big bang" (Cambrian explosion) that launched the diversification of two-sided animals on Earth ~0.6 billion years ago [287]. Remnants of that ancient regulatory hardware are still lurking in the genomes of modern bilaterians behind a cloud of more recent convergent adaptations that add a layer of noise to the signal. Some of those circuits have been unveiled here by comparing humans and flies.

Why have these circuits endured? Darwin addressed a comparable question long ago in a different context when he contemplated why arm and leg bones have persisted virtually unaltered throughout mammalian evolution. His explanation was that the underlying pattern (read "operating system") arose in the ancient common progenitor (read "urbilaterian") and persisted because it was robust enough to withstand any environmental changes that challenged the descendant lineages as they diverged from one another (boldface added):

> What can be more curious than that the hand of a man, formed for grasping, that of a mole for digging, the leg of the horse, the paddle of the porpoise, and the wing of the bat, should all be constructed on the same pattern, and should include the same bones, in the same relative positions? . . . The explanation is manifest on the theory of the natural selection of successive slight modifications . . . In changes of this nature, **there will be little or no tendency to modify the original pattern** . . . If we suppose that the ancient progenitor, the archetype as it may be called, of all mammals, had its limbs constructed on the existing general pattern, for whatever purpose they served, we can at once perceive the plain signification of the homologous construction of the limbs throughout the whole class. [463]

Humans and flies have come to inhabit very different worlds, in part because of the distinctive ecological forces that shaped our separate lineages over the eons, but also because of the colossal disparity in our body size [843].

> At the scale of a fly, viscosity, drag, and stickiness far outweigh the forces of gravity, mass, and inertia in importance; sounds are felt as much as heard, and flight is akin to swimming through syrup. To navigate this world, the fly relies on thousands of diverse sense organs to detect, transduce, and integrate different mechanical stimuli. Walking demands continual proprioceptive feedback to coordinate the action of six multijointed legs. Flight requires gyroscopes to detect changes in heading and attitude, and strain gauges to monitor each wing beat. [1143]

Yes, we tower over flies even more than Gulliver did over the Lilliputians, and we do live overtly different lives, but now, thanks to the facts uncovered by evo-devo, we are able to see beyond those banalities. From a genomic perspective, fruit flies are almost indistinguishable from us. Who could have imagined that these fairylike "gnats" would turn out to be so uncannily similar to us after all?

It has been nearly 100 years since Carter found Tut's tomb. Researchers have only recently entered the tomb, so to speak, of our urbilaterian ancestor, and we are still dazzled by the artifacts before us. It is up to the next generation of evo-devotees to survey these treasures and solve the puzzles that they pose.

References

1. Abbott, A., and other members of the *C. elegans* Sequencing Consortium (1998). Genome sequence of the nematode *C. elegans*: a platform for investigating biology. *Science* 282, 2012–2018.

2. Abbott, L.A., and Lindenmayer, A. (1981). Models for growth of clones in hexagonal cell arrangements: applications in *Drosophila* wing disc epithelia and plant epidermal tissues. *J. Theor. Biol.* 90, 495–544.

3. Abdus-Saboor, I., Al Nufal, M.J., Agha, M.V., Ruinart de Brimont, M., Fleischmann, A., and Shykind, B.M. (2016). An expression refinement process ensures singular odorant receptor gene choice. *Curr. Biol.* 26, 1083–1090.

4. Abelló, G., and Alsina, B. (2007). Establishment of a proneural field in the inner ear. *Int. J. Dev. Biol.* 51, 483–493.

5. Abmayr, S.M., and Pavlath, G.K. (2012). Myoblast fusion: lessons from flies and mice. *Development* 139, 641–656.

6. Aboitiz, F., and Montiel, J. (2003). One hundred million years of interhemispheric communication: the history of the corpus callosum. *Braz. J. Med. Biol. Res.* 36, 409–420.

7. Aboobaker, A.A., and Blaxter, M. (2010). The nematode story: *Hox* gene loss and rapid evolution. *In* J.S. Deutsch (ed.), *Hox Genes: Studies from the 20th to the 21st Century*. Landes Bioscience, Austin, TX, pp. 101–110.

8. Aboobaker, A.A., Tomancak, P., Patel, N., Rubin, G.M., and Lai, E.C. (2005). *Drosophila* microRNAs exhibit diverse spatial expression patterns during embryonic development. *PNAS* 102, #50, 18017–18022.

9. Abouheif, E. (1997). Developmental genetics and homology: a hierarchical approach. *Trends Ecol. Evol.* 12, 405–408.

10. Abouheif, E., Akam, M., Dickinson, W.J., Holland, P.W.H., Meyer, A., Patel, N.H., Raff, R.A., Roth, V.L., and Wray, G.A. (1997). Homology and developmental genes. *Trends Genet.* 13, 432–433.

11. Abramov, I., and Gordon, J. (1994). Color appearance: on seeing red – or yellow, or green, or blue. *Annu. Rev. Psychol.* 45, 451–485.

12. Abu-Shaar, M., and Mann, R.S. (1998). Generation of multiple antagonistic domains along the proximodistal axis during *Drosophila* leg development. *Development* 125, 3821–3830.

13. Abu-Shaar, M., Ryoo, H.D., and Mann, R.S. (1999). Control of the nuclear localization of Extradenticle by competing nuclear import and export signals. *Genes Dev.* 13, 935–945.

14. Abuin, L., Bargeton, B., Ulbrich, M.H., Isacoff, E.Y., Kellenberger, S., and Benton, R. (2011). Functional architecture of olfactory ionotropic glutamate receptors. *Neuron* 69, 44–60.

15. Acampora, D., Annino, A., Tuorto, F., Puelles, E., Lucchesi, W., Papalia, A., and Simeone, A. (2005). Otx genes in the evolution of the vertebrate brain. *Brain Res. Bull.* 66, 410–420.

16. Achim, K., and Arendt, D. (2014). Structural evolution of cell types by step-wise assembly of cellular modules. *Curr. Opin. Genet. Dev.* 27, 102–108.

17. Adachi, R., Sasaki, Y., Morita, K., Shirakawa, H., Goto, T., Furuyama, A., and Isono, K. (2012). Behavioral analysis of *Drosophila* transformants expressing human taste receptor genes in the gustatory receptor neurons. *J. Neurogenet.* 26, 198–205.

18. Adam, J., Myat, A., Le Roux, I., Eddison, M., Henrique, D., Ish-Horowicz, D., and Lewis, J. (1998). Cell fate choices and the expression of Notch, Delta and Serrate homologues in the chick inner ear: parallels with *Drosophila* sense-organ development. *Development* 125, 4645–4654.

19. Adams, D.S., Robinson, K.R., Fukumoto, T., Yuan, S., Albertson, R.C., Yelick, P., Kuo, L., McSweeney, M., and Levin, M. (2006). Early, H^+-V-ATPase-dependent proton flux is necessary for consistent left-right patterning of non-mammalian vertebrates. *Development* 133, 1657–1671.

20. Adler, P.N., and Nathans, J. (2016). The cellular compass. *Sci. Am.* 314, 66–71.

21. Affolter, M., Marty, T., and Vigano, M.A. (1999). Balancing import and export in development. *Genes Dev.* 13, 913–915.

22. Afzelius, B.A. (1976). A human syndrome caused by immobile cilia. *Science* 193, 317–319.

23. Afzelius, B.A. (1995). Situs inversus and ciliary abnormalities. What is the connection? *Int. J. Dev. Biol.* 39, 839–844.

24. Afzelius, B.A., and Stenram, U. (2006). Prevalence and genetics of immotile-cilia syndrome and left-handedness. *Int. J. Dev. Biol.* 50, 571–573.

25. Agi, E., Langen, M., Altschuler, S.J., Wu, L.F., Zimmermann, T., and Hiesinger, P.R. (2014). The evolution and development of neural superposition. *J. Neurogenet.* 28, 216–232.

26. Aguilar-Hidalgo, D., Domínguez-Cejudo, M.A., Amore, G., Brockmann, A., Lemos, M.C., Córdoba, A., and Casares, F. (2013). A Hh-driven gene network controls specification, pattern and size of the *Drosophila* simple eyes. *Development* 140, 82–92.

27. Aguilera, F., McDougall, C., and Degnan, B.M. (2013). Origin, evolution and classification of type-3 copper proteins: lineage-specific gene expansions and losses across the Metazoa. *BMC Evol. Biol.* 13, Article 96.

28. Ahn, Y., Mullan, H.E., and Krumlauf, R. (2014). Long-range regulation by shared retinoic acid response elements modulates dynamic expression of posterior *Hoxb* genes in CNS development. *Dev. Biol.* 388, 134–144.

29. Ahnelt, P.K., and Kolb, H. (2000). The mammalian photoreceptor: mosaic-adaptive design. *Prog. Retin. Eye Res.* 19, 711–777.

30. Akam, M. (1987). The molecular basis for metameric pattern in the *Drosophila* embryo. *Development* 101, 1–22.

31. Akam, M. (1989). *Hox* and HOM: homologous gene clusters in insects and vertebrates. *Cell* 57, 347–349.

32. Akam, M. (1998). *Hox* genes, homeosis and the evolution of segment identity: no need for hopeless monsters. *Int. J. Dev. Biol.* 42, 445–451.

33. Akam, M. (1998). *Hox* genes: from master genes to micromanagers. *Curr. Biol.* 8, R676–R678.

34. Akins, K.A., and Hahn, M. (2014). More than mere colouring: the role of spectral information in human vision. *Br. J. Philos. Sci.* 65, 125–171.

35. Akiyama-Oda, Y., and Oda, H. (2006). Axis specification in the spider embryo: *dpp* is required for radial-to-axial symmetry transformation and *sog* for ventral patterning. *Development* 133, 2347–2357.

36. Al-Qattan, M.M. (2003). Congenital duplication of the palm in a patient with multiple anomalies. *J. Hand Surg. Br.* 28B, 276–279.

37. Al Qattan, M.M. (2004). On the emerging evidence of a new category of duplication in the human hand: the dorsoventral duplication. *Plast. Reconstr. Surg.* 114, 1233–1237.

38. Ala-Laurila, P., and Rieke, F. (2014). Coincidence detection of single-photon responses in the inner retina at the sensitivity limit of vision. *Curr. Biol.* 24, 2888–2898.

39. Albalat, R. (2009). The retinoic acid machinery in invertebrates: ancestral elements and vertebrate innovations. *Mol. Cell. Endocrinol.* 313, 23–35.

40. Albalat, R. (2012). Evolution of the genetic machinery of the visual cycle: a novelty of the vertebrate eye? *Mol. Biol. Evol.* 29, 1461–1469.

41. Albalat, R., Brunet, F., Laudet, V., and Schubert, M. (2011). Evolution of retinoid and steroid signaling: vertebrate diversification from an amphioxus perspective. *Genome Biol. Evol.* 3, 985–1005.

42. Albert, J. (2011). Sensory transduction: the "swarm intelligence" of auditory hair bundles. *Curr. Biol.* 21, R632–R634.

43. Albert, J.T., and Göpfert, M.C. (2015). Hearing in *Drosophila*. *Curr. Opin. Neurobiol.* 34, 79–85.

44. Alberts, B., Johnson, A., Lewis, J., Raff, M., Roberts, K., and Walter, P. (2002). *Molecular Biology of the Cell*, 4th edn. Garland, New York, NY.

45. Alexander, T., Nolte, C., and Krumlauf, R. (2009). *Hox* genes and segmentation of the hindbrain and axial skeleton. *Annu. Rev. Cell Dev. Biol.* 25, 431–456.

46. Alfano, C., and Studer, M. (2013). Neocortical arealization: evolution, mechanisms, and open questions. *Dev. Neurobiol.* 73, 411–447.

47. Alié, A., and Manuel, M. (2010). The backbone of the post-synaptic density originated in a unicellular ancestor of choanoflagellates and metazoans. *BMC Evol. Biol.* 10, Article 34.

48. Allan, D.W., and Thor, S. (2015). Transcriptional selectors, masters, and combinatorial codes: regulatory principles of neural subtype specification. *WIREs Dev. Biol.* 4, 505–528.

49. Allen, A.E., Storchi, R., Martial, F.P., Petersen, R.S., Montemurro, M.A., Brown, T.M., and Lucas, R.J. (2014). Melanopsin-driven light adaptation in mouse vision. *Curr. Biol.* 24, 2481–2490.

50. Allen, J., and Chilton, J.K. (2009). The specific targeting of guidance receptors within neurons: who directs the directors? *Dev. Biol.* 327, 4–11.

51. Allison, W.T., Barthel, L.K., Skebo, K.M., Takechi, M., Kawamura, S., and Raymond, P.A. (2010). Ontogeny of cone photoreceptor mosaics in zebrafish. *J. Comp. Neurol.* 518, 4182–4195.

52. Alsina, B., Giraldez, F., and Pujades, C. (2009). Patterning and cell fate in ear development. *Int. J. Dev. Biol.* 53, 1503–1513.

53. Álvarez-Fernández, C., Tamirisa, S., Prada, F., Chernomoretz, A., Podhajcer, O., Blanco, E., and Martín-Blanco, E. (2015). Identification and functional analysis of healing regulators in *Drosophila*. *PLoS Genet.* 11, #2, e1004965.

54. Ambegaonkar, A.A., and Irvine, K.D. (2015). Coordination of planar cell polarity pathways through Spiny legs. *eLife* 4, e09946.

55. Ameisen, J.C. (2002). On the origin, evolution, and nature of programmed cell death: a timeline of four billion years. *Cell Death Differ.* 9, 367–393. [*See also* Green, D.R., and Fitzgerald, P. (2016). Just So stories about the evolution of apoptosis. *Curr. Biol.* 26, R620–R627.]

56. Amemiya, C.T., Miyake, T., and Rast, J.P. (2005). Echinoderms. *Curr. Biol.* 15, R944–R946.

57. Amemiya, C.T., Prohaska, S.J., Hill-Force, A., Cook, A., Wasserscheid, J., Ferrier, D.E.K., Pascual-Anaya, J., Garcia-Fernàndez, J., Dewar, K., and Stadler, P.F. (2008). The amphioxus *Hox* cluster: characterization, comparative genomics, and evolution. *J. Exp. Zool. B. Mol. Dev. Evol.* 310, 465–477.

58. Amodio, V., Tevy, M.F., Traina, C., Ghosh, T.K., and Capovilla, M. (2012). Transactivation in *Drosophila* of human enhancers by human transcription factors involved in congenital heart diseases. *Dev. Dyn.* 241, 190–199.

59. Amour, A., Bird, M., Chaudry, L., Deadman, J., Hayes, D., and Kay, C. (2004). General considerations for proteolytic cascades. *Biochem. Soc. Trans.* 32, 15–16.

60. Amrein, H., and Thorne, N. (2005). Gustatory perception and behavior in *Drosophila melanogaster*. *Curr. Biol.* 15, R673–R684.

61. Anderson, E., Peluso, S., Lettice, L.A., and Hill, R.E. (2012). Human limb abnormalities caused by disruption of hedgehog signaling. *Trends Genet.* 28, 364–373.

62. Anderson, J.M., Horton, P., Kim, E.-H., and Chow, W.S. (2012). Towards elucidation of dynamic structural changes of plant thylakoid architecture. *Philos. Trans. R. Soc. Lond. B* 367, 3515–3524.

63. Andl, T., Reddy, S.T., Gaddapara, T., and Millar, S.E. (2002). WNT signals are required for the initiation of hair follicle development. *Dev. Cell* 2, 643–653.

64. Andretic, R. (2015). Neurobiology: what drives flies to sleep? *Curr. Biol.* 25, R1086–R1088.

65. Andrews, G.L., Tanglao, S., Farmer, W.T., Morin, S., Brotman, S., Berberoglu, M.A., Price, H., Fernandez, G.C., Mastick, G.S., Charron, F., and Kidd, T. (2008). Dscam guides embryonic axons by Netrin-dependent and -independent functions. *Development* 135, 3839–3848.

66. Angelini, D.R., and Kaufman, T.C. (2005). Comparative developmental genetics and the evolution of arthropod body plans. *Annu. Rev. Genet.* 39, 95–119.

67. Angelini, D.R., and Kaufman, T.C. (2005). Insect appendages and comparative ontogenetics. *Dev. Biol.* 286, 57–77.

68. Angioy, A.M., Desogus, A., Barbarossa, I.T., Anderson, P., and Hansson, B.S. (2003). Extreme sensitivity in an olfactory system. *Chem. Senses* 28, 279–284.

69. Angueyra, J.M., Pulido, C., Malagón, G., Nasi, E., and Gomez, M.d.P. (2012). Melanopsin-expressing amphioxus photoreceptors transduce light via a phospholipase C signaling cascade. *PLoS ONE* 7, #1, e29813.

70. Annona, G., Holland, N.D., and D'Aniello, S. (2015). Evolution of the notochord. *EvoDevo* 6, Article 30.

71. Anonymous (2001). Little humans with wings? *Spectra: The Newsletter of the Carnegie Institution* Spring 2001, p. 2004.

72. Antic, D., Stubbs, J.L., Suyama, K., Kinter, C., Scott, M.P., and Axelrod, J. (2010). Planar cell polarity enables posterior localization of nodal cilia and left-right axis determination during mouse and *Xenopus* embryogenesis. *PLoS ONE* 5, #2, e8999.

73. Apostolopoulou, A.A., Rist, A., and Thum, A.S. (2015). Taste processing in *Drosophila* larvae. *Front. Integr. Neurosci.* 9, Article 50.

74. Appel, T.A. (1987). *The Cuvier–Geoffroy Debate*. Oxford University Press, New York, NY.

75. Applebury, M.L., Antoch, M.P., Baxter, L.C., Chun, L.L.Y., Falk, J.D., Farhangfar, F., Kage, K., Krzystolik, M.G., Lyass, L.A., and Robbins, J.T. (2000). The murine cone photoreceptor: a single cone type expresses both S and M opsins with retinal spatial patterning. *Neuron* 27, 513–523.

76. Araújo, S.J., and Tear, G. (2003). Axon guidance mechanisms and molecules: lessons from invertebrates. *Nat. Rev. Neurosci.* 4, 910–922.

77. Aravind, L., Dixit, V.M., and Koonin, E.V. (2001). Apoptotic molecular machinery: vastly increased complexity in vertebrates revealed by genome comparisons. *Science* 291, 1279–1284.

78. Arendt, D. (2003). Evolution of eyes and photoreceptor cell types. *Int. J. Dev. Biol.* 47, 563–571.

79. Arendt, D. (2005). Genes and homology in nervous system evolution: comparing gene functions, expression patterns, and cell type molecular fingerprints. *Theory Biosci.* 124, 185–197.

80. Arendt, D. (2008). The evolution of cell types in animals: emerging principles from molecular studies. *Nat. Rev. Genet.* 9, 868–882.

81. Arendt, D., Denes, A.S., Jékely, G., and Tessmar-Raible, K. (2008). The evolution of nervous system centralization. *Philos. Trans. R. Soc. Lond. B* 363, 1523–1528.

82. Arendt, D., Hausen, H., and Purschke, G. (2009). The "division of labour" model of eye evolution. *Philos. Trans. R. Soc. Lond. B* 364, 2809–2817.

83. Arendt, D., and Nübler-Jung, K. (1994). Inversion of dorsoventral axis? *Nature* 371, 26.

84. Arendt, D., and Nübler-Jung, K. (1996). Common ground plans in early brain development in mice and flies. *BioEssays* 18, 255–259.

85. Arendt, D., and Nübler-Jung, K. (1997). Dorsal or ventral: similarities in fate maps and gastrulation patterns in annelids, arthropods and chordates. *Mech. Dev.* 61, 7–21.

86. Arendt, D., and Nübler-Jung, K. (1999). Comparison of early nerve cord development in insects and vertebrates. *Development* 126, 2309–2325.

87. Arendt, D., and Nübler-Jung, K. (1999). Rearranging gastrulation in the name of yolk: evolution of gastrulation in yolk-rich amniote eggs. *Mech. Dev.* 81, 3–22.

88. Arendt, D., Tessmar-Raible, K., Snyman, H., Dorresteijn, A.W., and Wittbrodt, J. (2004). Ciliary photoreceptors with a vertebrate-type opsin in an invertebrate brain. *Science* 306, 869–871.

89. Arendt, D., Tosches, M.A., and Marlow, H. (2016). From nerve net to nerve ring, nerve cord and brain: evolution of the nervous system. *Nat. Rev. Neurosci.* 17, 61–72.

90. Arendt, D., and Wittbrodt, J. (2001). Reconstructing the eyes of Urbilateria. *Philos. Trans. R. Soc. Lond. B* 356, 1545–1563.

91. Armitage, S.A.O., Freiburg, R.Y., Kurtz, J., and Bravo, I.G. (2012). The evolution of *Dscam* genes across the arthropods. *BMC Evol. Biol.* 12, Article 53. [*See also* Leite, D.J., and McGregor, A.P. (2016). Arthropod evolution and development: recent insights from chelicerates and myriapods. *Curr. Opin. Genet. Dev.* 39, 93–100.]

92. Armitage, S.A.O., Peuss, R., and Kurtz, J. (2015). Dscam and pancrustacean immune memory: a review of the evidence. *Dev. Comp. Immunol.* 48, 315–323.

93. Armstrong, E.J., and Bischoff, J. (2004). Heart valve development: endothelial cell signaling and differentiation. *Circ. Res.* 95, 459–470.

94. Armstrong, J.R., and Ferguson, M.W.J. (1995). Ontogeny of the skin and the transition from scar-free to scarring phenotype during wound healing in the pouch young of a marsupial, *Monodelphis domestica. Dev. Biol.* 169, 242–260.
95. Arnheiter, H. (1998). Eyes viewed from the skin. *Nature* 391, 632–633.
96. Arnone, M.I., Rizzo, F., Annunciata, R., Cameron, R.A., Peterson, K.J., and Martínez, P. (2006). Genetic organization and embryonic expression of the ParaHox genes in the sea urchin *S. purpuratus*: insights into the relationship between clustering and colinearity. *Dev. Biol.* 300, 63–73.
97. Arnosti, D.N., and Kulkarni, M.M. (2005). Transcriptional enhancers: intelligent enhanceosomes or flexible billboards? *J. Cell. Biochem.* 94, 890–898.
98. Aronova, M.Z. (2009). Structural models of "simple" sense organs by the example of the first metazoa. *J. Evol. Biochem. Physiol.* 45, 179–196.
99. Aronowicz, J., and Lowe, C.J. (2006). *Hox* gene expression in the hemichordate *Saccoglossus kowalevskii* and the evolution of deuterostome nervous systems. *Integr. Comp. Biol.* 46, 890–901.
100. Arshavsky, V.Y. (2010). Vision: the retinoid cycle in *Drosophila. Curr. Biol.* 20, R96–R98.
101. Arthur, W. (2002). The emerging conceptual framework of evolutionary developmental biology. *Nature* 415, 757–764.
102. Arya, R., and White, K. (2015). Cell death in development: signaling pathways and core mechanisms. *Semin. Cell Dev. Biol.* 39, 12–19.
103. Arzi, A., and Sobel, N. (2011). Olfactory perception as a compass for olfactory neural maps. *Trends Cogn. Sci.* 15, 537–545.
104. Asadi-Pooya, A.A., Sharan, A., Nei, M., and Sperling, M.R. (2008). Corpus callosostomy. *Epilepsy Behav.* 13, 271–278.
105. Ashburner, M. (2006). *Won for All: How the Drosophila Genome Was Sequenced.* Cold Spring Harbor Laboratory Press, Plainview, NY.
106. Ashkenazi, A., and Salvesen, G. (2014). Regulated cell death: signaling and mechanisms. *Annu. Rev. Cell Dev. Biol.* 30, 337–356.
107. Aso, Y., Hattori, D., Yu, Y., Johnston, R.M., Iyer, N.A., Ngo, T.-T.B., Dionne, H., Abbott, L.F., Axel, R., Tanimoto, H., and Rubin, G.M. (2014). The neuronal architecture of the mushroom body provides a logic for associative learning. *eLife* 3, e04577.
108. Aso, Y., Sitaraman, D., Ichinose, T., Kaun, K.R., Vogt, K., Belliart-Guérin, G., Plaçais, P.-Y., Robie, A.A., Yamagata, N., Schnaitmann, C., Rowell, W.J., Johnston, R.M., Ngo, T.-T.B., Chen, N., Korff, W., Nitabach, M.N., Heberlein, U., Preat, T., Branson, K.M., Tanimoto, H., and Rubin, G.M. (2014). Mushroom body output neurons encode valence and guide memory-based action selection in *Drosophila. eLife* 3, e04580.
109. Asteriti, S., Grillner, S., and Cangiano, L. (2015). A Cambrian origin for vertebrate rods. *eLife* 4, e07166.
110. Atkinson, P.J., Najarro, E.H., Sayyid, Z.N., and Cheng, A.G. (2015). Sensory hair development and regeneration: similarities and differences. *Development* 142, 1561–1571.
111. Audibert, A., Simon, F., and Gho, M. (2005). Cell cycle diversity involves differential regulation of Cyclin E activity in the *Drosophila* bristle cell lineage. *Development* 132, 2287–2297.

112. Austin, J.R., II, and Staehelin, L.A. (2011). Three-dimensional architecture of grana and stroma thylakoids of higher plants as determined by electron tomography. *Plant Physiol.* 155, 1601–1611.

113. Avasthi, P., and Marshall, W.F. (2012). Stages of ciliogenesis and regulation of ciliary length. *Differentiation* 83, S30–S42.

114. Averof, M. (2002). Arthropod Hox genes: insights on the evolutionary forces that shape gene functions. *Curr. Opin. Genet. Dev.* 12, 386–392.

115. Aw, S., and Levin, M. (2008). What's left in asymmetry? *Dev. Dyn.* 237, 3453–3463.

116. Aw, S., and Levin, M. (2009). Is left-right asymmetry a form of planar cell polarity? *Development* 136, 355–366.

117. Axelrod, J.D., and Bergmann, D.C. (2014). Coordinating cell polarity: heading in the right direction? *Development* 141, 3298–3302.

118. Aylsworth, A.S. (2001). Clinical aspects of defects in the determination of laterality. *Am. J. Med. Genet.* 101, 345–355.

119. Azevedo, F.A.C., Carvalho, L.R.B., Grinberg, L.T., Farfel, J.M., Ferretti, R.E.L., Leite, R.E.P., Filho, W.J., Lent, R., and Herculano-Houzel, S. (2009). Equal numbers of neuronal and nonneuronal cells make the human brain an isometrically scaled-up primate brain. *J. Comp. Neurol.* 513, 532–541.

120. Babcock, L.E. (1993). The right and the sinister. *Nat. Hist.* 102, #7, 32–39.

121. Bachmann, A., and Knust, E. (1998). Dissection of *cis*-regulatory elements of the *Drosophila* gene *Serrate*. *Dev. Genes Evol.* 208, 346–351.

122. Bachmann, A., and Knust, E. (1998). Positive and negative control of *Serrate* expression during early development of the *Drosophila* wing. *Mech. Dev.* 76, 67–78.

123. Backfisch, B., Rajan, V.B.V., Fischer, R.M., Lohs, C., Arboleda, E., Tessmar-Raible, K., and Raible, F. (2013). Stable transgenesis in the marine annelid *Platynereis dumerilii* sheds new light on photoreceptor evolution. *PNAS* 110, #1, 193–198.

124. Badano, J.L., and Katsanis, N. (2006). Life without centrioles: cilia in the spotlight. *Cell* 125, 1228–1230.

125. Bae, B.-I., Jayaraman, D., and Walsh, C.A. (2015). Genetic changes shaping the human brain. *Dev. Cell* 32, 423–434.

126. Baek, J.H., Hatakeyama, J., Sakamoto, S., Ohtsuka, T., and Kageyama, R. (2006). Persistent and high levels of Hes1 expression regulate boundary formation in the developing central nervous system. *Development* 133, 2467–2476.

127. Baer, M.M., Chanut-Delalande, H., and Affolter, M. (2009). Cellular and molecular mechanisms underlying the formation of biological tubes. *Curr. Top. Dev. Biol.* 89, 137–162.

128. Baguñà, J., and Garcia-Fernàndez, J. (2003). Evo-devo: the long and winding road. *Int. J. Dev. Biol.* 47, 705–713.

129. Bailes, H.J., and Lucas, R.J. (2010). Melanopsin and inner retinal photoreception. *Cell. Mol. Life Sci.* 67, 99–111.

130. Bailly, X., Reichert, H., and Hartenstein, V. (2013). The urbilaterian brain revisited: novel insights into old questions from new flatworm clades. *Dev. Genes Evol.* 223, 149–157.

131. Bainbridge, D. (2008). *Beyond the Zonules of Zinn: A Fantastic Journey Through Your Brain*. Harvard University Press, Cambridge, MA.

132. Baird, I.L. (1974). Some aspects of the comparative anatomy and evolution of the inner ear in submammalian vertebrates. *Brain Behav. Evol.* 10, 11–36.

133. Bakalenko, N.I., Novikova, E.L., Nesterenko, A.Y., and Kulakova, M.A. (2013). *Hox* gene expression during postlarval development of the polychaete *Alitta virens*. *EvoDevo* 4, Article 13.

134. Baker, C.V.H., O'Neill, P., and McCole, R.B. (2008). Lateral line, otic and epibranchial placodes: developmental and evolutionary links? *J. Exp. Zool. B. Mol. Dev. Evol.* 310, 370–383.

135. Baker, N.E. (2001). Master regulatory genes; telling them what to do. *BioEssays* 23, 763–766.

136. Baker, N.E., and Firth, L.C. (2011). Retinal determination genes function along with cell-cell signals to regulate *Drosophila* eye development. *BioEssays* 33, 538–546.

137. Baker, R.E., Schnell, S., and Maini, P.K. (2009). Waves and patterning in developmental biology: vertebrate segmentation and feather bud formation as case studies. *Int. J. Dev. Biol.* 53, 783–794.

138. Balaban, E. (2005). Brain switching: studying evolutionary behavioral changes in the context of individual brain development. *Int. J. Dev. Biol.* 49, 117–124.

139. Balavoine, G. (2014). Segment formation in annelids: patterns, processes and evolution. *Int. J. Dev. Biol.* 58, 469–483.

140. Balavoine, G., and Adoutte, A. (2003). The segmented *Urbilateria*: a testable scenario. *Integr. Comp. Biol.* 43, 137–147.

141. Balavoine, G., de Rosa, R., and Adoutte, A. (2002). Hox clusters and bilaterian phylogeny. *Mol. Phylogenet. Evol.* 24, 366–373.

142. Baliga, B., and Kumar, S. (2003). Apaf-1/cytochrome *c* apoptosome: an essential initiator of caspase activation or just a sideshow? *Cell Death Differ.* 10, 16–18.

143. Baltimore, D. (2001). Our genome unveiled. *Nature* 409, 814–816.

144. Balzeau, A., Gilissen, E., and Grimaud-Hervé, D. (2102). Shared pattern of endocranial shape asymmetries among great apes, anatomically modern humans, and fossil hominins. *PLoS ONE* 7, #1, e29581.

145. Barendregt, M., Harvey, B.M., Rokers, B., and Dumoulin, S.O. (2015). Transformation from a retinal to a cyclopean representation in human visual cortex. *Curr. Biol.* 25, 1982–1987.

146. Bargmann, C.I. (2006). Comparative chemosensation from receptors to ecology. *Nature* 444, 295–301.

147. Barinaga, M. (1995). Focusing on the *eyeless* gene. *Science* 267, 1766–1767.

148. Barish, S., and Volkan, P.C. (2015). Mechanisms of olfactory receptor neuron specification in *Drosophila*. *WIREs Dev. Biol.* 4, 609–621.

149. Barnes, J. (ed.) (1984). *The Complete Works of Aristotle: The Revised Oxford Translation*. Princeton University Press, Princeton, NJ.

150. Barr-Gillespie, P.-G. (2015). Assembly of hair bundles, an amazing problem for cell biology. *Mol. Biol. Cell* 26, #15, 2727–2732. [*See also* Marshall, W.F. (2016). Cell geometry: how cells count and measure size. *Annu. Rev. Biophys.* 45, 49–64.]

151. Barretto, R.P.J., Gillis-Smith, S., Chandrashekar, J., Yarmolinsky, D.A., Schnitzer, M.J., Ryba, N.J.P., and Zuker, C.S. (2015). The neural representation of taste quality at the periphery. *Nature* 517, 373–376.

152. Barrios, N., González-Pérez, E., Hernández, R., and Campuzano, S. (2015). The homeodomain Iroquois proteins control cell cycle progression and regulate the size of developmental fields. *PLoS Genet.* 11, #8, e1005463.

153. Bartlett, H., Veenstra, G.J.C., and Weeks, D.L. (2010). Examining the cardiac NK-2 genes in early heart development. *Pediatr. Cardiol.* 31, 335–341.

154. Bartusiak, M. (2015). Rings, rings, rings: it's Saturn Giganticus! *Nat. Hist.* 123, #5, 10–11.

155. Barucca, M., Canapa, A., and Bicscotti, M.A. (2016). An overview of *Hox* genes in Lophotrochozoa: evolution and functionality. *J. Dev. Biol.* 4, Article 4010012.

156. Basch, M.L., Brown, R.M., II, Jen, H.-I., and Groves, A.K. (2016). Where hearing starts: the development of the mammalian cochlea. *J. Anat.* 228, 233–254.

157. Basto, R., Lau, J., Vinogradova, T., Gardiol, A., Woods, C.G., Khodjakov, A., and Raff, J.W. (2006). Flies without centrioles. *Cell* 125, 1375–1386.

158. Basu, B., and Brueckner, M. (2008). Cilia: multifunctional organelles at the center of vertebrate left-right asymmetry. *Curr. Top. Dev. Biol.* 85, 151–174.

159. Bates, C.J. (1995). Vitamin A. *Lancet* 345, 31–35.

160. Bateson, W. (1894). *Materials for the Study of Variation Treated with Especial Regard to Discontinuity in the Origin of Species.* MacMillan, London.

161. Battelle, B.-A., Kempler, K.E., Saraf, S.R., Marten, C.E., Dugger, D.R., Jr., Speiser, D.I., and Oakley, T.H. (2015). Opsins in Limulus eyes: characterization of three visible lightsensitive opsins unique to and co-expressed in median eye photoreceptors and a peropsin/RGR that is expressed in all eyes. *J. Exp. Biol.* 218, 466–479.

162. Baughman, K.W., McDougall, C., Cummins, S.F., Hall, M., Degnan, B.M., Satoh, N., and Shoguchi, E. (2014). Genomic organization of Hox and ParaHox clusters in the echinoderm, *Acanthaster planci. Genesis* 52, 952–958.

163. Bayascas, J.B., Castillo, E., and Saló, E. (1998). Platyhelminthes have a Hox code differently activated during regeneration, with genes closely related to those of spiralian protostomes. *Dev. Genes Evol.* 208, 467–473.

164. Baylies, M.K., Bate, M., and Ruiz Gomez, M. (1998). Myogenesis: a view from *Drosophila. Cell* 93, 921–927.

165. Bayramli, X., and Fuss, S.H. (2012). Born to run: patterning the *Drosophila* olfactory system. *Dev. Cell* 22, 240–241.

166. Bazin-Lopez, N., Valdivia, L.E., Wilson, S.W., and Gestri, G. (2015). Watching eyes take shape. *Curr. Opin. Genet. Dev.* 32, 73–79.

167. Bear, A., and Monteiro, A. (2013). Both cell-autonomous mechanisms and hormones contribute to sexual development in vertebrates and insects. *BioEssays* 35, 725–732.

168. Beatus, P., and Lendahl, U. (1998). Notch and neurogenesis. *J. Neurosci. Res.* 54, 125–136.

169. Beaulé, V., Tremblay, S., Lafleur, L.-P., Tremblay, S., Lassonde, M., Lepage, J.-F., and Théoret, H. (2015). Cortical thickness in adults with agensis of the corpus callosum. *Neuropsychologia* 77, 359–365.

170. Bechstedt, S., and Howard, J. (2007). Models of hair cell mechanotransduction. *Curr. Top. Membr.* 59, 399–424.

171. Bechstedt, S., and Howard, J. (2008). Hearing mechanics: a fly in your ear. *Curr. Biol.* 18, R869–R870.

172. Beckwith, E.J., and Yanovsky, M.J. (2014). Circadian regulation of gene expression: at the crossroads of transcriptional and post-transcriptional regulatory networks. *Curr. Opin. Genet. Dev.* 27, 35–42.

173. Beisel, K.W., Wang-Lundberg, Y., Maklad, A., and Fritzsch, B. (2005). Development and evolution of the vestibular sensory apparatus of the mammalian ear. *J. Vestib. Res.* 15, 225–241.

174. Bell, A. (2008). The pipe and the pinwheel: is pressure an effective stimulus for the 9 + 0 primary cilium? *Cell Biol. Int.* 32, 462–468.

175. Bell, A.T.A., and Niven, J.E. (2014). Individual-level, context-dependent handedness in the desert locust. *Curr. Biol.* 24, R382–R383.

176. Bell, M.L., Earl, J.B., and Britt, S.G. (2007). Two types of *Drosophila* R7 photoreceptor cells are arranged randomly: a model for stochastic cell-fate determination. *J. Comp. Neurol.* 502, 75–85.

177. Bell-Pedersen, D., Cassone, V.M., Earnest, D.J., Golden, S.S., Hardin, P.E., Thomas, T.L., and Zoran, M.J. (2005). Circadian rhythms from multiple oscillators: lessons from diverse organisms. *Nat. Rev. Genet.* 6, 544–556.

178. Bellardita, C., and Kiehn, O. (2015). Phenotypic characterization of speed-associated gait changes in mice reveals modular organization of locomotor networks. *Curr. Biol.* 25, 1426–1436.

179. Belle, M.D.C., and Piggins, H.D. (2012). Circadian time redoxed. *Science* 337, 805–806.

180. Bellen, H.J., and Yamamoto, S. (2015). Morgan's legacy: fruit flies and the functional annotation of conserved genes. *Cell* 163, 12–14.

181. Belluscio, L., Lodovichi, C., Feinstein, P., Mombaerts, P., and Katz, L.C. (2002). Odorant receptors instruct functional circuitry in the mouse olfactory bulb. *Nature* 419, 296–300.

182. Bengtson, S., Cunningham, J.A., Yin, C., and Donoghue, P.C.J. (2012). A merciful death for the "earliest bilaterian," *Vernanimalcula*. *Evol. Dev.* 14, 421–427.

183. Benito-Gutiérrez, È., and Arendt, D. (2009). CNS evolution: new insight from the mud. *Curr. Biol.* 19, R640–R642.

184. Benton, R., Sachse, S., Michnick, S.W., and Vosshall, L.B. (2006). Atypical membrane topology and heteromeric function of *Drosophila* odorant receptors in vivo. *PLoS Biol.* 4, #2, e20.

185. Benton, R., Vannice, K.S., Gomez-Diaz, C., and Vosshall, L.B. (2009). Variant ionotropic glutamate receptors as chemosensory receptors in *Drosophila*. *Cell* 136, 149–162. [*See also* Dey, S., and Stowers, L. (2016). Think you know how smell works? Sniff again. *Cell* 165, 1566–1567.]

186. Benzer, S. (1973). Genetic dissection of behavior. *Sci. Am.* 229, #6, 24–37.

187. Bernards, A., and Hariharan, I.K. (2001). Of flies and men: studying human disease in *Drosophila*. *Curr. Opin. Genet. Dev.* 11, 274–278.

188. Berry, J.A., Cervantes-Sandoval, I., Chakraborty, M., and Davis, R.L. (2015). Sleep facilitates memory by blocking dopamine neuron-mediated forgetting. *Cell* 161, 1656–1667.

189. Berson, D.M., Dunn, F.A., and Takao, M. (2002). Phototransduction by retinal ganglion cells that set the circadian clock. *Science* 295, 1070–1073.

190. Berthelsen, J., Kilstrup-Nielsen, C., Blasi, F., Mavilio, F., and Zappavigna, V. (1999). The subcellular localization of PBX1 and EXD proteins depends on nuclear import and export signals and is modulated by association with PREP1 and HTH. *Genes Dev.* 13, 946–953.

191. Bertrand, N., Castro, D.S., and Guillemot, F. (2002). Proneural genes and the specification of neural cell types. *Nat. Rev. Neurosci.* 3, 517–530.

192. Bertrand, S., and Escriva, H. (2011). Evolutionary crossroads in developmental biology: amphioxus. *Development* 138, 4819–4830.

193. Bertrand, V., Bisso, P., Poole, R.J., and Hobert, O. (2011). Notch-dependent induction of left/right asymmetry in *C. elegans* interneurons and motorneurons. *Curr. Biol.* 21, 1225–1231.

194. Bertucci, P., and Arendt, D. (2013). Somatic and visceral nervous systems – an ancient duality. *BMC Biol.* 11, Article 11.

195. Beshel, J., and Zhong, Y. (2013). Graded encoding of food odor value in the *Drosophila* brain. *J. Neurosci.* 33, #40, 15693–15704.

196. Besson, C., Bernard, F., Corson, F., Roualt, H., Reynaud, E., Keder, A., Mazouni, K., and Schweisguth, F. (2015). Planar cell polarity breaks the symmetry of PAR protein distribution prior to mitosis in *Drosophila* sensory organ precursor cells. *Curr. Biol.* 25, 1104–1110.

197. Bettex, D.A., Prêtre, R., and Chassot, P.-G. (2014). Is our heart a well-designed pump? The heart along animal evolution. *Eur. Heart J.* 35, 2322–2332.

198. Beurg, M., Xiong, W., Zhao, B., Müller, U., and Fettiplace, R. (2015). Subunit determination of the conductance of hair-cell mechanotransducer channels. *PNAS* 112, #5, 1589–1594.

199. Bharti, K., Gasper, M., Ou, J., Brucato, M., Clore-Gronenborn, K., Pickel, J., and Arnheiter, H. (2012). A regulatory loop involving PAX6, MITF, and WNT signaling controls retinal pigment epithelium development. *PLoS Genet.* 8, #7, e1002757.

200. Bhat, K.M. (2005). Slit-Roundabout signaling neutralizes Netrin-Frazzled-mediated attractant cue to specify the lateral positioning of longitudinal axon pathways. *Genetics* 170, 149–159.

201. Bhat, K.M., and Schedl, P. (1997). Requirement for *engrailed* and *invected* genes reveals novel regulatory interactions between *engrailed/invected*, *patched*, *gooseberry* and *wingless* during *Drosophila* neurogenesis. *Development* 124, 1675–1688.

202. Bhatia, S., Monahan, J., Ravi, V., Gautier, P., Murdoch, E., Brenner, S., Van Heyningen, V., Venkatesh, B., and Kleinjan, D.A. (2014). A survey of ancient conserved non-coding elements in the *PAX6* locus reveals a landscape of interdigitated *cis*-regulatory archipelagos. *Dev. Biol.* 387, 214–228.

203. Bier, E. (1997). Anti-neural-inhibition: a conserved mechanism for neural induction. *Cell* 89, 681–684.

204. Bier, E. (2005). *Drosophila*, the golden bug, emerges as a tool for human genetics. *Nat. Rev. Genet.* 6, 9–23.

205. Bier, E. (2011). Evolution of development: diversified dorsoventral patterning. *Curr. Biol.* 21, R591–R594.

206. Bier, E., and Bodmer, R. (2004). *Drosophila*, an emerging model for cardiac disease. *Gene* 342, 1–11.

207. Bier, E., and De Robertis, E.M. (2015). BMP gradients: a paradigm for morphogen-mediated developmental patterning. *Science* 348, aaa 5838.

208. Bilak, A., and Su, T.T. (2009). Regulation of *Drosophila melanogaster* pro-apoptotic gene *hid*. *Apoptosis* 14, 943–949.

209. Billeter, J.-C., and Levine, J.D. (2013). Who is he and what is he to you? Recognition in *Drosophila melanogaster*. *Curr. Opin. Neurobiol.* 23, 17–23.

210. Bisazza, A., Rogers, L.J., and Vallortigara, G. (1998). The origins of cerebral asymmetry: a review of evidence of behavioural and brain lateralization in fishes, reptiles and amphibians. *Neurosci. Biobehav. Rev.* 22, 411–426.

211. Bishop, S.A., Klein, T., Martinez Arias, A., and Couso, J.P. (1999). Composite signalling from *Serrate* and *Delta* establishes leg segments in *Drosophila* through *Notch*. *Development* 126, 2993–3003.

212. Bishopric, N.H. (2005). Evolution of the heart from bacteria to man. *Ann. N. Y. Acad. Sci.* 1047, 13–29.

213. Black, B.L., and Olson, E.N. (1998). Transcriptional control of muscle development by Myocyte enhancer factor-2 (Mef2) proteins. *Annu. Rev. Cell Dev. Biol.* 14, 167–196.

214. Black, D.L., and Zipursky, S.L. (2008). To cross or not to cross: alternately spliced forms of the Robo3 receptor regulate discrete steps in axonal midline crossing. *Neuron* 58, 297–298.

215. Blair, S.S. (2009). Segmentation in animals. *Curr. Biol.* 18, R991–R995.

216. Blair, S.S. (2014). Planar cell polarity: the importance of getting it backwards. *Curr. Biol.* 24, R835–R838.

217. Blanco, J., Girard, F., Kamachi, Y., Kondoh, H., and Gehring, W. (2005). Functional analysis of the chicken *d1-crystallin* enhancer activity in *Drosophila* reveals remarkable evolutionary conservation between chicken and fly. *Development* 132, 1895–1905.

218. Blanco, J., Pandey, R., Wasser, M., and Udolph, G. (2011). *Orthodenticle* is necessary for survival of a cluster of clonally related dopaminergic neurons in the *Drosophila* larval and adult brain. *Neural Dev.* 6, Article 34.

219. Blaxter, M., and Sunnucks, P. (2011). Velvet worms. *Curr. Biol.* 21, R238–R240.

220. Blum, M., Feistel, K., Thumberger, T., and Schweickert, A. (2014). The evolution and conservation of left-right patterning mechanisms. *Development* 141, 1603–1613.

221. Blum, M., Schweickert, A., Vick, P., Wright, C.V.E., and Danilchik, M.V. (2014). Symmetry breakage in the vertebrate embryo: when does it happen and how does it work? *Dev. Biol.* 393, 109–123.

222. Blumer, M.J.F. (1996). Alterations of the eyes during ontogenesis in *Aporrhais pespelecani* (Mollusca, Caenogastropoda). *Zoomorphology* 116, 123–131.

223. Bode, H. (2011). Axis formation in hydra. *Annu. Rev. Genet.* 45, 105–117.

224. Bodmer, R. (1993). The gene *tinman* is required for specification of the heart and visceral muscles in *Drosophila*. *Development* 118, 719–729.

225. Bodmer, R., and Venkatesh, T.V. (1998). Heart development in *Drosophila* and vertebrates: conservation of molecular mechanisms. *Dev. Genet.* 22, 181–186.

226. Boekhoff-Falk, G. (2005). Hearing in *Drosophila*: development of Johnston's organ and emerging parallels to vertebrate ear development. *Dev. Dyn.* 232, 550–558.

227. Boekhoff-Falk, G., and Eberl, D.F. (2014). The *Drosophila* auditory system. *WIREs Dev. Biol.* 3, 179–191.

228. Bohring, A., Stamm, T., Spaich, C., Haase, C., Spree, K., Hehr, U., Hoffmann, M., Ledig, S., Sel, S., Wieacker, P., and Röpke, A. (2009). *WNT10A* mutations are a frequent cause of a broad spectrum of ectodermal dysplasias with sex-biased manifestation pattern in heterozygotes. *Am. J. Hum. Genet.* 85, 97–105.

229. Bok, M.J., Porter, M.L., Place, A.R., and Cronin, T.W. (2014). Biological sunscreens tune polychromatic ultraviolet vision in mantis shrimp. *Curr. Biol.* 24, 1636–1642.

230. Bokolia, N.P., and Mishra, M. (2015). Hearing molecules, mechanism and transportation: modeled in *Drosophila melanogaster*. *Dev. Neurobiol.* 75, 109–130.

231. Boorman, C.J., and Shimeld, S.M. (2002). The evolution of left-right asymmetry in chordates. *BioEssays* 24, 1004–1011.

232. Boorman, C.J., and Shimeld, S.M. (2002). Pitx homeobox genes in *Ciona* and amphioxus show left-right asymmetry is a conserved chordate character and define the ascidian adenohypophysis. *Evol. Dev.* 4, 354–365.

233. Borges, R., Johnson, W.E., O'Brien, S.J., Vasconcelos, V., and Antunes, A. (2012). The role of gene duplication and unconstrained selective pressures in the melanopsin gene family evolution and vertebrate circadian rhythm regulation. *PLoS ONE* 7, #12, e52413.

234. Borst, A. (2009). *Drosophila*'s view on insect vision. *Curr. Biol.* 19, R36–R47.

235. Bouchard, M., de Caprona, D., Busslinger, M., Xu, P., and Fritzsch, B. (2010). Pax2 and Pax8 cooperate in mouse inner ear morphogenesis and innervation. *BMC Dev. Biol.* 10, Article 89.

236. Bowmaker, J.K. (2012). Evolution of the vertebrate eye. *In* O.F. Lazareva, T. Shimizu, and E.A. Wasserman (eds.), *How Animals See the World: Comparative Behavior, Biology, and Evolution of Vision.* Oxford University Press, New York, NY, pp. 441–472.

237. Bowmaker, J.K., and Hunt, D.M. (2006). Evolution of vertebrate visual pigments. *Curr. Biol.* 16, R484–R489.

238. Boyd, J.L., Skove, S.L., Rouanet, J.P., Pilaz, L.-J., Bepler, T., Gordan, R., Wray, G.A., and Silver, D.L. (2015). Human-chimpanzee differences in a *FZD8* enhancer alter cell-cycle dynamics in the developing neocortex. *Curr. Biol.* 25, 772–779.

239. Boyden, E.A. (1977). Development and growth of the airways. *In* W.A. Hodson (ed.), *Development of the Lung.* Marcel Dekker, New York, NY, pp. 3–35.

240. Bozza, T., Vassalli, A., Fuss, S.H., Zhang, J.-J., Weiland, B., Pacifico, R., Feinstein, P., and Mombaerts, P. (2009). Mapping of Class I and Class II odorant receptors to glomerular domains by two distinct types of olfactory sensory neurons in mice. *Neuron* 61, 220–233.

241. Brachmann, C.B., and Cagan, R.L. (2003). Patterning the fly eye: the role of apoptosis. *Trends Genet.* 19, 91–96.

242. Bradford, D.K., Cole, S.J., and Cooper, H.M. (2009). Netrin-1: diversity in development. *Int. J. Biochem. Cell Biol.* 41, 487–493.

243. Bradshaw, J.L., and Rogers, L.J. (1993). *The Evolution of Lateral Asymmetries, Language, Tool Use, and Intellect.* Academic Press, New York, NY.

244. Brainard, D.H., and Hurlbert, A.C. (2015). Colour vision: understanding #TheDress. *Curr. Biol.* 25, R551–R554.

245. Brand, T. (2003). Heart development: molecular insights into cardiac specification and early morphogenesis. *Dev. Biol.* 258, 1–19.

246. Brandler, W.M., Morris, A.P., Evans, D.M., Scerri, T.S., Kemp, J.P., Timpson, N.J., St. Pourcain, B., Smith, G.D., Ring, S.M., Stein, J.L., Monaco, A.P., Talcott, J.B., Fisher, S.E., Webber, C., and Paracchini, S. (2013). Common variants in left/right asymmetry genes and pathways are associated with relative hand skill. *PLoS Genet.* 9, #9, e1003751.

247. Brauckmann, S. (2012). Karl Ernst von Baer (1792–1876) and evolution. *Int. J. Dev. Biol.* 56, 653–660.

248. Bray, D. (1998). Signaling complexes: biophysical constraints on intracellular communication. *Annu. Rev. Biophys. Biomol. Struct.* 27, 59–75.

249. Brazeau, M.D., and Friedman, M. (2015). The origin and early phylogenetic history of jawed vertebrates. *Nature* 520, 490–497.

250. Breer, H., Fleischer, J., and Strotmann, J. (2006). The sense of smell: multiple olfactory subsystems. *Cell. Mol. Life Sci.* 63, 1465–1475.

251. Bremner, A.J., and van Velzen, J. (2015). Sensorimotor control: retuning the body-world interface. *Curr. Biol.* 25, R159–R161.

252. Brena, C., Chipman, A.D., Minelli, A., and Akam, M. (2006). Expression of trunk Hox genes in the centipede *Strigamia maritima*: sense and anti-sense transcripts. *Evol. Dev.* 8, 252–265.

253. Brennan, P.A., and Zufall, F. (2006). Pheromonal communication in vertebrates. *Nature* 444, 308–315.

254. Breslin, P.A.S., and Spector, A.C. (2008). Mammalian taste perception. *Curr. Biol.* 18, R148–R155.

255. Brierly, A.S. (2014). Diel vertical migration. *Curr. Biol.* 24, R1074–R1076.

256. Briscoe, A.D., and Chittka, L. (2001). The evolution of color vision in insects. *Annu. Rev. Entomol.* 46, 471–510.

257. Briscoe, J., and Small, S. (2015). Morphogen rules: design principles of gradient-mediated embryo patterning. *Development* 142, 3996–4009.

258. Brites, D., Brena, C., Ebert, D., and Du Pasquier, L. (2013). More than one way to produce protein diversity: duplication and limited alternative splicing of an adhesion molecule gene in basal arthropods. *Evolution* 67, 2999–3011.

259. Brockes, J.P., and Kumar, A. (2005). Appendage regeneration in adult vertebrates and implications for regenerative medicine. *Science* 310, 1919–1922. [*See also* Dall'Agnese, A., and Puri, P.L. (2016). Could we also be regenerative superheroes, like salamanders? *BioEssays* 38, 917–926.]

260. Broihier, H.T., Kuzin, A., Zhu, Y., Odenwald, W., and Skeath, J.B. (2004). *Drosophila* homeodomain protein Nkx6 coordinates motoneuron subtype identity and axonogenesis. *Development* 131, 5233–5242.

261. Bronowski, J. (1956). *Science and Human Values*. Harper & Row, New York, NY.

262. Brooke, N.M., Garcia-Fernàndez, J., and Holland, P.W.H. (1998). The ParaHox gene cluster is an evolutionary sister of the Hox gene cluster. *Nature* 392, 920–922.

263. Brookes, M. (2001). *Fly: The Unsung Hero of Twentieth-Century Science*. HarperCollins, New York, NY.

264. Brose, K., Bland, K.S., Wang, K.H., Arnott, D., Henzel, W., Goodman, C.S., Tessier-Lavigne, M., and Kidd, T. (1999). Slit proteins bind Robo receptors and have an evolutionarily conserved role in repulsive axon guidance. *Cell* 99, 795–806.

265. Brown, D.D., Martz, S.N., Binder, O., Goetz, S.C., Price, B.M.J., Smith, J.C., and Conlon, F.L. (2005). *Tbx5* and *Tbx20* act synergistically to control vertebrate heart morphogenesis. *Development* 132, 553–563.

266. Brown, N.A., and Lander, A. (1993). On the other hand . . . *Nature* 363, 303–304.

267. Brown, N.A., McCarthy, A., and Wolpert, L. (1990). The development of handed asymmetry in aggregation chimeras of *situs inversus* mutant and wild-type mouse embryo. *Development* 110, 949–954.

268. Brown, N.A., and Wolpert, L. (1990). The development of handedness in left/right asymmetry. *Development* 109, 1–9.

269. Brown, N.L., Patel, S., Brzezinski, J., and Glaser, T. (2001). *Math5* is required for retinal ganglion cell and optic nerve formation. *Development* 128, 2497–2508.

270. Brown, S.D.M., Hardisty-Hughes, R.E., and Mburu, P. (2008). Quiet as a mouse: dissecting the molecular and genetic basis of hearing. *Nat. Rev. Genet.* 9, 277–290.

271. Brown, T.M., Tsujimura, S.-i., Allen, A.E., Wynne, J., Bedford, R., Vickery, G., Vugler, A., and Lucas, R.J. (2012). Melanopsin-based brightness discrimination in mice and humans. *Curr. Biol.* 22, 1131–1141.

272. Brownstone, R.M., and Wilson, J.M. (2007). Strategies for delineating spinal locomotor rhythm-generating networks and the possible role of Hb9 interneurones in rhythmogenesis. *Brain Res. Rev.* 57, 64–76.

273. Brunet, I., Di Nardo, A.A., Sonnier, L., Beurdeley, M., and Prochiantz, A. (2007). The topological role of homeoproteins in the developing central nervous system. *Trends Neurosci.* 30, 260–267.

274. Brunet, T., Lauri, A., and Arendt, D. (2015). Did the notochord evolve from an ancient axial muscle? The axochord hypothesis. *BioEssays* 37, 836–850.

275. Bryant, D.A., and Frigaard, N.-U. (2006). Prokaryotic photosynthesis and phototropy illuminated. *Trends Microbiol.* 14, 488–496.

276. Bryant, P.J. (1993). The Polar Coordinate Model goes molecular. *Science* 259, 471–472.

277. Bryant, S.V., and Gardiner, D.M. (2016). The relationship between growth and pattern formation. *Regeneration* 3, 103–122.

278. Bryant, S.V., and Iten, L.E. (1976). Supernumerary limbs in amphibians: experimental production in *Notophthalmus viridescens* and a new interpretation of their formation. *Dev. Biol.* 50, 212–234.

279. Bryantsev, A.L., and Cripps, R.M. (2009). Cardiac gene regulatory networks in *Drosophila*. *Biochim. Biophys. Acta* 1789, 343–353.

280. Brzezinski, J.A., and Reh, T.A. (2015). Photoreceptor cell fate specification in vertebrates. *Development* 142, 3263–3273.

281. Bucher, G., Farzana, L., Brown, S.J., and Klingler, M. (2005). Anterior localization of maternal mRNAs in a short germ insect lacking *bicoid*. *Evol. Dev.* 7, 142–149.

282. Buchon, N., Osman, D., David, F.P.A., Fang, H.Y., Boquete, J.-P., Deplancke, B., and Lemaitre, B. (2013). Morphological and molecular characterization of adult midgut compartmentalization in *Drosophila*. *Cell Rep.* 3, 1725–1738.

283. Buckingham, M., Meilhac, S., and Zaffran, S. (2005). Building the mammalian heart from two sources of myocardial cells. *Nat. Rev. Genet.* 6, 826–835.

284. Budd, G.E. (2001). Why are arthropods segmented? *Evol. Dev.* 3, 332–342.

285. Budd, G.E. (2008). The earliest fossil record of the animals and its significance. *Philos. Trans. R. Soc. Lond. B* 363, 1425–1434.

286. Budd, G.E. (2012). Cambrian nervous wrecks. *Nature* 490, 180–181.

287. Budd, G.E. (2013). At the origin of animals: the revolutionary Cambrian fossil record. *Curr. Genomics* 14, 344–354.

288. Budd, G.E., and Jackson, I.S.C. (2016). Ecological innovations in the Cambrian and the origins of the crown group phyla. *Philos. Trans. R. Soc. Lond. B* 371, 20150287.

289. Buhr, E., and Van Gelder, R.N. (2014). The making of the master clock. *eLife* 3, e04014.

290. Bullock, T.H., Orkand, R., and Grinnell, A. (1977). *Introduction to Nervous Systems.* W.H. Freeman, San Francisco, CA.

291. Burke, A.C., Nelson, C.E., Morgan, B.A., and Tabin, C. (1995). *Hox* genes and the evolution of vertebrate axial morphology. *Development* 121, 333–346.

292. Burke, R.D. (2011). Deuterostome neuroanatomy and the body plan paradox. *Evol. Dev.* 13, 110–115.

293. Burmester, T., and Hankeln, T. (2007). The respiratory proteins of insects. *J. Insect Physiol.* 53, 285–294.

294. Burn, S.F., Boot, M.J., de Angelis, C., Doohan, R., Arques, C.G., Torres, M., and Hill, R.E. (2008). The dynamics of spleen morphogenesis. *Dev. Biol.* 318, 303–311.

295. Buschbeck, E.K., and Friedrich, M. (2008). Evolution of insect eyes: tales of ancient heritage, deconstruction, reconstruction, remodeling, and recycling. *Evol. Educ. Outreach* 1, 448–462.

296. Buschbeck, E.K., and Hauser, M. (2009). The visual system of male scale insects. *Naturwissenschaften* 96, 365–374.

297. Butler, S.J., and Tear, G. (2007). Getting axons onto the right path: the role of transcription factors in axon guidance. *Development* 134, 439–448.

298. Butts, T., Holland, P.W.H., and Ferrier, D.E.K. (2008). The Urbilaterian Super-Hox cluster. *Trends Genet.* 24, 259–262.

299. Butts, T., Holland, P.W.H., and Ferrier, D.E.K. (2010). Ancient homeobox gene loss and the evolution of chordate brain and pharynx development: deductions from amphioxus gene expression. *Proc. R. Soc. Lond. B* 277, 3381–3389.

300. Byrne, M., Martinez, P., and Morris, V. (2016). Evolution of a pentameral body plan was not linked to translocation of anterior *Hox* genes: the echinoderm HOX cluster revisited. *Evol. Dev.* 18, 137–143.

301. Cagan, R. (2009). Principles of *Drosophila* eye differentiation. *Curr. Top. Dev. Biol.* 89, 115–135.

302. Cagan, R.L., and Ready, D.F. (1989). The emergence of order in the *Drosophila* pupal retina. *Dev. Biol.* 136, 346–362.

303. Cajal, S.R., and Sánchez, D. (1915). Contribución al conocimiento de los centros nerviosos de los insectos. *Trab. Lab. Invest. Biol. Univ. Madrid* 13, 1–167 + 162 plates.

304. Callaerts, P., Halder, G., and Gehring, W.J. (1997). *Pax-6* in development and evolution. *Annu. Rev. Neurosci.* 20, 483–532.

305. Callander, D.C., Alcorn, M.R., Birsoy, B., and Rothman, J.H. (2014). Natural reversal of left-right gut/gonad asymmetry in *C. elegans* males is independent of embryonic chirality. *Genesis* 52, 581–587.

306. Cameron, R.A., Rowen, L., Nesbitt, R., Bloom, S., Rast, J.P., Berney, K., Arenas-Mena, C., Martinez, P., Lucas, S., Richardson, P.M., Davidson, E.H., Peterson, K.J., and Hood, L. (2006). Unusual gene order and organization of the sea urchin Hox cluster. *J. Exp. Zool. B. Mol. Dev. Evol.* 306, 45–58.

307. Campbell, G., and Tomlinson, A. (1995). Initiation of the proximodistal axis in insect legs. *Development* 121, 619–628.

308. Campbell, G., and Tomlinson, A. (1998). The roles of the homeobox genes *aristaless* and *Distal-less* in patterning the legs and wings of *Drosophila*. *Development* 125, 4483–4493.

309. Campbell, G., Weaver, T., and Tomlinson, A. (1993). Axis specification in the developing *Drosophila* appendage: the role of *wingless*, *decapentaplegic*, and the homeobox gene *aristaless*. *Cell* 74, 1113–1123.

310. Campo-Paysaa, F., Marlétaz, F., Laudet, V., and Schubert, M. (2008). Retinoic acid signaling in development: tissue-specific functions and evolutionary origins. *Genesis* 46, 640–656.

311. Cande, J., Prud'homme, B., and Gompel, N. (2013). Smells like evolution: the role of chemoreceptor evolution in behavioral change. *Curr. Opin. Neurobiol.* 23, 152–158.

312. Cañestro, C., Albalat, R., Irimia, M., and Garcia-Fernàndez, J. (2013). Impact of gene gains, losses and duplication modes on the origin and diversification of vertebrates. *Semin. Cell Dev. Biol.* 24, 83–94.

313. Cañestro, C., and Postlethwait, J.H. (2007). Development of a chordate anterior–posterior axis without classical retinoic acid signaling. *Dev. Biol.* 305, 522–538.

314. Cannon, J.T., Vellutini, B.C., Smith, J., III, Ronquist, F., Jondelius, U., and Hejnol, A. (2016). Xenacoelomorpha is the sister group to Nephrozoa. *Nature* 530, 89–93. [*See also* Hejnol, A., and Pang, K. (2016). Xenacoelomorpha's significance for understanding bilaterian evolution. *Curr. Opin. Genet. Dev.* 39, 48–54.]

315. Canto-Soler, M.V., and Adler, R. (2006). Optic cup and lens development requires Pax6 expression in the early optic vesicle during a narrow time window. *Dev. Biol.* 294, 119–132.

316. Cantore, E. (1977). *Scientific Man: The Humanistic Significance of Science*. Institute for Scientific Humanism, New York, NY.

317. Cao, N., Huang, Y., Zheng, J., Spencer, C.I., Zhang, Y., Fu, J.-D., Nie, B., Xie, M., Zhang, M., Wang, H., Ma, T., Xu, T., Shi, G., Srivastava, D., and Ding, S. (2016). Conversion of human fibroblasts into functional cardiomyocytes by small molecules. *Science* 352, 1216–1220.

318. Capdevila, J., Tsukui, T., Rodríguez Estaban, C., Zappavigna, V., and Izpisua Belmonte, J.C. (1999). Control of vertebrate limb outgrowth by the proximal factor Meis2 and distal antagonism of BMPs by Gremlin. *Mol. Cell* 4, 839–849.

319. Capellini, T.D., Zappavigna, V., and Selleri, L. (2011). Pbx homeodomain proteins: TALEnted regulators of limb patterning and outgrowth. *Dev. Dyn.* 240, 1063–1086.

320. Capilla, A., Johnson, R., Daniels, M., Benavente, M., Bray, S.J., and Galindo, M.I. (2012). Planar cell polarity controls directional Notch signaling in the *Drosophila* leg. *Development* 139, 2584–2593.

321. Capozzoli, N.J. (1995). Why are vertebrate nervous systems crossed? *Med. Hypotheses* 45, 471–475.

322. Capozzoli, N.J. (1999). Why do we speak with the left hemisphere? *Med. Hypotheses* 52, 497–503.

323. Carapuço, M., Nóvoa, A., Bobola, N., and Mallo, M. (2005). *Hox* genes specify vertebral types in the presomitic mesoderm. *Genes Dev.* 19, 2116–2121.

324. Carlson, B.M. (1994). *Human Embryology and Developmental Biology*. Mosby, St. Louis, MO.

325. Carmeliet, P., and Tessier-Lavigne, M. (2005). Common mechanisms of nerve and blood vessel wiring. *Nature* 436, 193–200.

326. Carroll, S.B. (1990). Zebra patterns in fly embryos: activation of stripes or repression of interstripes? *Cell* 60, 9–16.

327. Carroll, S.B. (2005). *Endless Forms Most Beautiful: The New Science of Evo Devo and the Making of the Animal Kingdom*. Norton, New York, NY.

328. Carroll, S.B. (2005). The origins of form. *Nat. Hist.* 114, #9, 58–63.

329. Carroll, S.B., DiNardo, S., O'Farrell, P.H., White, R.A.H., and Scott, M.P. (1988). Temporal and spatial relationships between segmentation and homeotic gene expression in *Drosophila* embryos: distributions of the *fushi tarazu*, *engrailed*, *Sex combs reduced*, *Antennapedia*, and *Ultrabithorax* proteins. *Genes Dev.* 2, 350–360.

330. Carroll, S.B., Grenier, J.K., and Weatherbee, S.D. (2005). *From DNA to Diversity: Molecular Genetics and the Evolution of Animal Design*, 2nd edn. Blackwell, Malden, MA.

331. Carson, H.L. (1983). Chromosomal sequences and interisland colonizations in Hawaiian *Drosophila*. *Genetics* 103, 465–482.

332. Casey, B., and Hackett, B.P. (2000). Left-right axis malformations in man and mouse. *Curr. Opin. Genet. Dev.* 10, 257–261.

333. Castelli-Gair, J. (1998). Implications of the spatial and temporal regulation of *Hox* genes on development and evolution. *Int. J. Dev. Biol.* 42, 437–444.

334. Catania, K.C. (2011). Natural-born killer. *Sci. Am.* 304, #4, 84–87.

335. Catela, C., Shin, M.M., and Dasen, J.S. (2015). Assembly and function of spinal circuits for motor control. *Annu. Rev. Cell Dev. Biol.* 31, 669–698.

336. Cattenoz, P.B., and Giangrande, A. (2015). New insights in the clockwork mechanism regulating lineage specification: lessons from the *Drosophila* nervous system. *Dev. Dyn.* 244, 332–341.

337. Caubit, X., Coré, N., Boned, A., Kerridge, S., Djabali, M., and Fasano, L. (2000). Vertebrate orthologues of the *Drosophila* region-specific patterning gene *teashirt*. *Mech. Dev.* 91, 445–448.

338. Cavodeassi, F., del Corral, R.D., Campuzano, S., and Domínguez, M. (1999). Compartments and organising boundaries in the *Drosophila* eye: the role of the homeodomain Iroquois proteins. *Development* 126, 4933–4942.

339. Cayouette, M., and Raff, M. (2002). Asymmetric segregation of Numb: a mechanism for neural specification from *Drosophila* to mammals. *Nat. Neurosci.* 5, 1265–1269.

340. Cehajic-Kapetanovic, J., Eleftheriou, C., Allen, A.E., Milosavljevic, N., Pienaar, A., Bedford, R., Davis, K.E., Bishop, P.N., and Lucas, R.J. (2015). Restoration of vision with ectopic expression of human rod opsin. *Curr. Biol.* 25, 2111–2122.

341. Certel, S.J., and Thor, S. (2004). Specification of *Drosophila* motoneuron identity by the combinatorial action of POU and LIM-HD factors. *Development* 131, 5429–5439.

342. Chacon-Heszele, M.F., and Chen, P. (2009). Mouse models for dissecting vertebrate planar cell polarity signaling in the inner ear. *Brain Res.* 1277, 130–140.

343. Challis, R.C., Tian, H., Wang, J., He, J., Jiang, J., Chen, X., Yin, W., Connelly, T., Ma, L., Yu, R., Pluznick, J.L., Storm, D.R., Huang, L., Zhao, K., and Ma, M. (2015). An olfactory cilia pattern in the mammalian nose ensures high sensitivity to odors. *Curr. Biol.* 25, 2503–2512.

344. Chan, Y.-H., and Marshall, W.F. (2012). How cells know the size of their organelles. *Science* 337, 1186–1189.

345. Chan, Y.-M., and Jan, Y.N. (1999). Conservation of neurogenic genes and mechanisms. *Curr. Opin. Neurobiol.* 9, 582–588.

346. Chandrashekar, J., Hoon, M.A., Ryba, N.J.P., and Zuker, C.S. (2006). The receptors and cells for mammalian taste. *Nature* 444, 288–294.

347. Chang, C.-P., Neilson, J.R., Bayle, J.H., Gestwicki, J.E., Kuo, A., Stankunas, K., Graef, I.A., and Crabtree, G.R. (2004). A field of myocardial-endocardial NFAT signaling underlies heart valve morphogenesis. *Cell* 118, 649–663.

348. Chang, D.C., and Reppert, S.M. (2001). The circadian clocks of mice and men. *Neuron* 29, 555–558.

349. Chang, H., Cahill, H., Smallwood, P.M., Wang, Y., and Nathans, J. (2015). Identification of *Astrotactin2* as a genetic modifier that regulates the global orientation of mammalian hair follicles. *PLoS Genet.* 11, #9, e1005532.

350. Chang, H.-Y., and Ready, D.F. (2000). Rescue of photoreceptor degeneration in rhodopsin-null *Drosophila* mutants by activated Rac1. *Science* 290, 1978–1980.

351. Chang, S., Johnston, R.J., Jr., and Hobert, O. (2003). A transcriptional regulatory cascade that controls left/right asymmetry in chemosensory neurons of *C. elegans*. *Genes Dev.* 17, 2123–2137.

352. Charité, J., de Graaff, W., Consten, D., Reijnen, M.J., Korving, J., and Deschamps, J. (1998). Transducing positional information to the *Hox* genes: critical interaction of *cdx* gene products with position-sensitive regulatory elements. *Development* 125, 4349–4358.

353. Charlton-Perkins, M., and Cook, T.A. (2010). Building a fly eye: terminal differentiation events of the retina, corneal lens, and pigmented epithelia. *Curr. Top. Dev. Biol.* 93, 129–173.

354. Charpentier, M.S., and Conlon, F.L. (2013). Cellular and molecular mechanisms underlying blood vessel lumen formation. *BioEssays* 36, 251–259.

355. Chatelin, L., Volovitch, M., Joliot, A.H., Perez, F., and Prochiantz, A. (1996). Transcription factor Hoxa-5 is taken up by cells in culture and conveyed to their nuclei. *Mech. Dev.* 55, 111–117.

356. Chea, H.K., Wright, C.V., and Swalla, B.J. (2005). Nodal signaling and the evolution of deuterostome gastrulation. *Dev. Dyn.* 234, 269–278.

357. Cheatle Jarvela, A.M., and Pick, L. (2016). Evo-devo: discovery of diverse mechanisms regulating development. *Curr. Top. Dev. Biol.* 117, 253–274.

358. Chédotal, A. (2011). Further tales of the midline. *Curr. Opin. Neurobiol.* 21, 68–75.

359. Cheesman, S.E., Layden, M.J., Ohlen, T.V., Doe, C.Q., and Eisen, J.S. (2004). Zebrafish and fly Nkx6 proteins have similar CNS expression patterns and regulate motoneuron formation. *Development* 131, 5221–5232.

360. Chen, C.-K., Woodruff, M.L., Chen, F.S., Shim, H., Cilluffo, M.C., and Fain, G.L. (2010). Replacing the rod with the cone transducin α subunit decreases sensitivity and accelerates response decay. *J. Physiol.* 588, 3231–3241.

361. Chen, H., Lun, Y., Ovchinnikov, D., Kokubo, H., Oberg, K.C., Pepicelli, C.V., Gan, L., Lee, B., and Johnson, R.L. (1998). Limb and kidney defects in *Lmx1b* mutant mice suggest an involvement of *LMX1B* in human nail patella syndrome. *Nat. Genet.* 19, 51–55.

362. Chen, H., Xu, Z., Mei, C., Yu, D., and Small, S. (2012). A system of repressor gradients spatially organizes the boundaries of Bicoid-dependent target genes. *Cell* 149, 618–629.

363. Chen, J.-Y. (2011). The origins and key innovations of vertebrates and arthropods. *Paleoworld* 20, 257–278.

364. Chen, S.-K., Badea, T.C., and Hattar, S. (2011). Photoentrainment and pupillary light reflex are mediated by distinct populations of ipRGCs. *Nature* 476, 92–95.

365. Chen, Z., Gore, B.B., Long, H., Ma, L., and Tessier-Lavigne, M. (2008). Alternative splicing of the Robo3 axon guidance receptor governs the midline switch from attraction to repulsion. *Neuron* 58, 325–332.

366. Chen, Z., Zhu, J.-y., Fu, Y., Richman, A., and Han, Z. (2016). Wnt4 is required for ostia development in the *Drosophila* heart. *Dev. Biol.* 413, 188–198.

367. Cheng, N., Tsunenari, T., and Yau, K.-W. (2009). Intrinsic light response of retinal horizontal cells of teleosts. *Nature* 460, 899–903.

368. Cherry, S., Jin, E.J., Özel, M.N., Lu, Z., Agi, E., Wang, D., Jung, W.-H., Epstein, D., Meinertzhagen, I.A., Chan, C.-C., and Hiesinger, P.R. (2013). Charcot–Marie–Tooth 2B mutations in *rab7* cause dosage-dependent neurodegeneration due to partial loss of function. *eLife* 2, e01064.

369. Chesler, A., and Firestein, S. (2008). Current views on odour receptors. *Nature* 452, 944.

370. Chiang, A.S., Lin, C.Y., Chuang, C.C., Chang, H.M., and Hsieh, C.-H. (2011). Three-dimensional reconstruction of brain-wide wiring networks in *Drosophila* at single-cell resolution. *Curr. Biol.* 21, 1–11.

371. Chien, C.-B. (1998). Why does the growth cone cross the road? *Neuron* 20, 3–6.

372. Chien, Y.-H., Keller, R., Kintner, C., and Shook, D.R. (2015). Mechanical strain determines the axis of planar polarity in ciliated epithelia. *Curr. Biol.* 25, 2774–2784.

373. Chilton, J.K. (2006). Molecular mechanisms of axon guidance. *Dev. Biol.* 292, 13–24.

374. Chintapalli, V.R., Terhzaz, S., Wang, J., Al Bratty, M., Watson, D.G., Herzyk, P., Davies, S.A., and Dow, J.A.T. (2012). Functional correlates of positional and gender-specific renal asymmetry in *Drosophila*. *PLoS ONE* 7, #4, e32577.

375. Chipman, A.D. (2008). Annelids step forward. *Evol. Dev.* 10, 141–142.

376. Chipman, A.D. (2010). Parallel evolution of segmentation by co-option of ancestral gene regulatory networks. *BioEssays* 32, 60–70.

377. Chipman, A.D., Ferrier, D.E.K., Brena, C., Qu, J., Hughes, D.S.T., Schröder, R., Torres-Oliva, M., Znassi, N., Jiang, H., Almeida, F.C., Alonso, C.R., Apostolou, Z., Aqrawi, P., Arthur, W., Barna, J.C.J., Blankenburg, K.P., Brites, D., Capella-Gutiérrez, S., Coyle, M., Dearden, P.K., DuPasquier, L., Duncan, E.J., Ebert, D., Eibner, C., Erikson, G., Evans, P.D., Extavour, C.G., Francisco, L., Gabaldón, T., Gillis, W.J., Goodwin-Horn, E.A., Green, J.E., Griffiths-Jones, S., Grimmelikhuijzen, C.J.P., Gubbala, S., Guigó, R., Han, Y., Hauser, F., Havlak, P., Hayden, L., Helbing, S., Holder, M., Hui, J.H.L., Hunn, J.P., Hunnekuhl, V.S., Jackson, L., Javaid, M., Jhangiani, S.N., Jiggins, F.M., Jones, T.E., Kaiser, T.S., Kalra, D., Kenny, N.J., Korchina, V., Kovar, C.L., Kraus, F.B., Lapraz, F., Lee, S.L., Lv, J., Mandapat, C., Manning, G., Mariotti, M., Mata, R., Mathew, T., Neumann, T., Newsham, I., Ngo, D.N., Ninova, M., Okwuonu, G., Ongeri, F., Palmer, W.J., Patil, S., Patraquim, P., Pham, C., Pu, L.-L., Putnam, N.H., Rabouille, C., Ramos, O.M., Rhodes, A.C., Robertson, H.E., Robertson, H.M., Ronshaugen, M., Rozas, J., Saada, N., Sánchez-Gracia, A., Scherer, S.E., Schurko, A.M., Siggens, K.W., Simmons, D., Stief, A., Stolle, E., Telford, M.J., Tessmar-Raible, K., Thornton, R., van der Zee, M., von Haeseler, A., Williams, J.M., Willis, J.H., Wu, Y., Zou, X., Lawson, D., Muzny, D.M., Worley, K.C., Gibbs, R.A., Akam, M., and Richards, S. (2014). The first myriapod genome sequence reveals conservative arthropod gene content and genome organisation in the centipede *Strigamia maritima*. *PLoS Biol.* 12, #11, e1002005.

378. Chisholm, A., and Tessier-Lavigne, M. (1999). Conservation and divergence of axon guidance mechanisms. *Curr. Opin. Neurobiol.* 9, 603–615.

379. Chisholm, A.D., and Horvitz, H.R. (1995). Patterning of the *Caenorhabditis elegans* head region by the *Pax-6* family member *vab-3*. *Nature* 377, 52–55.

380. Choksi, S.P., Southall, T.D., Bossing, T., Edoff, K., de Wit, E., Fischer, B.E., van Steensel, B., Micklem, G., and Brand, A.H. (2006). Prospero acts as a binary switch between self-renewal and differentiation in *Drosophila* neural stem cells. *Dev. Cell* 11, 775–789.

381. Chouhan, N.S., Wolf, R., Helfrich-Förster, C., and Heisenberg, M. (2015). Flies remember the time of day. *Curr. Biol.* 25, 1619–1624.

382. Chourrout, D., Delsuc, F., Chourrout, P., Edvardsen, R.B., Rentzsch, F., Renfer, E., Jensen, M.F., Zhu, B., de Jong, P., Steele, R.E., and Technau, U. (2006). Minimal ProtoHox cluster inferred from bilaterian and cnidarian Hox complements. *Nature* 442, 684–687.

383. Chow, R.L., Altmann, C.R., Lang, R.A., and Hemmati-Brivanlou, A. (1999). Pax6 induces ectopic eyes in a vertebrate. *Development* 126, 4213–4222.

384. Christiaen, L., Jaszczyszyn, Y., Kerfant, M., Kano, S., Thermes, V., and Joly, J.-S. (2007). Evolutionary modification of mouth position in deuterostomes. *Semin. Cell Dev. Biol.* 18, 502–511.

385. Christian, J.L. (2012). Morphogen gradients in development: from form to function. *WIREs Dev. Biol.* 1, 3–15.

386. Christodoulou, F., Raible, F., Tomer, R., Simakov, O., Trachana, K., Klaus, S., Snyman, H., Hannon, G.J., Bork, P., and Arendt, D. (2010). Ancient animal micro-RNAs and the evolution of tissue identity. *Nature* 463, 1084–1088.

387. Christoffels, V.M., Smits, G.J., Kispert, A., and Moorman, A.F.M. (2010). Development of the pacemaker tissues of the heart. *Circ. Res.* 106, 240–254.

388. Chuang, C.-F., VanHoven, M.K., Fetter, R.D., Verselis, V.K., and Bargmann, C.I. (2007). An innexin-dependent cell network establishes left-right neuronal asymmetry in *C. elegans. Cell* 129, 787–799.

389. Chuang, J.-Z., Zhao, Y., and Sung, C.-H. (2007). SARA-regulated vesicular targeting underlies formation of the light-sensing organelle in mammalian rods. *Cell* 130, 535–547.

390. Ciechanska, E., Dansereau, D.A., Svendsen, P.C., Heslip, T.R., and Brook, W.J. (2007). *dAP-2* and *defective proventriculus* regulate *Serrate* and *Delta* expression in the tarsus of *Drosophila melanogaster. Genome* 50, 693–705.

391. Clandinin, T.R., and Giocomo, L.M. (2015). Internal compass puts flies in their place. *Nature* 521, 165–166.

392. Clark, K.L., Yutzey, K.E., and Benson, D.W. (2006). Transcription factors and congenital heart defects. *Annu. Rev. Physiol.* 68, 97–121.

393. Clarke, P.G.H. (1981). Chance, repetition, and error in the development of normal nervous systems. *Perspect. Biol. Med.* 25, 2–19.

394. Clarke, S.L., VanderMeer, J.E., Wenger, A.M., Schaar, B.T., Ahituv, N., and Bejerano, G. (2012). Human developmental enhancers conserved between deuterostomes and protostomes. *PLoS Genet.* 8, #8, e1002852.

395. Clayton, J.D., Kyriacou, C.P., and Reppert, S.M. (2001). Keeping time with the human genome. *Nature* 409, 829–831.

396. Cloney, R.A. (1982). Ascidian larvae and the events of metamorphosis. *Am. Zool.* 22, 817–826.

397. Clowney, E.J., LeGros, M.A., Mosley, C.P., Clowney, F.G., Markenskoff-Papadimitriou, E.C., Myllys, M., Barnea, G., Larabell, C.A., and Lomvardas, S. (2012). Nuclear aggregation of olfactory receptor genes governs their monogenic expression. *Cell* 151, 724–737.

398. Cogan, G.B., Thesen, T., Carlson, C., Doyle, W., Devinsky, O., and Pesaran, B. (2014). Sensory-motor transformations for speech occur bilaterally. *Nature* 507, 94–98.

399. Cohen, B., Simcox, A.A., and Cohen, S.M. (1993). Allocation of the thoracic imaginal primordia in the *Drosophila* embryo. *Development* 117, 597–608.

400. Cohen, E.D., and Morrisey, E.E. (2008). A house with many rooms: how the heart got its chambers with *foxn4. Genes Dev.* 22, 706–710.

401. Cohen, S.M., Brönner, G., Küttner, F., Jürgens, G., and Jäckle, H. (1989). *Distal-less* encodes a homoeodomain protein required for limb development in *Drosophila. Nature* 338, 432–434.

402. Collett, T.S. (2002). Insect vision: controlling actions through optic flow. *Curr. Biol.* 12, R615–R617.

403. Colley, N.J. (2000). Actin' up with Rac1. *Science* 290, 1902–1903.

404. Collin, S.P., Knight, M.A., Davies, W.L., Potter, I.C., Hunt, D.M., and Trezise, A.E.O. (2003). Ancient color vision: multiple opsin genes in the ancestral vertebrates. *Curr. Biol.* 13, R864–R865.

405. Collins, A.G., and Valentine, J.W. (2001). Defining phyla: evolutionary pathways to metazoan body plans. *Evol. Dev.* 3, 432–442.

406. Collins, M.M., Baumholtz, A.I., Simard, A., Gregory, M., Cyr, D.G., and Ryan, A.K. (2015). Claudin-10 is required for relay of left–right patterning cues from Hensen's node to the lateral plate mesoderm. *Dev. Biol.* 401, 236–248.

407. Collins, M.M., and Ryan, A.K. (2014). Are there conserved roles for the extracellular matrix, cilia, and junctional complexes in left-right patterning? *Genesis* 52, 488–502.

408. Collu, G.M., and Mlodzik, M. (2015). Planar polarity: converting a morphogen gradient into cellular polarity. *Curr. Biol.* 25, R372–R374.

409. Comai, G., and Tajbakhsh, S. (2014). Molecular and cellular regulation of skeletal myogenesis. *Curr. Top. Dev. Biol.* 110, 1–73.

410. Concha, M.L., Russell, C., Regan, J.C., Tawk, M., Sidi, S., Gilmour, D.T., Kapsimali, M., Sumoy, L., Goldstone, K., Amaya, E., Kimelman, D., Nicolson, T., Gründer, S., Gomperts, M., Clarke, J.D.W., and Wilson, S.W. (2003). Local tissue interactions across the dorsal midline of the forebrain establish CNS laterality. *Neuron* 39, 423–438.

411. Conradt, B. (2009). Genetic control of programmed cell death during animal development. *Annu. Rev. Genet.* 43, 493–523.

412. Cook, C.E., Chenevert, J., Larsson, T.A., Arendt, D., Houliston, E., and Lénárt, P. (2016). Old knowledge and new technologies allow rapid development of model organisms. *Mol. Biol. Cell* 27, 882–887.

413. Cook, J.E. (1996). Spatial properties of retinal mosaics: an empirical evaluation of some existing measures. *Vis. Neurosci.* 13, 15–30.

414. Cook, T. (2003). Cell diversity in the retina: more than meets the eye. *BioEssays* 25, 921–925.

415. Cook, T., and Desplan, C. (2001). Photoreceptor subtype specification: from flies to humans. *Semin. Cell Dev. Biol.* 12, 509–518.

416. Cooke, J. (2004). Developmental mechanism and evolutionary origin of vertebrate left/right asymmetries. *Biol. Rev.* 79, 377–407.

417. Cooke, J. (2004). The evolutionary origins and significance of vertebrate left-right organisation. *BioEssays* 26, 413–421.

418. Copeland, J.W.R., Nasiadka, A., Dietrich, B.H., and Krause, H.M. (1996). Patterning of the *Drosophila* embryo by a homeodomain-deleted Ftz polypeptide. *Nature* 379, 162–165.

419. Copf, T., Schröder, R., and Averof, M. (2004). Ancestral role of *caudal* genes in axis elongation and segmentation. *PNAS* 101, 17711–17715.

420. Corballis, M.C. (1991). *The Lopsided Ape: Evolution of the Generative Mind*. Oxford University Press, New York, NY.

421. Corballis, M.C. (2014). Left brain, right brain: facts and fantasies. *PLoS Biol.* 12, #1, e1001767.

422. Corballis, M.C., and Morgan, M.J. (1978). On the biological basis of human laterality. I. Evidence for a maturational left-right gradient. *Behav. Brain Sci.* 2, 261–269.

423. Corballis, M.C., and Morgan, M.J. (1978). On the biological basis of human laterality. II. The mechanisms of inheritance. *Behav. Brain Sci.* 2, 270–336.

424. Cordes, R., Schuster-Gossler, K., Serth, K., and Gossier, A. (2004). Specification of vertebral identity is coupled to Notch signalling and the segmentation clock. *Development* 131, 1221–1233.

425. Córdoba, S., and Estella, C. (2014). The bHLH-PAS transcription factor Dysfusion regulates tarsal joint formation in response to Notch activity during *Drosophila* leg development. *PLoS Genet.* 10, #10, e1004621.

426. Coren, S., and Porac, C. (1977). Fifty centuries of right-handedness: the historical record. *Science* 198, 631–632.
427. Corless, J.M. (2012). Cone outer segments: a biophysical model of membrane dynamics, shape retention, and lamella formation. *Biophys. J.* 102, 2697–2705.
428. Costa, A., Sanchez-Guardado, L., Juniat, S., Gale, J.E., Daudet, N., and Henrique, D. (2015). Generation of sensory hair cells by genetic programming with a combination of transcription factors. *Development* 142, 1948–1959.
429. Coulter, D.E., Swaykus, E.A., Beran-Koehn, M.A., Goldberg, D., Wieschaus, E., and Schedl, P. (1990). Molecular analysis of *odd-skipped*, a zinc finger encoding segmentation gene with a novel pair-rule expression pattern. *EMBO J.* 8, 3795–3804.
430. Couso, J.P. (2009). Segmentation, metamerism and the Cambrian explosion. *Int. J. Dev. Biol.* 53, 1305–1316.
431. Coutelis, J.-B., Géminard, C., Spéder, P., Suzanne, M., Petzoldt, A.G., and Noselli, S. (2013). *Drosophila* left/right asymmetry establishment is controlled by the Hox gene Abdominal-B. *Dev. Cell* 24, 89–97.
432. Coutelis, J.-B., González-Morales, N., Géminard, C., and Noselli, S. (2014). Diversity and convergence in the mechanisms establishing L/R asymmetry in metazoa. *EMBO Rep.* 15, 926–937.
433. Coutelis, J.B., Petzoldt, A.G., Spéder, P., Suzanne, M., and Noselli, S. (2008). Left-right asymmetry in *Drosophila*. *Semin. Cell Dev. Biol.* 19, 252–262.
434. Couto, A., Alenius, M., and Dickson, B.J. (2005). Molecular, anatomical, and functional organization of the *Drosophila* olfactory system. *Curr. Biol.* 15, 1535–1547.
435. Couturier, L., Vodovar, N., and Schweisguth, F. (2012). Endocytosis by Numb breaks Notch symmetry at cytokinesis. *Nat. Cell Biol.* 14, 131–139.
436. Cowan, W.M., Fawcett, J.W., O'Leary, D.D.M., and Stanfield, B.B. (1984). Regressive events in neurogenesis. *Science* 225, 1258–1265.
437. Craig, D.A., and Mary-Sasal, N. (2013). A detailed description of *Simulium* (*Meilloniellum*) *adersi* (Pomeroy) from Mayotte, Comoro islands, with comments on bionomics and biogeography (Diptera: Simuliidae). *Zootaxa* 3641, 129–148.
438. Cripps, R.M., and Olson, E.N. (2002). Control of cardiac development by an evolutionarily conserved transcriptional network. *Dev. Biol.* 246, 14–28.
439. Crocker, J., Tamori, Y., and Erives, A. (2008). Evolution acts on enhancer organization to fine-tune gradient threshold readouts. *PLoS Biol.* 6, #11, e263.
440. Cronin, T.W., Johnsen, S., Marshall, N.J., and Warrant, E.J. (2014). *Visual Ecology.* Princeton University Press, Princeton, NJ.
441. Cronin, T.W., and Porter, M.L. (2014). The evolution of invertebrate photopigments and photoreceptors. *In* D.M. Hunt, M.W. Hankins, S.P. Collin, and N.J. Marshall (eds.), *Evolution of Visual and Non-Visual Pigments.* Springer, New York, NY, pp. 105–135.
442. Croset, V., Rytz, R., Cummins, S.F., Budd, A., Brawand, D., Kaessmann, H., Gibson, T.J., and Benton, R. (2010). Ancient protostome origin of chemosensory ionotropic glutamate receptors and the evolution of insect taste and olfaction. *PLoS Genet.* 6, #8, e1001064.
443. Crouzet, S.M., Busch, N.A., and Ohla, K. (2015). Taste quality decoding parallels taste sensations. *Curr. Biol.* 25, 890–896.
444. Crow, J.F. (2001). Shannon's brief foray into genetics. *Genetics* 159, 915–917.
445. Cummins, M., Pueyo, J.I., Greig, S.A., and Couso, J.P. (2003). Comparative analysis of leg and antenna development in wild-type and homeotic *Drosophila melanogaster*. *Dev. Genes Evol.* 213, 319–327.

446. Curcio, C.A., Sloan, K.R., Jr., Packer, O., Hendrickson, A.E., and Kalina, R.E. (1987). Distribution of cones in human and monkey retina: individual variability and radial asymmetry. *Science* 236, 579–582.

447. Currie, K.W., Brown, D.D.R., Zhu, S., Xu, C.J., Voisin, V., Bader, G.D., and Pearson, B.J. (2016). HOX gene complement and expression in the planarian *Schmidtea mediterranea. EvoDevo* 7, Article 7.

448. Curto, G.G., Gard, C., and Ribes, V. (2015). Structures and properties of PAX linked regulatory networks architecting and pacing the emergence of neuronal diversity. *Semin. Cell Dev. Biol.* 44, 75–86.

449. Cveki, A., and Piatigorsky, J. (1996). Lens development and crystallin gene expression: many roles for Pax-6. *BioEssays* 18, 621–630.

450. da Fonseca, R.N., Lynch, J.A., and Roth, S. (2009). Evolution of axis formation: mRNA localization, regulatory circuits and posterior specification in non-model arthropods. *Curr. Opin. Genet. Dev.* 19, 404–411.

451. Dacey, D.M., Liao, H.-W., Peterson, B.B., Robinson, F.R., Smith, V.C., Pokorny, J., Yau, K.-W., and Gamlin, P.D. (2005). Melanopsin-expressing ganglion cells in primate retina signal colour and irradiance and project to the LGN. *Nature* 433, 749–754.

452. Dahmann, C., Oates, A.C., and Brand, M. (2011). Boundary formation and maintenance in tissue development. *Nat. Rev. Genet.* 12, 43–55.

453. Dallos, P. (2008). Cochlear amplification, outer hair cells and prestin. *Curr. Opin. Neurobiol.* 18, 370–376.

454. Dallos, P., and Fakler, B. (2002). Prestin, a new type of motor protein. *Nat. Rev. Mol. Cell Biol.* 3, 104–111.

455. Dallos, P., Wu, X., Cheatham, M.A., Gao, J., Zheng, J., Anderson, C.T., Jia, S., Wang, X., Cheng, W.H.Y., Sengupta, S., He, D.Z.Z., and Zuo, J. (2008). Prestin-based outer hair cell motility is necessary for mammalian cochlear amplification. *Neuron* 58, 333–339.

456. Dalton, R.P., and Lomvardas, S. (2015). Chemosensory receptor specificity and regulation. *Annu. Rev. Neurosci.* 38, 331–349.

457. Dambly-Chaudière, C., Sapède, D., Soubiran, F., Decorde, K., Gompel, N., and Ghysen, A. (2003). The lateral line of zebrafish: a model system for the analysis of morphogenesis and neural development in vertebrates. *Biol. Cell* 95, 579–587.

458. Damen, W.G.M. (2002). *fushi tarazu*: a Hox gene changes its role. *BioEssays* 24, 992–995.

459. Damen, W.G.M. (2007). Evolutionary conservation and divergence of the segmentation process in arthropods. *Dev. Dyn.* 236, 1379–1391. [*See also* Leite, D.J., and McGregor, A.P. (2016). Arthropod evolution and development: recent insights from chelicerates and myriapods. *Curr. Opin. Genet. Dev.* 39, 93–100.]

460. Danchin, E.G.J., and Pontarotti, P. (2004). Stastical evidence for a more than 800-million-year-old evolutionarily conserved genomic region in our genome. *J. Mol. Evol.* 59, 587–597.

461. Daneman, R., and Barres, B.A. (2005). The blood–brain barrier: lessons from moody flies. *Cell* 123, 9–12.

462. Darras, S., and Nishida, H. (2001). The BMP signaling pathway is required together with the FGF pathway for notochord induction in the ascidian embryo. *Development* 128, 2629–2638.

463. Darwin, C. (1859). *On the Origin of Species by Means of Natural Selection, or the Preservation of Favoured Races in the Struggle for Life.* John Murray, London.

464. Dasen, J.S., and Jessell, T.M. (2009). Hox networks and the origins of motor neuron diversity. *Curr. Top. Dev. Biol.* 88, 169–200.

465. Dasen, J.S., Tice, B.C., Brenner-Morton, S., and Jessell, T.M. (2005). A Hox regulatory network establishes motor neuron pool identity and target-muscle connectivity. *Cell* 123, 477–491.

466. Datta, R.R., Cruickshank, T., and Kumar, J.P. (2011). Differential selection within the *Drosophila* retinal determination network and evidence for functional divergence between paralog pairs. *Evol. Dev.* 13, 58–71.

467. Davidson, B.P., and Tam, P.P.L. (2000). The node of the mouse embryo. *Curr. Biol.* 10, R617–R619.

468. Davidson, E.H. (2001). *Genomic Regulatory Systems: Development and Evolution.* Academic Press, New York, NY.

469. Davidson, E.H. (2006). *The Regulatory Genome: Gene Regulatory Networks in Development and Evolution.* Academic Press, New York, NY.

470. Davies, W.I.L., Collin, S.P., and Hunt, D.M. (2012). Molecular ecology and adaptation of visual pigments in craniates. *Mol. Ecol.* 21, 3121–3158.

471. Davis, G.K., and Patel, N.H. (1999). The origin and evolution of segmentation. *Trends Genet.* 9, #12, M68–M72.

472. Davis, G.K., and Patel, N.H. (2002). Short, long, and beyond: molecular and embryological approaches to insect segmentation. *Annu. Rev. Entomol.* 47, 669–699. [*See also* Schmidt-Ott, U., and Lynch, J.A. (2016). Emerging developmental genetic model systems in holometabolous insects. *Curr. Opin. Genet. Dev.* 39, 116–128.]

473. Davis, G.K., and Patel, N.H. (2003). Playing by pair-rules? *BioEssays* 25, 425–429.

474. Davis, R.H. (2004). The age of model organisms. *Nat. Rev. Genet.* 5, 69–76.

475. Davison, A., McDowell, G.S., Holden, J.M., Johnson, H.F., Koutsovoulos, G.D., Liu, M.M., Hulpiau, P., Van Roy, F., Wade, C.M., Banerjee, R., Yang, F., Chiba, S., Davey, J.W., Jackson, D.J., Levin, M., and Blaxter, M.L. (2016). Formin is associated with left-right asymmetry in the pond snail and the frog. *Curr. Biol.* 26, 654–660.

476. Dawkins, R. (1998). *Unweaving the Rainbow: Science, Delusion and the Appetite for Wonder.* Houghton Mifflin, New York, NY.

477. de Brito Sanchez, G., and Giurfa, M. (2011). A comparative analysis of neural taste processing in animals. *Philos. Trans. R. Soc. Lond. B* 366, 2171–2180.

478. de Bruyne, M., and Warr, C.G. (2005). Molecular and cellular organization of insect chemosensory neurons. *BioEssays* 28, 23–34.

479. de Celis, J.F., and Barrio, R. (2009). Regulation and function of Spalt proteins during animal development. *Int. J. Dev. Biol.* 53, 1385–1398.

480. de Ibarra, N.H., Vorobyev, M., and Menzel, R. (2014). Mechanisms, functions and ecology of colour vision in the honeybee. *J. Comp. Physiol. A* 200, 411–433.

481. de Melo, J., Peng, G.-H., Chen, S., and Blackshaw, S. (2011). The Spalt family transcription factor Sall3 regulates the development of cone photoreceptors and retinal horizontal interneurons. *Development* 138, 2325–2336.

482. de Mendoza, A., Sebé-Pedrós, A., Sestak, M.S., Matejcic, M., Torruella, G., Domazet-Loso, T., and Ruiz-Trillo, I. (2013). Transcription factor evolution in eukaryotes and the assembly of the regulatory toolkit in multicellular lineages. *PNAS* 110, E4858–E4866.

483. de Monasterio, F.M., Shein, S.J., and McCrane, E.P. (1981). Staining of blue-sensitive cones of the macaque retina by a fluorescent dye. *Science* 231, 1278–1281.

484. De Robertis, E.M. (2008). Evo-devo: variations on ancestral themes. *Cell* 132, 185–195.

485. De Robertis, E.M. (2008). The molecular ancestry of segmentation mechanisms. *PNAS* 105, #43, 16411–16412.

486. De Robertis, E.M., and Sasai, Y. (1996). A common plan for dorsoventral patterning in Bilateria. *Nature* 380, 37–40.

487. de Rosa, R., Grenier, J.K., Andreeva, T., Cook, C.E., Adoutte, A., Akam, M., Carroll, S.B., and Balavoine, G. (1999). Hox genes in brachiopods and priapulids and protostome evolution. *Nature* 399, 772–776.

488. de Rosa, R., Prud'homme, B., and Balavoine, G. (2005). *caudal* and *even-skipped* in the annelid *Platynereis dumerilii* and the ancestry of posterior growth. *Evol. Dev.* 7, 574–587.

489. de Velasco, B., Erclik, T., Shy, D., Sclafani, J., Lipshitz, H., McInnes, R., and Hartenstein, V. (2007). Specification and development of the pars intercerebralis and pars lateralis, neuroendocrine command centers in the *Drosophila* brain. *Development* 302, 309–323.

490. de Visser, J.A.G.M., and Krug, J. (2014). Empirical fitness landscapes and the predictability of evolution. *Nat. Rev. Genet.* 15, 480–490.

491. Deans, M.R. (2013). A balance of form and function: planar polarity and development of the vestibular maculae. *Semin. Cell Dev. Biol.* 24, 490–498.

492. Dearden, P., and Akam, M. (1999). Developmental evolution: axial patterning in insects. *Curr. Biol.* 9, R591–R594.

493. Decker, R.S., Koyama, E., and Pacifici, M. (2014). Genesis and morphogenesis of limb synovial joints and articular cartilage. *Matrix Biol.* 39, 5–10.

494. Degnan, B.M., Vervoort, M., Larroux, C., and Richards, G.S. (2009). Early evolution of metazoan transcription factors. *Curr. Opin. Genet. Dev.* 19, 591–599.

495. Dekkers, M.P.J., Nikoletopoulou, V., and Barde, Y.-A. (2013). Death of developing neurons: new insights and implications for connectivity. *J. Cell Biol.* 203, 385–393. [*See also* Yaron, A., and Schuldiner, O. (2016). Common and divergent mechanisms in developmental neuronal remodeling and dying back neurodegeneration. *Curr. Biol.* 26, R628–R639.]

496. Delgado, I., and Torres, M. (2016). Gradients, waves and timers, an overview of limb patterning models. *Semin. Cell Dev. Biol.* 49, 109–115.

497. Delidakis, C., and Artavanis-Tsakonas, S. (1992). The Enhancer of split [E(spl)] locus of *Drosophila* encodes seven independent helix-loop-helix proteins. *PNAS* 89, 8731–8735.

498. Delling, M., Indzhykulian, A.A., Liu, X., Li, Y., Xie, T., Corey, D.P., and Clapham, D.E. (2016). Primary cilia are not calcium-responsive mechanosensors. *Nature* 531, 656–660.

499. Delmas, P., Hao, J., and Rodat-Despoix, L. (2011). Molecular mechanisms of mechanotransduction in mammalian sensory neurons. *Nat. Rev. Neurosci.* 12, 139–153.

500. Delsuc, F., Brinkmann, H., Chourrout, D., and Philippe, H. (2006). Tunicates and not cephalochordates are the closest relatives of vertebrates. *Nature* 439, 965–968.

501. Delventhal, R., and Carlson, J.R. (2016). Bitter taste receptors confer diverse functions to neurons. *eLife* 5, e11181.

502. DeMaria, S., Berke, A.P., Van Name, E., Heravian, A., Ferreira, T., and Ngai, J. (2013). Role of a ubiquitously expressed receptor in the vertebrate olfactory system. *J. Neurosci.* 33, #38, 15235–15247.

503. DeMaria, S., and Ngai, J. (2010). The cell biology of smell. *J. Cell Biol.* 191, 443–452.

504. Demuth, J.P., and Wade, M.J. (2007). Maternal expression increases the rate of *bicoid* evolution by relaxing selective constraint. *Genetica* 129, 37–43.

505. Denes, A.S., Jékely, G., Steinmetz, P.R.H., Raible, F., Snyman, H., Prud'homme, B., Ferrier, D.E.K., Balavoine, G., and Arendt, D. (2007). Molecular architecture of annelid nerve cord supports common origin of nervous system centralization in Bilateria. *Cell* 129, 277–288.

506. Deng, H., Tan, T., and Yuan, L. (2015). Advances in the molecular genetics of non-syndromic polydactyly. *Expert Rev. Mol. Med.* 17, e18.

507. Denver, R.J. (2008). Chordate metamorphosis: ancient control by iodothyronines. *Curr. Biol.* 18, R567–R569.

508. Depetris-Chauvin, A., Galagovsky, D., and Grosjean, Y. (2015). Chemicals and chemoreceptors: ecologically relevant signals driving behavior in *Drosophila*. *Front. Ecol. Evol.* 3, Article 41.

509. Derelle, R., Lopez, P., Le Guyader, H., and Manuel, M. (2007). Homeodomain proteins belong to the ancestral molecular toolkit of eukaryotes. *Evol. Dev.* 9, 212–219.

510. Deschamps, J. (2007). Ancestral and recently recruited global control of the *Hox* genes in development. *Curr. Opin. Genet. Dev.* 17, 422–427.

511. Deschamps, J., and van Nes, J. (2005). Developmental regulation of the Hox genes during axial morphogenesis in the mouse. *Development* 132, 2931–2942.

512. Deutsch, J.S. (2004). Segments and parasegments in arthropods: a functional perspective. *BioEssays* 26, 1117–1125.

513. Deutsch, J.S. (2010). Homeosis and beyond. What is the function of the Hox genes? *In* J.S. Deutsch (ed.), *Hox Genes: Studies from the 20th to the 21st Century*. Landes Bioscience, Austin, TX, pp. 155–165.

514. Devenport, D. (2014). The cell biology of planar cell polarity. *J. Cell Biol.* 207, 171–179.

515. Devenport, D., and Fuchs, E. (2008). Planar polarization in embryonic epidermis orchestrates global asymmetric morphogenesis of hair follicles. *Nature Cell Biol.* 10, 1257–1268.

516. Dewey, E.B., Taylor, D.T., and Johnston, C.A. (2015). Cell fate decision making through oriented cell division. *J. Dev. Biol.* 3, 129–157.

517. Dhouailly, D., Olivera-Martinez, I., Fliniaux, I., Missier, S., Viallet, J.P., and Thelu, J. (2004). Skin field formation: morphogenetic events. *Int. J. Dev. Biol.* 48, 85–91.

518. Diaz-Benjumea, F.J., and Cohen, S.M. (1993). Interaction between dorsal and ventral cells in the imaginal disc directs wing development in *Drosophila*. *Cell* 75, 741–752.

519. Dicke, R.H., Peebles, P.J.E., Roll, P.G., and Wilkinson, D.T. (1965). Cosmic black-body radiation. *Astrophys. J.* 142, 414–419.

520. Dickinson, M.H. (1999). Haltere-mediated equilibrium reflexes of the fruit fly, *Drosophila melanogaster*. *Philos. Trans. R. Soc. Lond. B* 354, 903–916.

521. Dickinson, M.H. (2015). Motor control: how dragonflies catch their prey. *Curr. Biol.* 25, R232–R234.

522. Dickinson, M.H., Farley, C.T., Full, R.J., Koehl, M.A.R., Kram, R., and Lehman, S. (2000). How animals move: an integrative view. *Science* 288, 100–106.

523. Dickson, B. (2001). Moving on. *Science* 291, 1910–1911.

524. Dickson, B.J., and Gilestro, G.F. (2006). Regulation of commissural axon pathfinding by Slit and its Robo receptors. *Annu. Rev. Cell Dev. Biol.* 22, 651–675.

525. Dimiccoli, M., Girard, B., Berthoz, A., and Bennequin, D. (2013). Striola magica: a functional explanation of otolith geometry. *J. Comput. Neurosci.* 35, 125–154.

526. Diogo, R., Smith, C.M., and Ziermann, J.M. (2015). Evolutionary developmental pathology and anthropology: a new field linking development, comparative anatomy, human evolution, morphological variations and defects, and medicine. *Dev. Dyn.* 244, 1357–1374.

527. Dissel, S., Angadi, V., Kirszenblat, L., Suzuki, Y., Donlea, J., Klose, M., Koch, Z., English, D., Winsky-Sommerer, R., van Swinderen, B., and Shaw, P.J. (2015). Sleep restores behavioral plasticity to *Drosophila* mutants. *Curr. Biol.* 25, 1270–1281.

528. Dissel, S., Hansen, C.N., Özkaya, Ö., Hemsley, M., Kyriacou, C.P., and Rosato, E. (2014). The logic of circadian organization in *Drosophila*. *Curr. Biol.* 24, 2257–2266.

529. Do, M.T.H., and Yau, K.-W. (2010). Intrinsically photosensitive retinal ganglion cells. *Physiol. Rev.* 90, 1547–1581.

530. Domigan, C.K., Ziyad, S., and Iruela-Arispe, M.L. (2014). Canonical and noncanonical vascular endothelial growth factor pathways: new developments in biology and signal transduction. *Arterioscler. Thromb. Vasc. Biol.* 35, 30–39.

531. Domínguez, L., González, A., and Moreno, N. (2015). Patterns of hypothalamic regionalization in amphibians and reptiles: common traits revealed by a genoarchitectonic approach. *Front. Neuroanat.* 9, Article 3.

532. Domyan, E.T., Kronenberg, Z., Infante, C.R., Vickrey, A.I., Stringham, S.A., Bruders, R., Guernsey, M.W., Park, S., Payne, J., Beckstead, R.B., Kardon, G., Menke, D.B., Yandell, M., and Shapiro, M.D. (2016). Molecular shifts in limb identity underlie development of feathered feet in two domestic avian species. *eLife* 5, e12115.

533. Donelson, N.C., and Sanyal, S. (2015). Use of *Drosophila* in the investigation of sleep disorders. *Exp. Neurol.* 274, 72–79.

534. Dong, Y., Cirimotich, C.M., Pike, A., Chandra, R., and Dimopoulos, G. (2012). *Anopheles* NF-kB-regulated splicing factors direct pathogen-specific repertoires of the hypervariable pattern recognition receptor AgDscam. *Cell Host Microbe* 12, 521–530.

535. Dong, Y., Taylor, H.E., and Dimopoulos, G. (2006). AgDscam, a hypervariable immunoglobulin domain-containing receptor of the *Anopheles gambiae* innate immune system. *PLoS Biol.* 4, #7, e229.

536. Donner, A.L., and Maas, R.L. (2004). Conservation and non-conservation of genetic pathways in eye specification. *Int. J. Dev. Biol.* 48, 743–753.

537. Döring, C., Gosda, J., Tessmar-Raible, K., Hausen, H., Arendt, D., and Purschke, G. (2013). Evolution of clitellate phaosomes from rhabdomeric photoreceptor cells of polychaetes: a study in the leech *Helobdella robusta* (Annelida, Sedentaria, Clitellata). *Front. Zool.* 10, Article 52.

538. dos Reis, M., Thawornwattana, Y., Angelis, K., Telford, M.J., Donoghue, P.C.J., and Yang, Z. (2015). Uncertainty in the timing of origin of animals and the limits of precision in molecular timescales. *Curr. Biol.* 25, 2939–2950.

539. Douglas, R.H., Partridge, J.C., Dulai, K., Hunt, D., Mullineaux, C.W., Tauber, A.Y., and Hynninen, P.H. (1998). Dragon fish see using chlorophyll. *Nature* 393, 423–424.

540. Doupé, D.P., and Jones, P.H. (2013). Cycling progenitors maintain epithelia while diverse cell types contribute to repair. *BioEssays* 35, 443–451.

541. Dowling, J.E. (2012). *The Retina: An Approachable Part of the Brain*, 2nd edn. Harvard University Press, Cambridge, MA.

542. Downey, G., and Lende, D.H. (2012). Evolution and the brain. *In* D.H. Lende and G. Downey (eds.), *The Encultured Brain: An Introduction to Neuroanthropology*. MIT Press, Cambridge, MA., pp. 103–137.

543. Driever, W. (2004). The Bicoid morphogen papers (II): account from Wolfgang Driever. *Cell* S116, S7–S9.

544. Driever, W., and Nüsslein-Volhard, C. (1988). The *bicoid* protein determines position in the *Drosophila* embryo in a concentration-dependent manner. *Cell* 54, 95–104.

545. Driever, W., and Nüsslein-Volhard, C. (1988). A gradient of *bicoid* protein in *Drosophila* embryos. *Cell* 54, 83–93.

546. Driver, E.C., Sillers, L., Coate, T.M., Rose, M.F., and Kelley, M.W. (2013). The *Atoh1*-lineage gives rise to hair cells and supporting cells within the mammalian cochlea. *Dev. Biol.* 376, 86–98.

547. Dror, A.A., and Avraham, K.B. (2010). Hearing impairment: a panoply of genes and functions. *Neuron* 68, 293–308.

548. Duboc, V., Dufourcq, P., Blader, P., and Roussigné, M. (2015). Asymmetry of the brain: development and implications. *Annu. Rev. Genet.* 49, 647–672.

549. Duboc, V., and Lepage, T. (2006). A conserved role for the nodal signaling pathway in the establishment of dorso-ventral and left-right axes in deuterostomes. *J. Exp. Zool. B. Mol. Dev. Evol.* 310, 41–53.

550. Duboc, V., Röttinger, E., Lapraz, F., Besnardeau, L., and Lepage, T. (2005). Left-right asymmetry in the sea urchin embryo is regulated by Nodal signaling on the right side. *Dev. Cell* 9, 147–158.

551. Duboule, D. (2007). The rise and fall of Hox gene clusters. *Development* 134, 2549–2560.

552. Duboule, D., and Dollé, P. (1989). The structural and functional organization of the murine HOX gene family resembles that of *Drosophila* homeotic genes. *EMBO J.* 8, 1497–1505.

553. Dubowy, C.M., and Cavanaugh, D.J. (2014). Sleep: a neuropeptidergic wake-up call for flies. *Curr. Biol.* 24, R1092–R1094.

554. Dubrulle, J., and Pourquié, O. (2004). Coupling segmentation to axis formation. *Development* 131, 5783–5793.

555. DuBuc, T.Q., Ryan, J.F., Shinzato, C., Satoh, N., and Martindale, M.Q. (2012). Coral comparative genomics reveal expanded *Hox* cluster in the cnidarian-bilaterian ancestor. *Integr. Comp. Biol.* 52, 835–841.

556. Dudley, R., and Yanoviak, S.P. (2011). Animal aloft: the origins of aerial behavior and flight. *Integr. Comp. Biol.* 51, 926–936.

557. Duelli, P. (1978). An insect retina without microvilli in the male scale insect, *Eriococcus* sp. (Eriococcidae, Homoptera). *Cell Tissue Res.* 187, 417–427.

558. Dugas-Ford, J., and Ragsdale, C.W. (2015). Levels of homology and the problem of the neocortex. *Annu. Rev. Neurosci.* 38, 351–368.

559. Dulac, C. (2006). Charting olfactory maps. *Science* 314, 606–607.

560. Dulai, K.S., von Dornum, M., Mollon, J.D., and Hunt, D.M. (1999). The evolution of trichromatic color vision by opsin gene duplication in New World and Old World primates. *Genome Res.* 9, 629–638.

561. Duncan, J.S., and Fritzsch, B. (2012). Evolution of sound and balance perception: innovations that aggregate single hair cells into the ear and transform a gravistatic sensor into the organ of Corti. *Anat. Rec.* 295, 1760–1774.

562. Dunn, C.W., Giribet, G., Edgecombe, G.D., and Hejnol, A. (2014). Animal phylogeny and its evolutionary implications. *Annu. Rev. Ecol. Evol. Syst.* 45, 371–395.

563. Dunn, C.W., and Ryan, J.F. (2015). The evolution of animal genomes. *Curr. Opin. Genet. Dev.* 35, 25–32.

564. Dupé, V., and Lumsden, A. (2001). Hindbrain patterning involves graded responses to retinoic acid signalling. *Development* 128, 2199–2208.

565. Durston, A.J. (2012). Global posterior prevalence is unique to vertebrates: a dance to the music of time? *Dev. Dyn.* 241, 1799–1807.

566. Durston, A.J., Wacker, S., Bardine, N., and Jansen, H.J. (2012). Time space translation: a *Hox* mechanism for vertebrate A-P patterning. *Curr. Genomics* 13, 300–307.

567. Duttke, S.H.C., Doolittle, R.F., Wang, Y.-L., and Kadonga, J.T. (2014). TRF2 and the evolution of the bilateria. *Genes Dev.* 28, 2071–2076.

568. Duverger, O., and Morasso, M.I. (2014). To grow or not to grow: hair morphogenesis and human genetic hair disorders. *Semin. Cell Dev. Biol.* 25–26, 22–23.

569. Duysens, J., and Van de Crommert, H.W.A.A. (1998). Neural control of locomotion. Part 1: The central pattern generator from cats to humans. *Gait Posture* 7, 131–141.

570. Dyer, M.A., Livesey, F.J., Cepko, C.L., and Oliver, G. (2003). Prox1 function controls progenitor cell proliferation and horizontal cell genesis in the mammalian retina. *Nat. Genet.* 34, 53–58.

571. Eakin, R.M. (1965). Evolution of photoreceptors. *Cold Spring Harb. Symp. Quant. Biol.* 30, 363–370.

572. Eakin, R.M. (1979). Evolutionary significance of photoreceptors: in retrospect. *Am. Zool.* 19, 647–653.

573. Eakin, R.M., and Brandenberger, J.L. (1980). Unique eye of probable evolutionary significance. *Science* 211, 1189–1190.

574. Eakin, R.M., and Westfall, J.A. (1962). Fine structure of photoreceptors in amphioxus. *J. Ultrastruct. Res.* 6, 531–539.

575. Eatock, R.A., and Hurley, K.M. (2003). Functional development of hair cells. *Curr. Top. Dev. Biol.* 57, 389–447.

576. Eatock, R.A., and Songer, J.E. (2011). Vestibular hair cells and afferents: two channels for head motion signals. *Annu. Rev. Neurosci.* 34, 501–534.

577. Eberl, D.F., and Boekhoff-Falk, G. (2007). Development of Johnston's organ in *Drosophila*. *Int. J. Dev. Biol.* 51, 679–687.

578. Eberl, D.F., Hardy, R.W., and Kernan, M.J. (2000). Genetically similar transduction mechanisms for touch and hearing in *Drosophila*. *J. Neurosci.* 20, #16, 5981–5988.

579. Ecker, J.L., Dumitrescu, O.N., Wong, K.Y., Alam, N.M., Chen, S.-K., LeGates, T., Renna, J.M., Prusky, G.T., Berson, D.M., and Hattar, S. (2010). Melanopsin-expressing retinal ganglion-cell photoreceptors: cellular diversity and role in pattern vision. *Neuron* 67, 49–60.

580. Eddison, M., Le Roux, I., and Lewis, J. (2000). Notch signaling in the development of the inner ear: lessons from *Drosophila*. *PNAS* 97, 11692–11699.

581. Edelman, G.M. (1993). Neural Darwinism: selection and reentrant signaling in higher brain function. *Neuron* 10, 115–125.

582. Edgar, L.G., Carr, S., Wang, H., and Wood, W.B. (2001). Zygotic expression of the *caudal* homolog *pal-1* is required for posterior patterning in *Caenorhabditis elegans* embryogenesis. *Dev. Biol.* 229, 71–88.

583. Edgar, R.S., Green, E.W., Zhao, Y., van Ooijen, G., Olmedo, M., Qin, X., Xu, Y., Pan, M., Valekunja, U.K., Feeney, K.A., Maywood, E.S., Hastings, M.H., Baliga, N.S., Merrow, M., Millar, A.J., Johnson, C.H., Kyriacou, C.P., O'Neill, J.S., and Reddy, A.B. (2012). Peroxiredoxins are conserved markers of circadian rhythms. *Nature* 485, 459–464.

584. Effertz, T., Scharr, A.L., and Ricci, A.J. (2015). The how and why of identifying the hair cell mechano-electrical transduction channel. *Pflugers Arch.* 467, 73–84.

585. Effertz, T., Wiek, R., and Göpfert, M.C. (2011). NompC TRP channel is essential for *Drosophila* sound receptor function. *Curr. Biol.* 21, 592–597.

586. Egger, B., Chell, J.M., and Brand, A.H. (2008). Insights into neural stem cell biology from flies. *Philos. Trans. R. Soc. Lond. B* 363, 39–56.

587. Eichmann, A., and Thomas, J.-L. (2013). Molecular parallels between neural and vascular development. *Cold Spring Harb. Perspect. Med.* 3, a006551.

588. Elliott, D.A., Solloway, M.J., Wise, N., Biben, C., Costa, M.W., Furtado, M.B., Lange, M., Dunwoodie, S., and Harvey, R.P. (2006). A tyrosine-rich domain within homeodomain transcription factor Nkx2-5 is an essential element in the early cardiac transcriptional regulatory machinery. *Development* 133, 1311–1322.

589. Elofsson, R., and Dahl, E. (1970). The optic neuropiles and chiasmata of Crustacea. *Z. Zellforsch. Mikrosk Anat.* 107, 343–360.

590. Elsaesser, R., and Paysan, J. (2007). The sense of smell, its signalling pathways, and the dichotomy of cilia and microvilli in olfactory sensory cells. *BMC Neurosci.* 8 (Suppl. 3), Article S1.

591. Enard, W. (2015). Human evolution: enhancing the brain. *Curr. Biol.* 25, R409–R430.

592. Enjin, A., Zaharieva, E.E., Frank, D.D., Mansourian, S., Suh, G.B., Gallio, M., and Stensmyr, M.C. (2016). Humidity sensing in *Drosophila*. *Curr. Biol.* 26, 1352–1358.

593. Epstein, M., Pillemer, G., Yelin, R., Yisraeli, J.K., and Fainsod, A. (1997). Patterning of the embryo along the anterior-posterior axis: the role of the *caudal* genes. *Development* 124, 3805–3814.

594. Erclik, T., Hartenstein, V., Lipshitz, H.D., and McInnes, R.R. (2008). Conserved role of the *Vsx* genes supports a monophyletic origin for bilaterian visual systems. *Curr. Biol.* 18, 1278–1287.

595. Erclik, T., Hartenstein, V., McInnes, R.R., and Lipshitz, H.D. (2009). Eye evolution at high resolution: the neuron as a unit of homology. *Dev. Biol.* 332, 70–79.

596. Erickson, T., French, C.R., and Waskiewicz, A.J. (2010). Meis1 specifies positional information in the retina and tectum to organize the zebrafish visual system. *Neural Dev.* 5, Article 22.

597. Eriksson, B.J., Larson, E.T., Thörnqvist, P.-O., Tait, N.N., and Budd, G.E. (2005). Expression of *engrailed* in the developing brain and appendages of the ony-chophoran *Euperipatoides kanangrensis* (Reid). *J. Exp. Zool. B. Mol. Dev. Evol.* 304, 220–228.

598. Ernsberger, U. (2015). Can the "neuron theory" be complemented by a universal mechanism for generic neuronal differentiation? *Cell Tissue Res.* 359, 343–384.

599. Ernst, O.P., Lodowski, D.T., Elstner, M., Hegemann, P., Brown, L.S., and Kandori, H. (2014). Microbial and animal rhodopsins: structures, functions, and molecular mechanisms. *Chem. Rev.* 114, 126–163.

600. Erskine, L., and Herrera, E. (2007). The retinal ganglion cell axon's journey: insights into molecular mechanims of axon guidance. *Dev. Biol.* 308, 1–14.

601. Erwin, D.H., and Davidson, E.H. (2002). The last common bilaterian ancestor. *Development* 129, 3021–3032.

602. Esteves, F.F., Springhorn, A., Kague, E., Taylor, E., Pyrowolakis, G., Fisher, S., and Bier, E. (2014). BMPs regulate *msx* gene expression in the dorsal neuroectoderm of *Drosophila* and vertebrates by distinct mechanisms. *PLoS Genet.* 10, #9, e1004625.

603. Estévez-Calvar, N., Romero, A., Figueras, A., and Novoa, B. (2013). Genes of the mitochondrial apoptotic pathway in *Mytilus galloprovincialis*. *PLoS ONE* 8, #4, e61502.

604. Etchberger, J.F., Flowers, E.B., Poole, R.J., Bashllari, E., and Hobert, O. (2009). *Cis*-regulatory mechanisms of left/right asymmetric neuron-subtype specification in *C. elegans*. *Development* 136, 147–160.

605. Evans, C.J., Hartenstein, V., and Banerjee, U. (2003). Thicker than blood: conserved mechanisms in *Drosophila* and vertebrate hematopoiesis. *Dev. Cell* 5, 673–690.

606. Evans, C.J., Sinenko, S.A., Mandal, L., Martinez-Agosto, J.A., Hartenstein, V., and Benerjee, U. (2008). Genetic dissection of hematopoiesis using *Drosophila* as a model system. *Adv. Dev. Biol.* 18, 259–299.

607. Evans, S.M. (1999). Vertebrate tinman homologues and cardiac differentiation. *Semin. Cell Dev. Biol.* 10, 73–83.

608. Evans, T.A., and Bashaw, G.J. (2010). Axon guidance at the midline: of mice and flies. *Curr. Opin. Neurobiol.* 20, 79–85.

609. Evans, T.A., Santiago, C., Arbeille, E., and Bashaw, G.J. (2015). Robo2 acts in trans to inhibit Slit-Robo1 repulsion in pre-crossing commissural axons. *eLife* 4, e08407.

610. Ewing, T. (1993). Genetic "master switch" for left-right symmetry found. *Science* 260, 624–625.

611. Extavour, C.G., and Akam, M. (2003). Mechanisms of germ cell specification across the metazoans: epigenesis and preformation. *Development* 130, 5869–5884.

612. Extavour, C.G.M. (2007). Evolution of the bilaterian germ line: lineage origin and modulation of specification mechanisms. *Integr. Comp. Biol.* 47, 770–785.

613. Extavour, C.G.M. (2008). Urbisexuality: the evolution of bilaterian germ cell specification and reproductive systems. *In* A. Minelli and G. Fusco (eds.), *Evolving Pathways: Key Themes in Evolutionary Developmental Biology*. Cambridge University Press, New York, NY, pp. 321–342.

614. Ezan, J., and Montcouquiol, M. (2013). Revisiting planar cell polarity in the inner ear. *Semin. Cell Dev. Biol.* 24, 499–506. [*See also* Faubel, R., *et al.* (2016). Cilia-based flow network in the brain ventricles. *Science* 353, 176–178.]

615. Fabri, M., and Polonara, G. (2013). Functional topography of human corpus callosum: an fMRI mapping study. *Neural Plast.* 2013, Article 251308.

616. Fähling, M., Mrowka, R., Steege, A., Kirschner, K.M., Benko, E., Förstera, B., Persson, P.B., Thiele, B.J., Meier, J.C., and Scholz, H. (2009). Translational regulation of the human achaete-scute homologue-1 by fragile X mental retardation protein. *J. Biol. Chem.* 284, #7, 4255–4266.

617. Fain, G.L. (2003). *Sensory Transduction*. Sinauer, Sunderland, MA.

618. Fain, G.L. (2016). Phototransduction: making the chromophore to see through the murk. *Curr. Biol.* 25, R1126–R1142.

619. Fain, G.L., Hardie, R., and Laughlin, S.B. (2010). Phototransduction and the evolution of photoreceptors. *Curr. Biol.* 20, R114–R124.

620. Fan, J.-Y., Preuss, F., Muskus, M.J., Bjes, E.S., and Price, J.L. (2009). *Drosophila* and vertebrate Casein Kinase Iδ exhibits evolutionary conservation of circadian function. *Genetics* 181, 139–152.

621. Fanto, M., and McNeill, H. (2004). Planar polarity from flies to vertebrates. *J. Cell Sci.* 117, 527–533.

622. Farah, M.J. (2015). An ethics toolbox for neurotechnology. *Neuron* 86, 34–37.

623. Farfán, C., Shigeno, S., Nödl, M.-T., and de Couet, H.G. (2009). Developmental expression of *apterous/Lhx2/9* in the sepiolid squid *Euprymna scolopes* supports an ancestral role in neural development. *Evol. Dev.* 11, 354–362.

624. Farnum, C.E., and Wilsman, N.J. (2011). Axonemal positioning and orientation in three-dimensional space for primary cilia: what is known, what is assumed, and what needs clarification. *Dev. Dyn.* 240, 2405–2431.

625. Farrar, N.R., and Spencer, G.E. (2008). Pursuing a "turning point" in growth cone research. *Dev. Biol.* 318, 102–111.

626. Farris, S.M. (2015). Evolution of brain elaboration. *Philos. Trans. R. Soc. Lond. B* 370, 20150054.

627. Faurie, C., and Raymond, M. (2004). Handedness frequency over more than ten thousand years. *Proc. R. Soc. Lond. B* 271, S43–S45.

628. Fears, S.C., Scheibel, K., Abaryan, Z., Lee, C., Service, S.K., Jorgensen, M.J., Fairbanks, L.A., Cantor, R.M., Freimer, N.B., and Woods, R.P. (2011). Anatomic brain asymmetry in vervet monkeys. *PLoS ONE* 6, #12, e28243.

629. Feiler, R., Bjornson, R., Kirschfeld, K., Mismer, D., Rubin, G.M., Smith, D.P., Socolich, M., and Zuker, C.S. (1992). Ectopic expression of ultraviolet-rhodopsins in the blue photoreceptor cells of *Drosophila*: visual physiology and photochemistry of transgenic animals. *J. Neurosci.* 12, #10, 3862–3868.

630. Fekete, D.M., and Wu, D.K. (2002). Revisiting cell fate specification in the inner ear. *Curr. Opin. Neurobiol.* 12, 35–42.

631. Ferguson, L., Marlétaz, F., Carter, J.-M., Taylor, W.R., Gibbs, M., Breuker, C.J., and Holland, P.W.H. (2014). Ancient expansion of the Hox cluster in Lepidoptera generated four homeobox genes implicated in extra-embryonic tissue formation. *PLoS Genet.* 10, #10, e1004698.

632. Fernald, R.D. (2000). Evolution of eyes. *Curr. Opin. Neurobiol.* 10, 444–450.

633. Fernald, R.D. (2004). Evolving eyes. *Int. J. Dev. Biol.* 48, 701–705.

634. Fernald, R.D. (2006). Casting a genetic light on the evolution of eyes. *Science* 313, 1914–1918.

635. Ferreira, L.M.R., and Mostajo-Radji, M.A. (2013). How induced pluripotent stem cells are redefining personalized medicine. *Gene* 520, 1–6.

636. Ferreira, T., Wilson, S.R., Choi, Y.G., Risso, D., Dudoit, S., Speed, T.P., and Ngai, J. (2014). Silencing of odorant receptor genes by G protein βγ signaling ensures the expression of one odorant receptor per olfactory sensory neuron. *Neuron* 81, 847–859.

637. Ferrier, D.E.K. (2012). Evolutionary crossroads in developmental biology: annelids. *Development* 139, 2543–2653.

638. Ferrier, D.E.K. (2016). The origin of the Hox/ParaHox genes, the Ghost Locus hypothesis and the complexity of the first animal. *Brief. Funct. Genomics* 15, 1–9 (doi 10.1093).

639. Ferrier, D.E.K., and Minguillón, C. (2003). Evolution of the Hox/ParaHox gene clusters. *Int. J. Dev. Biol.* 47, 605–611.

640. Fettiplace, R., and Kim, K.X. (2014). The physiology of mechanoelectrical transduction channels in hearing. *Physiol. Rev.* 94, 951–986.

641. Feuda, R., Hamilton, S.C., McInerney, J.O., and Pisani, D. (2012). Metazoan opsin evolution reveals a simple route to animal vision. *PNAS* 109, #46, 18868–18872.

642. Finlay, B.L. (2008). The developing and evolving retina: using time to organize form. *Brain Res.* 1192, 5–16.

643. Fiore, V.G., Dolan, R.J., Strausfeld, N.J., and Hirth, F. (2015). Evolutionarily conserved mechanisms for the selection and maintenance of behavioural activity. *Philos. Trans. R. Soc. Lond. B* 370, 20150053.

644. Firestein, S. (2001). How the olfactory system makes sense of scents. *Nature* 413, 211–218.

645. Fishman, M.C., and Olson, E.N. (1997). Parsing the heart: genetic modules for organ assembly. *Cell* 91, 153–156.

646. Fitch, D.H.A., and Sudhaus, W. (2002). One small step for worms, one giant leap for "Bauplan"? *Evol. Dev.* 4, 243–246.

647. FitzGerald, B.J., Richardson, K., and Wesson, D.W. (2014). Olfactory tubercle stimulation alters odor preference behavior and recruits forebrain reward and motivational centers. *Front. Behav. Neurosci.* 8, Article 81.

648. Flanagan, J.G., and Van Vactor, D. (1998). Through the looking glass: axon guidance at the midline choice point. *Cell* 92, 429–432.

649. Fleming, A., Kishida, M.G., Kimmel, C.B., and Keynes, R.J. (2015). Building the backbone: the development and evolution of vertebral patterning. *Development* 142, 1733–1744.

650. Florio, M., Albert, M., Taverna, E., Namba, T., Brandl, H., Lewitus, E., Haffner, C., Sykes, A., Wong, F.K., Peters, J., Guhr, E., Klemroth, S., Prüfer, K., Kelso, J., Naumann, R., Nüsslein, I., Dahl, A., Lachmann, R., Pääbo, S., and Huttner, W.B. (2015). Human-specific gene *ARHGAP11B* promotes basal progenitor amplification and neocortex expansion. *Science* 347, 1465–1470.

651. Foltys, H., Krings, T., Meister, I.G., Sparing, R., Boroojerdi, B., Thron, A., and Topper, R. (2003). Motor representation in patients rapidly recovering after stroke: a functional magnetic resonance imaging and transcranial magnetic stimulation study. *Clin. Neurophysiol.* 114, 2404–2415.

652. Fomenou, M.D., Scaal, M., Stockdale, F.E., Christ, B., and Huang, R. (2005). Cells of all somitic compartments are determined with respect to segmental identity. *Dev. Dyn.* 233, 1386–1393.

653. Fontdevila, A. (2011). *The Dynamic Genome: A Darwinian Approach*. Oxford University Press, New York, NY.

654. Formosa-Jordan, P., and Ibañes, M. (2014). Competition in Notch signaling with cis enriches cell fate decisions. *PLoS ONE* 9, #4, e95744.

655. Foronda, D., Martin, P., and Sánchez-Herrero, E. (2012). *Drosophila* Hox and sex-determination genes control segment elimination through EGFR and *extramacrochaete* activity. *PLoS Genet.* 8, #8, e1002874.

656. Forouhar, A.S., Liebling, M., Hickerson, A., Nasiraei-Moghaddam, A., Tsai, H.-J., Hove, J.R., Fraser, S.E., Dickenson, M.E., and Gharib, M. (2006). The embryonic vertebrate heart tube is a dynamic suction pump. *Science* 312, 751–753.

657. Fortey, R. (2012). *Horseshoe Crabs and Velvet Worms: The Story of the Animals and Plants That Time Has Left Behind*. Knopf, New York, NY.

658. Fortini, M.E. (2009). Notch signaling: the core pathway and its posttranslational regulation. *Dev. Cell* 16, 633–647.

659. Fortunato, S.A.V., Adamski, M., Ramos, O.M., Leininger, S., Liu, J., Ferrier, D.E.K., and Adamska, M. (2014). Calcisponges have a ParaHox gene and dynamic expression of dispersed NK homeobox genes. *Nature* 514, 620–623.

660. Fortunato, S.A.V., Leininger, S., and Adamski, M. (2014). Evolution of the Pax-Six-Eya-Dach network: the calcisponge case study. *EvoDevo* 5, Article 23.

661. Foster, K.W. (2009). Eye evolution: two eyes can be better than one. *Curr. Biol.* 19, R208–R210.

662. Foster, R.G., and Hankins, M.W. (2007). Circadian vision. *Curr. Biol.* 17, R746–R751.

663. Franchini, L.F., and Pollard, K.S. (2015). Can a few non-coding mutations make a human brain? *BioEssays* 37, 1054–1061.

664. Frankel, N., Davis, G.K., Vargas, D., Wang, S., Payre, F., and Stern, D.L. (2010). Phenotypic robustness conferred by apparently redundant transcriptional enhancers. *Nature* 466, 490–493.

665. Franz, E.A., Chiaroni-Clarke, R., Woodrow, S., Glendining, K.A., Jasoni, C.L., Robertson, S.P., Gardner, R.J.M., and Markie, D. (2015). Congenital mirror movements: phenotypes associated with *DCC* and *RAD51* mutations. *J. Neurol. Sci.* 351, 140–145.

666. Franze, K. (2013). The mechanical control of nervous system development. *Development* 140, 3069–3077.

667. Frasnelli, E., Vallortigara, G., and Rogers, L.J. (2012). Left-right asymmetries of behaviour and nervous system in invertebrates. *Neurosci. Biobehav. Rev.* 36, 1273–1291.

668. Frawley, L.E., and Orr-Weaver, T.L. (2015). Polyploidy. *Curr. Biol.* 25, R353–R357.

669. Freeman, G., and Lundelius, J.W. (1982). The developmental genetics of dextrality and sinistrality in the gastropod *Lymnaea peregra*. *W. Roux's Arch.* 191, 69–83.

670. Freeman, M.R., and Doherty, J. (2006). Glial cell biology in *Drosophila* and vertebrates. *Trends Neurosci.* 29, 82–90.

671. Freeman, R., Ikuta, T., Wu, M., Koyanagi, R., Kawashima, T., Tagawa, K., Humphreys, T., Fang, G.-C., Fujiyama, A., Saiga, H., Lowe, C., Worley, K., Jenkins, J., Schmutz, J., Kirschner, M., Rokhsar, D., Satoh, N., and Gerhart, J. (2012). Identical genomic organization of two hemichordate Hox clusters. *Curr. Biol.* 22, 2053–2058.

672. Freibaum, B.D., Lu, Y., Lopez-Gonzalez, R., Kim, N.C., Almeida, S., Lee, K.-H., Badders, N., Valentine, M., Miller, B.L., Wong, P.C., Petrucelli, L., Kim, H.J., Gao, F.-B., and Taylor, J.P. (2015). GGGGCC repeat expansion in *C9orf72* compromises nucleocytoplasmic transport. *Nature* 525, 129–133.

673. French, V., Bryant, P.J., and Bryant, S.V. (1976). Pattern regulation in epimorphic fields. *Science* 193, 969–981.

674. Frentiu, F.D., and Briscoe, A.D. (2008). A butterfly eye's view of birds. *BioEssays* 30, 1151–1162.

675. Frenzel, H., Bohlender, J., Pinsker, K., Wohlleben, B., Tank, J., Lechner, S.G., Schiska, D., Jaijo, T., Rüshendorf, F., Saar, K., Jordan, J., Millán, J.M., Gross, M., and Lewin, G.R. (2012). A genetic basis for mechanosensory traits in humans. *PLoS Biol.* 10, #5, e1001318.

676. Friedmann, D., Hoagland, A., Berlin, S., and Isacoff, E.Y. (2015). A spinal opsin controls early neural activity and drives a behavioral light response. *Curr. Biol.* 25, 69–74.

677. Friedrich, F. (2006). Ancient mechanisms of visual sense organ development based on comparison of the gene networks controlling larval eye, ocellus, and compound eye specification in *Drosophila*. *Arthropod Struct. Dev.* 35, 357–378.

678. Friedrich, M. (2015). Evo-devo gene toolkit update: at least seven Pax transcription factor subfamilies in the last common ancestor of bilaterian animals. *Evol. Dev.* 17, 255–257.

679. Frigon, A. (2012). Central pattern generators of the mammalian spinal cord. *Neuroscientist* 18, 56–69.

680. Fritsch, M., Wollesen, T., de Oliveira, A.L., and Wanninger, A. (2015). Unexpected co-linearity of Hox gene expression in an aculiferan mollusk. *BMC Evol. Biol.* 15, Article 151.

681. Fritzsch, B., and Beisel, K.W. (2001). Evolution and development of the vertebrate ear. *Brain Res. Bull.* 55, 711–721.

682. Fritzsch, B., and Beisel, K.W. (2003). Molecular conservation and novelties in vertebrate ear development. *Curr. Top. Dev. Biol.* 57, 1–44.

683. Fritzsch, B., and Beisel, K.W. (2004). Keeping sensory cells and evolving neurons to connect them to the brain: molecular conservation and novelties in vertebrate ear development. *Brain Behav. Evol.* 64, 182–197.

684. Fritzsch, B., Beisel, K.W., Pauley, S., and Soukup, G. (2007). Molecular evolution of the vertebrate mechanosensory cell and ear. *Int. J. Dev. Biol.* 51, 663–678.

685. Fritzsch, B., Eberl, D.F., and Beisel, K.W. (2010). The role of bHLH genes in ear development and evolution: revisiting a 10-year-old hypothesis. *Cell. Mol. Life Sci.* 67, 3089–3099.

686. Fritzsch, B., Jahan, I., Pan, N., Kersigo, J., Duncan, J., and Kopecky, B. (2011). Dissecting the molecular basis of organ of Corti development: where are we now? *Hear. Res.* 276, 16–26.

687. Fritzsch, B., Pauley, S., and Beisel, K.W. (2006). Cells, molecules and morphogenesis: the making of the vertebrate ear. *Brain Res.* 1091, 151–171.

688. Fritzsch, B., and Piatigorsky, J. (2005). Ancestry of photic and mechanic sensation? *Science* 308, 1113–1114.

689. Fritzsch, B., and Straka, H. (2014). Evolution of vertebrate mechanosensory hair cells and inner ears: toward identifying stimuli that select mutation driven altered morphologies. *J. Comp. Physiol. A* 200, 5–18.

690. Fröbius, A.C., Matus, D.Q., and Seaver, E.C. (2008). Genomic organization and expression demonstrate spatial and temporal *Hox* gene colinearity in the lophotrochozoan *Capitella* sp. I. *PLoS ONE* 3, #12, e4004.

691. Frohns, A., Frohns, F., Naumann, S.C., Layer, P.G., and Löbrich, M. (2014). Inefficient double-strand break repair in murine rod photoreceptors with inverted heterochromatin organization. *Curr. Biol.* 24, 1080–1090.

692. Fu, X., Yan, Y., Xu, P.S., Geerlof-Vidavsky, I., Chong, W., Gross, M.L., and Holy, T.E. (2015). A molecular code for identity in the vomeronasal system. *Cell* 163, 313–323.

693. Fu, Y., Kefalov, V., Luo, D.-G., Xue, T., and Yau, K.-W. (2008). Quantal noise from human red cone pigment. *Nat. Neurosci.* 11, 565–571.

694. Fuchs, E. (2007). Scratching the surface of skin development. *Nature* 445, 834–842.

695. Fuchs, P.A. (2015). How many proteins does it take to gate hair cell mechanotransduction? *PNAS* 112, #5, 1254–1255.

696. Fuchs, Y., and Steller, H. (2011). Programmed cell death in animal development and disease. *Cell* 147, 742–758.

697. Fuerst, P.G., Bruce, F., Rounds, R.P., Erskine, L., and Burgess, R.W. (2012). Cell autonomy of DSCAM function in retinal development. *Dev. Biol.* 361, 326–337.

698. Fuerst, P.G., Koizumi, A., Masland, R.H., and Burgess, R.W. (2008). Neurite arborization and mosaic spacing in the mouse retina require DSCAM. *Nature* 451, 470–474.

699. Fujii, S., Yavuz, A., Slone, J., Jagge, C., Song, X., and Amrein, H. (2015). *Drosophila* sugar receptors in sweet taste perception, olfaction, and internal nutrient sensing. *Curr. Biol.* 25, 621–627.

700. Fukushige, T., Brodigan, T.M., Schriefer, L.A., Waterston, R.H., and Krause, M. (2006). Defining the transcriptional redundancy of early bodywall muscle development in *C. elegans*: evidence for a unified theory of animal muscle development. *Genes Dev.* 20, 3395–3406.

701. Furlong, R.F., and Holland, P.W.H. (2002). Bayesian phylogenetic analysis supports monophyly of Ambulacraria and of cyclostomes. *Zool. Sci.* 19, 593–599.

702. Fusco, M.A., Wajsenzon, I.J.R., de Carvalho, S.L., da Silva, R.T., Einicker-Lamas, M., Cavalcante, L.A., and Allodi, S. (2014). Vascular endothelial growth factor-like and its receptor in a crustacean optic ganglia: a role in neuronal differentiation? *Biochem. Biophys. Res. Comm.* 447, 299–303.

703. Fuss, S.H., and Ray, A. (2009). Mechanisms of odorant receptor gene choice in *Drosophila* and vertebrates. *Mol. Cell. Neurosci.* 41, 101–112.

704. Gaillard, D., Xu, M., Liu, F., Millar, S.E., and Barlow, L.A. (2015). β-Catenin signaling biases multipotent lingual epithelial progenitors to differentiate and acquire specific taste cell fates. *PLoS Genet.* 11, #5, e1005208.

705. Galic, M., and Matis, M. (2015). Polarized trafficking provides spatial cues for planar cell polarization within a tissue. *BioEssays* 37, 678–686.

706. Galindo, M.I., Fernández-Garza, D., Phillips, R., and Couso, J.P. (2011). Control of *Distal-less* expression in the *Drosophila* appendages by functional 3' enhancers. *Dev. Biol.* 353, 396–410.

707. Galizia, C.G., and Rössler, W. (2010). Parallel olfactory systems in insects: anatomy and function. *Annu. Rev. Entomol.* 55, 399–420.

708. Galli-Resta, L. (1998). Patterning the vertebrate retina: the early appearance of retinal mosaics. *Semin. Cell Dev. Biol.* 9, 279–284.

709. Gallo, G. (2013). Mechanisms underlying the initiation and dynamics of neuronal filopodia: from neurite formation to synaptogenesis. *Int. Rev. Cell Mol. Biol.* 301, 95–156.

710. Ganmor, E., Segev, R., and Schneidman, E. (2015). A thesaurus for a neural population code. *eLife* 4, e06134.

711. Gao, Q., Yuan, B., and Chess, A. (2000). Convergent projections of *Drosophila* olfactory neurons to specific glomeruli in the antennal lobe. *Nat. Neurosci.* 3, 780–785. [*See also* Berck, M.E., *et al.* (2016). The wiring diagram of a glomerular olfactory system. *eLife* 5, e14859.]

712. Gao, Y., Lan, Y., Liu, H., and Jiang, R. (2011). The zinc finger transcription factors Osr1 and Osr2 control synovial joint formation. *Dev. Biol.* 352, 83–91.

713. Gao, Y., Lan, Y., Ovitt, C.E., and Jiang, R. (2009). Functional equivalence of the zinc finger transcription factors Osr1 and Osr2 in mouse development. *Dev. Biol.* 328, 200–209.

714. Garcia-Fernàndez, J. (2005). The genesis and evolution of homeobox gene clusters. *Nat. Rev. Genet.* 6, 881–892.

715. Garcia-Fernàndez, J. (2005). Hox, ParaHox, ProtoHox: facts and guesses. *Heredity* 94, 145–152.

716. Garcia-Fernàndez, J., and Benito-Gutiérrez, É. (2009). It's a long way from amphioxus: descendants of the earliest chordate. *BioEssays* 31, 665–675.

717. García-Frigola, C., Carreres, M.I., Vegar, C., Mason, C., and Herrera, E. (2008). Zic2 promotes axonal divergence at the optic chiasm midline by EphB1-dependent and -independent mechanisms. *Development* 135, 1833–1841.

718. Gardner, K.H., and Correa, F. (2012). How plants see the invisible. *Science* 335, 1451–1452.

719. Garm, A., and Nilsson, D.-E. (2014). Visual navigation in starfish: first evidence for the use of vision and eyes in starfish. *Proc. R. Soc. Lond. B* 281, 20132011.

720. Garm, A., Oskarsson, M., and Nilsson, D.-E. (2011). Box jellyfish use terrestrial visual cues for navigation. *Curr. Biol.* 21, 798–803.

721. Garrett, A.M., and Burgess, R.W. (2015). Self-awareness in the retina. *eLife* 4, e10233.

722. Garrity, P.A. (2010). Feel the light. *Nature* 468, 900–901.

723. Garry, D.J., and Olson, E.N. (2006). A common progenitor at the heart of development. *Cell* 127, 1101–1104.

724. Garstang, M., and Ferrier, D.E.K. (2013). Time is of the essence for ParaHox homeobox gene clustering. *BMC Biol.* 11, Article 72.

725. Garvie, C.W., and Wolberger, C. (2001). Recognition of specific DNA sequences. *Mol. Cell* 8, 937–946.

726. Gaunt, S.J. (1997). Chick limbs, fly wings and homology at the fringe. *Nature* 386, 324–325.

727. Gaunt, S.J., Drage, D., and Trubshaw, R.C. (2008). Increased Cdx protein dose effects upon axial patterning in transgenic lines of mice. *Development* 135, 2511–2520.

728. Gaviño, M.A., and Reddien, P.W. (2011). A Bmp/Admp regulatory circuit controls maintenance and regeneration of dorsal-ventral polarity in planarians. *Curr. Biol.* 21, 294–299.

729. Gee, H. (2008). The amphioxus unleashed. *Nature* 453, 999–1000.

730. Gehring, W., and Rosbash, M. (2003). The coevolution of blue-light photoreception and circadian rhythms. *J. Mol. Evol.* 57, S286–S289.

731. Gehring, W.J. (1985). Homeotic genes, the homeo box, and the genetic control of development. *Cold Spring Harb. Symp. Quant. Biol.* 50, 243–251.

732. Gehring, W.J. (1998). *Master Control Genes in Development and Evolution: The Homeobox Story*. Yale University Press, New Haven, CT.

733. Gehring, W.J. (2002). The genetic control of eye development and its implications for the evolution of the various eye-types. *Int. J. Dev. Biol.* 46, 65–73.

734. Gehring, W.J. (2012). The animal body plan, the prototypic body segment, and eye evolution. *Evol. Dev.* 14, 34–46.

735. Gehring, W.J. (2014). The evolution of vision. *WIREs Dev. Biol.* 3, 1–40.

736. Gehring, W.J., and Ikeo, K. (1999). *Pax6*: mastering eye morphogenesis and eye evolution. *Trends Genet.* 15, 371–377.

737. Géléoc, G.S.G., and Holt, J.R. (2014). Sound strategies for hearing restoration. *Science* 344, 596.

738. Géminard, C., González-Morales, N., Coutelis, J.B., and Noselli, S. (2014). The Myosin ID pathway and left-right asymmetry in *Drosophila*. *Genesis* 52, 471–480.

739. Genikhovich, G., Fried, P., Prünster, M.M., Schinko, J.B., Gilles, A.F., Fredman, D., Meier, K., Iber, D., and Technau, U. (2015). Axis patterning by BMPs: cnidarian network reveals evolutionary constraints. *Cell Rep.* 10, 1646–1654.

740. Geoffroy St.-Hilaire, E. (1822). Considérations générales sur la vertèbre. *Mém. Mus. Hist. Nat.* 9, 89–119 + Planches V–VII (ff).

741. Gerdes, J.M., Davis, E.E., and Katsanis, N. (2009). The vertebrate primary cilium in development, homeostasis, and disease. *Cell* 137, 32–45.

742. Gerhart, J., and Kirschner, M. (1997). *Cells, Embryos, and Evolution*. Blackwell Science, Malden, MA.

743. Gerhart, J., Lowe, C., and Kirschner, M. (2005). Hemichordates and the origin of chordates. *Curr. Opin. Genet. Dev.* 15, 461–467.

744. Gerkema, M.P., Davies, W.I.L., Foster, R.G., Menaker, M., and Hut, R.A. (2013). The nocturnal bottleneck and the evolution of activity patterns in mammals. *Proc. R. Soc. Lond. B* 280, 20130508. [*See also* Kim, J.-W., *et al.* (2016). Recruitment of rod photoreceptors from short-wavelength-sensitive cones during the evolution of nocturnal vision in mammals. *Dev. Cell* 37, 520–532.]

745. Gerstner, J.R., and Yin, J.C.P. (2010). Circadian rhythms and memory formation. *Nat. Rev. Neurosci.* 11, 577–588.

746. Ghysen, A. (1990). Origins of segment periodicity. *Nature* 344, 297–298.

747. Ghysen, A. (2003). The origin and evolution of the nervous system. *Int. J. Dev. Biol.* 47, 555–562.

748. Gibbons, B. (1986). The intimate sense of smell. *Natl. Geogr.* 170, #3, 324–361.

749. Gibson, G. (2000). Evolution: *Hox* genes and the cellared wine principle. *Curr. Biol.* 10, R452–R455.

750. Giessel, A.J., and Datta, S.R. (2014). Olfactory maps, circuits and computations. *Curr. Opin. Neurobiol.* 24, 120–132.

751. Gilbert, A.N., and Firestein, S. (2002). Dollars and scents: commercial opportunities in olfaction and taste. *Nat. Neurosci.* 5 (Suppl.), 1043–1045.

752. Gilbert, J.-M. (2002). The evolution of *engrailed* genes after duplication and speciation events. *Dev. Genes Evol.* 212, 307–318.

753. Gilbert, S.F., and Bolker, J.A. (2001). Homologies of process and modular elements of embryonic construction. *J. Exp. Zool.* 291, 1–12.

754. Gilbert, S.L., Dobyns, W.B., and Lahn, B.T. (2005). Genetic links between brain development and brain evolution. *Nat. Rev. Genet.* 6, 581–590.

755. Gilestro, G.F. (2008). Redundant mechanisms for regulation of midline crossing in *Drosophila. PLoS ONE* 3, #11, e3798.

756. Giljov, A., Karenina, K., Ingram, J., and Malashichev, Y. (2015). Parallel emergence of true handedness in the evolution of marsupials and placentals. *Curr. Biol.* 25, 1878–1884.

757. Gillespie, P.G., and Walker, R.G. (2001). Molecular basis of mechanosensory transduction. *Nature* 413, 194–202.

758. Gillies, T.E., and Cabernard, C. (2011). Cell division orientation in animals. *Curr. Biol.* 21, R599–R609.

759. Giorgianni, M., and Patel, N.H. (2005). Conquering land, air and water: the evolution and development of arthropod appendages. *In* D.E.G. Briggs (ed.), *Evolving Form and Function: Fossils and Development.* Yale University Peabody Museum of Natural History, New Haven, CT., pp. 159–180.

760. Gitton, Y., Dahmane, N., Balk, S., Ruiz i Altaba, A., Neidhardt, L., Scholze, M., Herrmann, B.G., Kanlem, P., Benkahla, A., Schrinner, S., Yildirimman, R., Herwig, R., Lehrach, H., and Yaspo, M.-L. (2002). A gene expression map of human chromosome 21 orthologues in the mouse. *Nature* 420, 586–590.

761. Godwin, J. (2014). The promise of perfect adult tissue repair and regeneration in mammals: learning from regenerative amphibians and fish. *BioEssays* 36, 861–871.

762. Goetz, J.J., Farris, C., Chowdhury, R., and Trimarchi, J.M. (2014). Making of a retinal cell: insights into retinal cell-fate determination. *Int. Rev. Cell Mol. Biol.* 308, 273–321.

763. Goldbeter, A., Gonze, D., and Pourquié, O. (2007). Sharp developmental thresholds defined through bistability by antagonistic gradients of retinoic acid and FGF signaling. *Dev. Dyn.* 236, 1495–1508.

764. Goldman, A.L., van der Goes van Naters, W., Lessing, D., Warr, C.G., and Carlson, J.R. (2005). Coexpression of two functional odor receptors in one neuron. *Neuron* 45, 661–666.

765. Goldsmith, T.H. (1990). Optimization, constraint, and history in the evolution of eyes. *Q. Rev. Biol.* 65, 281–322.

766. Goldsmith, T.H. (2013). Evolutionary tinkering with visual photoreception. *Vis. Neurosci.* 30, 21–37.

767. Goltsev, Y., Fuse, N., Frasch, M., Zinzen, R.P., Lanzaro, G., and Levine, M. (2007). Evolution of the dorsal-ventral patterning network in the mosquito, *Anopheles gambiae*. *Development* 134, 2415–2424.

768. Gómez-Skarmeta, J.L., and Modolell, J. (2002). *Iroquois* genes: genomic organization and function in vertebrate neural development. *Curr. Opin. Genet. Dev.* 12, 403–408.

769. Gómez-Skarmeta, J.L., Rodríguez, I., Martínez, C., Culí, J., Ferrés-Marcó, D., Beamonte, D., and Modolell, J. (1995). *Cis*-regulation of *achaete* and *scute*: shared enhancer-like elements drive their coexpression in proneural clusters of the imaginal discs. *Genes Dev.* 9, 1869–1882.

770. González, J., Ranz, J.M., and Ruiz, A. (2002). Chromosomal elements evolve at different rates in the *Drosophila* genome. *Genetics* 161, 1137–1154.

771. González-Morales, N., Géminard, C., Lebreton, G., Cerezo, D., Coutelis, J.-B., and Noselli, S. (2015). The atypical cadherin Dachsous controls left-right asymmetry in *Drosophila*. *Dev. Cell* 33, 675–689.

772. Goode, D.K., Callaway, H.A., Cerda, G.A., Lewis, K.E., and Elgar, G. (2011). Minor change, major difference: divergent functions of highly conserved cis-regulatory elements subsequent to whole genome duplication events. *Development* 138, 879–884.

773. Goode, D.K., and Elgar, G. (2009). The *PAX258* gene subfamily: a comparative perspective. *Dev. Dyn.* 238, 2951–2974.

774. Goodrich, L.V., and Strutt, D. (2011). Principles of planar polarity in animal development. *Development* 138, 1877–1892.

775. Gooijers, J., and Swinnen, S.P. (2014). Interactions between brain structure and behavior: the corpus callosum and bimanual coordination. *Neurosci. Biobehav. Rev.* 43, 1–19.

776. Göpfert, M.C., and Hennig, R.M. (2016). Hearing in insects. *Annu. Rev. Entomol.* 61, 257–276.

777. Göpfert, M.C., and Robert, D. (2002). The mechanical basis of *Drosophila* audition. *J. Exp. Biol.* 205, 1199–1208.

778. Göpfert, M.C., Stocker, H., and Robert, D. (2002). *atonal* is required for exoskeletal joint formation in the *Drosophila* auditory system. *Dev. Dyn.* 225, 106–109.

779. Gorbunov, D., Sturlese, M., Nies, F., Kluge, M., Bellanda, M., Battistutta, R., and Oliver, D. (2014). Molecular architecture and the structural basis for anion interaction in prestin and SLC26 transporters. *Nat. Commun.* 5, Article 3622.

780. Gorfinkiel, N., Morata, G., and Guerrero, I. (1997). The homeobox gene *Distal-less* induces ventral appendage development in *Drosophila*. *Genes Dev.* 11, 2259–2271.

781. Gosse, N.J., Nevin, L.M., and Baier, H. (2008). Retinotopic order in the absence of axon competition. *Nature* 452, 892–895.

782. Goto, S., and Hayashi, S. (1997). Cell migration within the embryonic limb primordium of *Drosophila* as revealed by a novel fluorescence method to visualize mRNA and protein. *Dev. Genes Evol.* 207, 194–198.

783. Goto, S., and Hayashi, S. (1997). Specification of the embryonic limb primordium by graded activity of Decapentaplegic. *Development* 124, 125–132.

784. Gottardo, M., Pollarolo, G., Llamazares, S., Reina, J., Riparbelli, M.G., Callaini, G., and Gonzalez, C. (2015). Loss of centrobin enables daughter centrioles to form sensory cilia in *Drosophila*. *Curr. Biol.* 25, 2319–2324.

785. Gould, S.J. (1977). *Ontogeny and Phylogeny*. Harvard University Press, Cambridge, MA.

786. Gould, S.J. (1994). Common pathways of illumination. *Nat. Hist.* 103, #12, 10–20.

787. Gould, S.J. (1997). As the worm turns. *Nat. Hist.* 106, #1, 24–27, 68–73.

788. Gould, S.J. (2002). *The Structure of Evolutionary Theory*. Harvard University Press, Cambridge, MA.

789. Graham, A., Butts, T., Lumsden, A., and Kiecker, C. (2014). What can vertebrates tell us about segmentation? *EvoDevo* 5, Article 24.

790. Graham, A., Papalopulu, N., and Krumlauf, R. (1989). The murine and *Drosophila* homeobox gene complexes have common features of organization and expression. *Cell* 57, 367–378.

791. Graham, D.M., Wong, K.Y., Shapiro, P., Frederick, C., Pattabiraman, K., and Berson, D.M. (2008). Melanopsin ganglion cells use a membrane-associated rhabdomeric phototransduction cascade. *J. Neurophysiol.* 99, 2522–2532.

792. Graham, T.G.W., Tabei, S.M.A., Dinner, A.R., and Rebay, I. (2010). Modeling bistable cell-fate choices in the *Drosophila* eye: qualitative and quantitative perspectives. *Development* 137, 2265–2278.

793. Grande, C., and Patel, N.H. (2009). Nodal signalling is involved in left–right asymmetry in snails. *Nature* 457, 1007–1011.

794. Grati, M., and Kachar, B. (2011). Myosin VIIa and sans localization at stereocilia upper tip-link density implicates these Usher syndrome proteins in mechanotransduction. *PNAS* 108, #28, 11476–11481.

795. Graur, D., Zheng, Y., Price, N., Azevedo, R.B.R., Zufall, R.A., and Elhaik, E. (2013). On the immortality of television sets: "function" in the human genome according to the evolution-free gospel of ENCODE. *Genome Biol.* 5, 578–590.

796. Graveley, B.R. (2005). Mutually exclusive splicing of the insect Dscam pre-mRNA directed by competing intronic RNA secondary structures. *Cell* 123, 65–73.

797. Graveley, B.R., Kaur, A., Gunning, D., Zipursky, S.L., Rowen, L., and Clemens, J.C. (2004). The organization and evolution of the dipteran and hymenopteran *Down syndrome cell adhesion molecule* (*Dscam*) genes. *RNA* 10, 1499–1506.

798. Graw, J. (2010). Eye development. *Curr. Top. Dev. Biol.* 90, 343–386.

799. Graziussi, D.F., Suga, H., Schmid, V., and Gehring, W.J. (2012). The "*eyes absent*" (*eya*) gene in the eye-bearing hydrozoan jellyfish *Cladonema radiatum*: conservation of the retinal determination network. *J. Exp. Zool. B. Mol. Dev. Evol.* 318, 257–267.

800. Green, J., and Akam, M. (2013). Evolution of the pair rule gene network: insights from a centipede. *Dev. Biol.* 382, 235–245.

801. Green, M.M. (2002). It really is not a fruit fly. *Genetics* 162, 1–3.

802. Greenberg, L., and Hatini, V. (2009). Essential roles for *lines* in mediating leg and antennal proximodistal patterning and generating a stable Notch signaling interface at segment borders. *Dev. Biol.* 330, 93–104.

803. Greenberg, L., and Hatini, V. (2011). Systematic expression and loss-of-function analysis defines spatially restricted requirements for *Drosophila RhoGEFs* and *RhoGAPs* in leg morphogenesis. *Mech. Dev.* 128, 5–17.

804. Greenow, K., and Clarke, A.R. (2012). Controlling the stem cell compartment and regeneration in vivo: the role of pluripotency pathways. *Physiol. Rev.* 92, 75–99.

805. Greenspan, R.J., and Dierick, H.A. (2004). "Am not I a fly like thee?" From genes in fruit flies to behavior in humans. *Hum. Mol. Genet.* 13, R267–R273.

806. Greenspan, R.J., and van Swinderen, B. (2004). Cognitive consonance: complex brain functions in the fruit fly and its relatives. *Trends Neurosci.* 27, 707–711.

807. Greer, J.M., Puetz, J., Thomas, K.R., and Capecchi, M.R. (2000). Maintenance of functional equivalence during paralogous Hox gene evolution. *Nature* 403, 661–665.

808. Gregor, T., McGregor, A.P., and Wieschaus, E.F. (2008). Shape and function of the Bicoid morphogen gradient in dipteran species with different sized embryos. *Dev. Biol.* 316, 350–358.

809. Gregory, T.R. (2008). The evolution of complex organs. *Evol. Educ. Outreach* 1, 358–389.

810. Greven, H. (2007). Comments on the eyes of tardigrades. *Arthropod Struct. Dev.* 36, 401–407.

811. Griffin, C., Kleinjan, D.A., Doe, B., and van Heyningen, V. (2002). New 3′ elements control *Pax6* expression in the developing pretectum, neural retina and olfactory region. *Mech. Dev.* 112, 89–100.

812. Grigorian, M., and Hartenstein, V. (2013). Hematopoiesis and hematopoietic organs in arthropods. *Dev. Genes Evol.* 223, 103–115.

813. Grill, S.W. (2010). Forced to be unequal. *Science* 330, 597–598.

814. Grimes, A.C., and Kirby, M.L. (2009). The outflow tract of the heart in fishes: anatomy, genes and evolution. *J. Fish Biol.* 74, 983–1036.

815. Grindley, J.C., Davidson, D.R., and Hill, R.E. (1995). The role of *Pax-6* in eye and nasal development. *Development* 121, 1433–1442.

816. Gros, J., Feistel, K., Viebahn, C., Blum, M., and Tabin, C. (2009). Cell movements at Hensen's node establish left/right asymmetric gene expression in the chick. *Science* 324, 941–944.

817. Groves, A.K., and Fekete, D.M. (2012). Shaping sound in space: the regulation of inner ear patterning. *Development* 139, 245–257.

818. Grünbaum, B., and Shephard, G.C. (1987). *Tilings and Patterns.* W.H. Freeman, New York, NY.

819. Grus, W.E., and Zhang, J. (2006). Origin and evolution of the vertebrate vomeronasal system viewed through system-specific genes. *BioEssays* 28, 709–718.

820. Grus, W.E., and Zhang, J. (2009). Origin of the genetic components of the vomeronasal system in the common ancestor of all extant vertebrates. *Mol. Biol. Evol.* 26, 407–419.

821. Gruss, M., Bushell, T.J., Bright, D.P., Lieb, W.R., Mathie, A., and Franks, N.P. (2004). Two-pore-domain K+ channels are a novel target for the anesthetic gases xenon, nitrous oxide, and cyclopropane. *Mol. Pharmacol.* 65, 443–452.

822. Guarner, A., Manjón, C., Edwards, K., Steller, H., Suzanne, M., and Sánchez-Herrero, E. (2014). The *zinc finger homeodomain-2* gene of *Drosophila* controls *Notch* targets and regulates apoptosis in the tarsal segments. *Dev. Biol.* 385, 350–365.

823. Guarnieri, D.J., and Heberlein, U. (2003). *Drosophila melanogaster*, a genetic model system for alcohol research. *Int. Rev. Neurobiol.* 54, 199–228.

824. Gubb, D., and García-Bellido, A. (1982). A genetic analysis of the determination of cuticular polarity during development in *Drosophila melanogaster*. *J. Embryol. Exp. Morphol.* 68, 37–57.

825. Gubb, D., Green, C., Huen, D., Coulson, D., Johnson, G., Tree, D., Collier, S., and Roote, J. (1999). The balance between isoforms of the Prickle LIM domain protein is critical for planar polarity in *Drosophila* imaginal discs. *Genes Dev.* 13, 2315–2327.

826. Guerin, M.B., McKernan, D.P., O'Brien, C.J., and Cotter, T.G. (2006). Retinal ganglion cells: dying to survive. *Int. J. Dev. Biol.* 50, 665–674.

827. Guertin, P.A. (2009). The mammalian central pattern generator for locomotion. *Brain Res. Rev.* 62, 45–56.

828. Guichard, C., Harricane, M.-C., Lafitte, J.-J., Godard, P., Zaegel, M., Tack, V., Lalau, G., and Bouvagnet, P. (2001). Axonemal dynein intermediate-chain gene (*DNAI1*) mutations result in situs inversus and primary ciliary diskinesia (Kartagener syndrome). *Am. J. Hum. Genet.* 68, 1030–1035.

829. Guillemot, F., Lo, L.-C., Johnson, J.E., Auerbach, A., Anderson, D.J., and Joyner, A.L. (1993). Mammalian *achaete-scute* homolog 1 is required for the early development of olfactory and autonomic neurons. *Cell* 75, 463–476.

830. Guillery, R.W. (1994). No crossing at the chiasm. *Nature* 367, 597–598.

831. Gulino, A., Di Marcotullio, L., and Screpanti, I. (2010). The multiple functions of Numb. *Exp. Cell Res.* 316, 900–906.

832. Güntürkün, O. (2012). Brain asymmetry in vertebrates. In O.F. Lazareva, T. Shimizu, and E.A. Wasserman (eds.), *How Animals See the World: Comparative Behavior, Biology, and Evolution of Vision.* Oxford University Press, Oxford, pp. 501–519.

833. Guo, N., Hawkins, C., and Nathans, J. (2004). Frizzled6 controls hair patterning in mice. *PNAS* 101, #25, 9277–9281.

834. Gurtan, A.M., and Sharp, P.A. (2013). The role of miRNAs in regulating gene expression networks. *J. Mol. Biol.* 425, 3582–3600.

835. Guthrie, S. (1999). Axon guidance: starting and stopping with Slit. *Curr. Biol.* 9, R432–R435.

836. Gutierrez-Mazariegos, J., Schubert, M., and Laudet, V. (2014). Evolution of retinoic acid receptors and retinoic acid signaling. In M.A. Asson-Batres and C. Rochette-Egly (eds.), *The Biochemistry of Retinoic Acid Receptors I: Structure, Activation, and Function at the Molecular Level.* Springer, New York, NY, pp. 55–73.

837. Gutierrez-Mazariegos, J., Theodosiou, M., Campo-Paysaa, F., and Schubert, M. (2011). Vitamin A: a multifunctional tool for development. *Semin. Cell Dev. Biol.* 22, 603–610.

838. Ha, T.S., and Smith, D.P. (2009). Odorant and pheromone receptors in insects. *Front. Cell. Neurosci.* 3, Article 10.

839. Häder, T., La Rosée, A., Ziebold, U., Busch, M., Taubert, H., Jäckle, H., and Rivera-Pomar, R. (1998). Activation of posterior pair-rule stripe expression in response to maternal *caudal* and zygotic *knirps* activities. *Mech. Dev.* 71, 177–186.

840. Haeussler, M., Jaszczyszyn, Y., Christiaen, L., and Joly, J.-S. (2010). A *cis*-regulatory signature for chordate anterior neuroectodermal genes. *PLoS Genet.* 6, #4, e1000912.

841. Hafen, E., Kuroiwa, A., and Gehring, W.J. (1984). Spatial distribution of transcripts from the segmentation gene *fushi tarazu* during *Drosophila* embryonic development. *Cell* 37, 833–841.

842. Haithcock, J., Billington, N., Choi, K., Fordham, J., Sellers, J.R., Stafford, W.F., White, H., and Forgacs, E. (2011). The kinetic mechanism of mouse myosin VIIA. *J. Biol. Chem.* 286, 8819–8828.

843. Haldane, J.B.S. (1928). On being the right size. In *Possible Worlds and Other Papers.* Harper & Bros., New York, NY, pp. 20–28.

844. Halder, G., Callaerts, P., and Gehring, W.J. (1995). Induction of ectopic eyes by targeted expression of the *eyeless* gene in *Drosophila*. *Science* 267, 1788–1792.

845. Hale, R., and Strutt, D. (2015). Conservation of planar polarity pathway function across the animal kingdom. *Annu. Rev. Genet.* 49, 529–551.

846. Hall, B.K. (1984). *Homology: The Hierarchical Basis of Comparative Biology*. Academic Press, San Diego, CA.

847. Hall, B.K. (2012). Parallelism, deep homology, and evo-devo. *Evol. Dev.* 14, 29–33.

848. Hall, B.K., and Olson, W.M. (eds.) (2003). *Keywords and Concepts in Evolutionary Developmental Biology*. Harvard University Press, Cambridge, MA.

849. Hall, J.C. (2003). A neurogeneticist's manifesto. *J. Neurogenet.* 17, 1–90.

850. Hallem, E.A., and Carlson, J.R. (2004). The odor coding system of *Drosophila*. *Trends Genet.* 20, 453–459.

851. Hallem, E.A., and Carlson, J.R. (2006). Coding of odors by a receptor repertoire. *Cell* 125, 143–160.

852. Hallsson, J.H., Haflidadóttir, B.S., Stivers, C., Odenwald, W., Arnheiter, H., Pignoni, F., and Steingrímsson, E. (2004). The basic helix-loop-helix leucine zipper transcription factor *Mitf* is conserved in *Drosophila* and functions in eye development. *Genetics* 167, 233–241.

853. Halpern, M.E., Hobert, O., and Wright, C.V.E. (2014). Left-right asymmetry: advances and enigmas. *Genesis* 52, 451–454.

854. Hamada, H. (2015). Role of physical forces in embryonic development. *Semin. Cell Dev. Biol.* 47–48, 88–91.

855. Hamilton, C.R., and Vermeire, B.A. (1988). Complementary hemispheric specialization in monkeys. *Science* 242, 1691–1694.

856. Hamilton, E.E., and Kay, S.A. (2008). Snapshot: circadian clock proteins. *Cell* 135, 368.

857. Hammond, K.L., Baxendale, S., McCauley, D., Ingham, P.W., and Whitfield, T.T. (2009). Expression of *patched*, *prdm1* and *engrailed* in the lamprey somite reveals conserved responses to Hedgehog signaling. *Evol. Dev.* 11, 27–40.

858. Hammond, K.L., and Whitfield, T.T. (2006). The developing lamprey ear closely resembles the zebrafish otic vesicle: *otx1* expression can account for all major patterning differences. *Development* 133, 1347–1357.

859. Hampel, S., Franconville, R., Simpson, J.H., and Seeds, A.M. (2015). A neural command circuit for grooming movement control. *eLife* 4, e308758.

860. Hanchate, N.K., Kondoh, K., Lu, Z., Kuang, D., Ye, X., Qui, X., Pachter, L., Trapnell, C., and Buck, L.B. (2015). Single-cell transcriptomics reveals receptor transformations during olfactory neurogenesis. *Science* 350, 1251–1255.

861. Hankins, M.W., and Hughes, S. (2014). Vision: melanopsin as a novel irradiance detector at the heart of vision. *Curr. Biol.* 24, R1055–R1057.

862. Hannibal, R.L., and Patel, N.H. (2013). What is a segment? *EvoDevo* 4, Article 35.

863. Hanson, I.M. (2001). Mammalian homologues of the *Drosophila* eye specification genes. *Semin. Cell Dev. Biol.* 12, 475–484.

864. Hansson, B.S., and Stensmyr, M.C. (2011). Evolution of insect olfaction. *Neuron* 72, 698–711.

865. Hao, I., Green, R.B., Dunaevsky, O., Lengyel, J.A., and Rauskolb, C. (2003). The *odd-skipped* family of zinc finger genes promotes *Drosophila* leg segmentation. *Dev. Biol.* 263, 282–295. [*See also* Suzanne, M. (2016). Molecular and cellular mechanisms involved in leg joint morphogenesis. *Semin. Cell Dev. Biol.* 55, 131–138.]

866. Hara, Y., Yamaguchi, M., Akasaka, K., Nakano, H., Nonaka, M., and Amemiya, S. (2006). Expression patterns of *Hox* genes in larvae of the sea lily *Metacrinus rotundus*. *Dev. Genes Evol.* 216, 797–809.

867. Harada, T., Harada, C., and Parada, L.F. (2007). Molecular regulation of visual system development: more than meets the eye. *Genes Dev.* 21, 367–378.

868. Hardie, R., and Raghu, P. (2001). Visual transduction in *Drosophila*. *Nature* 413, 186–193.

869. Hardie, R.C. (1985). Functional organization of the fly retina. *In* D. Ottoson (ed.), *Progress in Sensory Physiology*. Springer-Verlag, Berlin, pp. 1–79.

870. Hardie, R.C. (1986). The photoreceptor array of the dipteran retina. *Trends Neurosci.* 9, 419–423.

871. Hardie, R.C. (2012). Polarization vision: *Drosophila* enters the arena. *Curr. Biol.* 22, R12–R14.

872. Hardie, R.C., and Franze, K. (2012). Photomechanical responses in *Drosophila* photoreceptors. *Science* 338, 260–263.

873. Hardin, P.E. (2000). From biological clock to biological rhythms. *Genome Biol.* 1, #4, reviews1023.

874. Hardin, P.E. (2005). The circadian timekeeping system of *Drosophila*. *Curr. Biol.* 15, R714–R722.

875. Hardin, P.E. (2011). Molecular genetic analysis of circadian timekeeping in *Drosophila*. *Adv. Genet.* 74, 141–173.

876. Harding, K., and Levine, M. (1988). Gap genes define the limits of Antennapedia and Bithorax gene expression during early development in *Drosophila*. *EMBO J.* 7, 205–214.

877. Hardison, R.C. (2008). Globin genes on the move. *J. Biol.* 7, Article 35.

878. Hardison, R.C. (2012). Evolution of hemoglobin and its genes. *Cold Spring Harb. Perspect. Med.* 2, a011627.

879. Harris, W.A. (1997). *Pax-6*: where to be conserved is not conservative. *PNAS* 94, 2098–2100.

880. Harrison, R.G. (1917). Transplantation of limbs. *PNAS* 3, #4, 245–251.

881. Hartenstein, V. (1993). *Atlas of Drosophila Development*. Cold Spring Harbor Laboratory Press, Plainview, NY.

882. Hartenstein, V. (2006). The neuroendocrine system of invertebrates: a developmental and evolutionary perspective. *J. Endocrinology* 190, 555–570.

883. Hartenstein, V., and Mandal, L. (2006). The blood/vascular system in a phylogenetic perspective. *BioEssays* 28, 1203–1210.

884. Hartenstein, V., and Stollewerk, A. (2015). The evolution of early neurogenesis. *Dev. Cell* 32, 390–407.

885. Hartenstein, V., Takashima, S., and Adams, K.L. (2010). Conserved genetic pathways controlling the development of the diffuse endocrine system in vertebrates and *Drosophila*. *Gen. Comp. Endocrinol.* 166, 462–469.

886. Hartenstein, V., Tepass, U., and Gruszynski-Defeo, E. (1994). Embryonic development of the stomatogastric nervous system in *Drosophila*. *J. Comp. Neurol.* 350, 367–381.

887. Hartline, D.K. (2011). The evolutionary origins of glia. *Glia* 59, 1215–1236.

888. Hartman, M.A., Finan, D., Sivaramakrishnan, S., and Spudich, J.A. (2011). Principles of unconventional myosin function and tarteting. *Annu. Rev. Cell Dev. Biol.* 27, 133–155.

889. Harvey, R.P. (1996). *NK-2* homeobox genes and heart development. *Dev. Biol.* 178, 203–216.

890. Harvey, R.P. (2002). Patterning the vertebrate heart. *Nat. Rev. Genet.* 3, 544–556.

891. Harzsch, S. (2002). The phylogenetic significance of crustacean optic neuropils and chiasmata: a re-examination. *J. Comp. Neurol.* 453, 10–21.

892. Harzsch, S. (2004). The tritocerebrum of Euarthropoda: a "non-*drosophilo*centric" perspective. *Evol. Dev.* 6, 303–309.

893. Harzsch, S., Vilpoux, K., Blackburn, D.C., Platchetzki, D., Brown, N.L., Melzer, R., Kempler, K.E., and Battelle, B.A. (2006). Evolution of arthropod visual systems: development of the eyes and central visual pathways in the horseshoe crab *Limulus polyphemus* Linnaeus, 1758 (Chelicerata, Xiphosura). *Dev. Dyn.* 235, 2641–2655.

894. Haskel-Ittah, M., Ben-Zvi, D., Branski-Arieli, M., Schejter, E.D., Shilo, B.-Z., and Barkai, N. (2012). Self-organized shuttling: generating sharp dorsoventral polarity in the early *Drosophila* embryo. *Cell* 150, 1016–1028.

895. Hassan, B.A., and Hiesinger, P.R. (2015). Beyond molecular codes: simple rules to wire complex brains. *Cell* 163, 285–291. [For a paean to the wonders of science in general and neuroscience in particular, *see* Hassan, B.A. (2016). The I in scientist. *Cell* 166, 790–793.]

896. Hattar, S., Liao, H.-W., Takao, M., Berson, D.M., and Yau, K.-W. (2002). Melanopsin-containing retinal ganglion cells: architecture, projections, and intrinsic photosensitivity. *Science* 295, 1065–1070.

897. Hattori, D., Chen, Y., Matthews, B.J., Salwinski, L., Sabatti, C., Grueber, W.B., and Zipursky, S.L. (2009). Robust discrimination between self and non-self neurites requires thousands of Dscam1 isoforms. *Nature* 461, 644–648.

898. Hattori, D., Demir, E., Kim, H.W., Viragh, E., Zipursky, S.L., and Dickson, B. (2007). Dscam diversity is essential for neuronal wiring and self-recognition. *Nature* 449, 223–227.

899. Hattori, D., Millard, S.S., Wojtowicz, W.M., and Zipursky, S.L. (2008). Dscam-mediated cell recognition regulates neural circuit formation. *Annu. Rev. Cell Dev. Biol.* 24, 597–620.

900. Haun, C., Alexander, J., Stainier, D.Y., and Okkema, P.G. (1998). Rescue of *Caenorhabditis elegans* pharyngeal development by a vertebrate heart specification gene. *PNAS* 95, 5072–5075.

901. Hauser, F.E., van Hazel, I., and Chang, B.S.W. (2014). Spectral tuning in vertebrate short wavelength-sensitive 1 (SWS1) visual pigments: can wavelength sensitivity be inferred from sequence data? *J. Exp. Zool. B. Mol. Dev. Evol.* 322, 529–539.

902. Hayashi, T., and Murakami, R. (2001). Left-right asymmetry in *Drosophila melanogaster* gut development. *Dev. Growth Differ.* 43, 239–246.

903. Haynes, P.R., Christmann, B.L., and Griffith, L.C. (2015). A single pair of neurons links sleep to memory consolidation in *Drosophila melanogaster*. *eLife* 4, e03868.

904. Hayward, A.G., II, Joshi, P., and Skromne, I. (2015). Spatiotemporal analysis of zebrafish *hox* gene regulation by Cdx4. *Dev. Dyn.* 244, 1564–1573.

905. He, D.Z.Z., Zheng, J., Kalinec, F., Kakehata, S., and Santos-Sacchi, J. (2006). Tuning in to the amazing outer hair cell: membrane wizardry with a twist and shout. *J. Membr. Biol.* 209, 119–134.

906. He, H., Kise, Y., Izadifar, A., Urwyler, O., Ayaz, D., Parthasarthy, A., Yan, B., Erfurth, M.-L., Dascenco, D., and Schmucker, D. (2014). Cell-intrinsic requirement of Dscam1 isoform diversity for axon collateral formation. *Science* 344, 1182–1886.

907. He, S., Dong, W., Deng, Q., Weng, S., and Sun, W. (2003). Seeing more clearly: recent advances in understanding retinal circuitry. *Science* 302, 408–411.

908. Heckman, C.A., and Plummer, H.K., III (2013). Filopodia as sensors. *Cell. Signal.* 25, 2298–2311.

909. Heed, T. (2010). Touch perception: how we know where we are touched. *Curr. Biol.* 20, R604–R606.

910. Heffer, A., Löhr, U., and Pick, L. (2011). *ftz* evolution: findings, hypotheses and speculations. *BioEssays* 33, 910–918.

911. Heffer, A., and Pick, L. (2013). Conservation and variation in *Hox* genes: how insect models pioneered the evo-devo field. *Annu. Rev. Entomol.* 58, 161–179.

912. Heffer, A., Shultz, J.W., and Pick, L. (2010). Surprising flexibility in a conserved Hox transcription factor over 550 million years of evolution. *PNAS* 107, 18040–18045.

913. Heimberg, A., and McGlinn, E. (2012). Building a robust A-P axis. *Curr. Genomics* 13, 278–288.

914. Heine, P., Dohle, E., Bumsted-O'Brien, K., Engelkamp, D., and Schulte, D. (2008). Evidence for an evolutionary conserved role of *homothorax/Meis1/2* during vertebrate retina development. *Development* 135, 805–811.

915. Hejnol, A., and Dunn, C.W. (2016). Animal evolution: are phyla real? *Curr. Biol.* 26, R424–R426.

916. Hejnol, A., and Lowe, C.J. (2014). Animal evolution: stiff or squishy notochord origins? *Curr. Biol.* 24, R1131–R1133.

917. Hejnol, A., and Martindale, M.Q. (2009). Coordinated spatial and temporal expression of *Hox* genes during embryogenesis in the acoel *Convolutriloba longifissura*. *BMC Biol.* 7, Article 65.

918. Hejnol, A., and Rentzsch, F. (2015). Neural nets. *Curr. Biol.* 25, R782–R786.

919. Hejnol, A., and Scholtz, G. (2004). Clonal analysis of Distal-less and engrailed expression patterns during early morphogenesis of uniramous and biramous crustacean limbs. *Dev. Genes Evol.* 214, 473–485.

920. Held, L.I., Jr. (1977). Analysis of Bristle-Pattern Formation in *Drosophila*. PhD thesis, Department of Molecular Biology, University of California, Berkeley, CA.

921. Held, L.I., Jr. (1991). Bristle patterning in *Drosophila*. *BioEssays* 13, 633–640.

922. Held, L.I., Jr. (1992). *Models for Embryonic Periodicity*. Karger, Basel.

923. Held, L.I., Jr. (2002). *Imaginal Discs: The Genetic and Cellular Logic of Pattern Formation*. Cambridge University Press, New York, NY.

924. Held, L.I., Jr. (2005). Suppressing apoptosis fails to cure "extra-joint syndrome" or to stop sex-comb rotation. *Dros. Inf. Serv.* 88, 9–10.

925. Held, L.I., Jr. (2009). *Quirks of Human Anatomy: An Evo-Devo Look at the Human Body*. Cambridge University Press, New York, NY.

926. Held, L.I., Jr. (2010). The evo-devo puzzle of human hair patterning. *Evol. Biol.* 37, 113–122.

927. Held, L.I., Jr. (2010). The evolutionary geometry of human anatomy: discovering our inner fly. *Evol. Anthropol.* 19, 227–235.

928. Held, L.I., Jr. (2014). *How the Snake Lost Its Legs: Curious Tales from the Frontier of Evo-Devo*. Cambridge University Press, New York, NY. [*See also* Guerreiro, I., *et al.* (2016). Reorganisation of *Hoxd* regulatory landscapes during the evolution of a snake-like body plan. *eLife* 5, e16087.]

929. Held, L.I., Jr., Duarte, C.M., and Derakhshanian, K. (1986). Extra joints and misoriented bristles on *Drosophila* legs. *In* H. Slavkin (ed.), *Progress in Developmental Biology (Part A)*. Alan R. Liss, New York, NY, pp. 293–296.

930. Held, L.I., Jr., Duarte, C.M., and Derakhshanian, K. (1986). Extra tarsal joints and abnormal cuticular polarities in various mutants of *Drosophila melanogaster*. *Roux's Arch. Dev. Biol.* 195, 145–157.

931. Helenius, I.T., and Beitel, G.J. (2008). The first "Slit" is the deepest: the secret to a hollow heart. *J. Cell Biol.* 182, 221–223.

932. Helfrich-Förster, C. (2004). The circadian clock in the brain: a structural and functional comparison between mammals and insects. *J. Comp. Physiol. A* 190, 601–613.

933. Hellige, J.B. (1993). *Hemispheric Asymmetry: What's Right and What's Left*. Harvard University Press, Cambridge, MA.

934. Helmstädter, M., Lüthy, K., Gödel, M., Simons, M., Ashish, Nihalani, D., Rensing, S.A., Fischbach, K.-F., and Huber, T.B. (2012). Functional study of mammalian Neph proteins in *Drosophila melanogaster*. *PLoS ONE* 7, #7, e40300.

935. Hemani, Y., and Soller, M. (2012). Mechanisms of *Drosophila Dscam* mutually exclusive splicing regulation. *Biochem. Soc. Trans.* 40, 804–809.

936. Hering, L., Henze, M.J., Kohler, M., Kelber, A., Bleidorn, C., Leschke, M., Nickel, B., Meyer, M., Kircher, M., Sunnucks, P., and Mayer, G. (2012). Opsins in Onychophora (velvet worms) suggest a single origin and subsequent diversification of visual pigments in arthropods. *Mol. Biol. Evol.* 29, 3451–3458.

937. Herranz, H., Eichenlaub, T., and Cohen, S.M. (2016). Cancer in *Drosophila*: imaginal discs as a model for epithelial tumor formation. *Curr. Top. Dev. Biol.* 116, 181–199.

938. Herrera, E., and Garcia-Frigola, C. (2008). Genetics and development of the optic chiasm. *Frontiers Biosci.* 13, 1646–1653.

939. Herron, J.C., Freeman, S., Hodin, J., Miner, B., and Sidor, C. (2014). *Evolutionary Analysis*, 5th edn. Pearson, Upper Saddle River, NJ.

940. Heyn, P., Kalinka, A.T., Tomancak, P., and Neugebauer, K.M. (2014). Introns and gene expression: cellular constraints, transcriptional regulation, and evolutionary consequences. *BioEssays* 37, 148–154.

941. Hibino, T., Ishii, Y., Levin, M., and Nishino, A. (2006). Ion flow regulates left-right asymmetry in sea urchin development. *Dev. Genes Evol.* 216, 265–276.

942. Hibino, T., Nishino, A., and Amemiya, S. (2006). Phylogenetic correspondence of the body axes in bliaterians is revealed by the right-sided expression of *Pitx* genes in echinoderm larvae. *Dev. Growth Differ.* 48, 587–595.

943. Hilbrant, M., Almudi, I., Leite, D.J., Kuncheria, L., Posnien, N., Nunes, M.D.S., and McGregor, A.P. (2014). Sexual dimorphism and natural variation within and among species in the *Drosophila* retinal mosaic. *BMC Evol. Biol.* 14, Article 240.

944. Hildebrand, J.G., and Shepherd, G.M. (1997). Mechanisms of olfactory discrimination: converging evidence for common principles across phyla. *Annu. Rev. Neurosci.* 20, 595–631.

945. Himanen, J.-P., Saha, N., and Nikolov, D.B. (2007). Cell-cell signaling via Eph receptors and ephrins. *Curr. Opin. Cell Biol.* 19, 534–542.

946. Hires, S.A., Pammer, L., Svoboda, K., and Golomb, D. (2013). Tapered whiskers are required for active tactile sensation. *eLife* 2, e01350.

947. Hiromi, Y., and Gehring, W.J. (1987). Regulation and function of the *Drosophila* segmentation gene *fushi tarazu*. *Cell* 50, 963–974.

948. Hirth, F. (2010). On the origin and evolution of the tripartite brain. *Brain Behav. Evol.* 76, 3–10.

949. Hirth, F., Kammermeier, L., Frei, E., Walldorf, U., Noll, M., and Reichert, H. (2003). An urbilaterian origin of the tripartite brain: developmental genetic insights from *Drosophila*. *Development* 130, 2365–2373.
950. Hirth, F., and Reichert, H. (1999). Conserved genetic programs in insect and mammalian brain development. *BioEssays* 21, 677–684.
951. Hisatomi, O., and Tokunaga, F. (2002). Molecular evolution of proteins involved in vertebrate phototransduction. *Comp. Biochem. Physiol. B* 133, 509–522.
952. Hoare, D.J., McCrohan, C.R., and Cobb, M. (2008). Precise and fuzzy coding by olfactory sensory neurons. *J. Neurosci.* 28, #39, 9710–9722.
953. Hodin, J. (2000). Plasticity and constraints in development and evolution. *J. Exp. Zool.* 288, 1–20.
954. Hofer, H., Carroll, J., Neitz, J., Neitz, M., and Williams, D.R. (2005). Organization of the human trichromatic cone mosaic. *J. Neurosci.* 25, #42, 9669–9679.
955. Hofmeyer, K., and Treisman, J. (2008). Sensory systems: seeing the world in a new light. *Curr. Biol.* 18, R919–R921.
956. Hogan, B.L.M. (1995). Upside-down ideas vindicated. *Nature* 376, 210–211.
957. Holland, L.Z. (2000). Body-plan evolution in the Bilateria: early antero-posterior patterning and the deuterostome-protostome dichotomy. *Curr. Opin. Genet. Dev.* 10, 434–442.
958. Holland, L.Z. (2005). Non-neural ectoderm is really neural: evolution of developmental patterning mechanisms in the non-neural ectoderm of chordates and the problem of sensory cell homologies. *J. Exp. Zool. B. Mol. Dev. Evol.* 304, 304–323.
959. Holland, L.Z. (2014). Genomics, evolution and development of amphioxus and tunicates: the Goldilocks Principle. *J. Exp. Zool. B. Mol. Dev. Evol.* 324, 342–352.
960. Holland, L.Z. (2015). Evolution of basal deuterostome nervous systems. *J. Exp. Biol.* 218, 637–645.
961. Holland, L.Z. (2015). The origin and evolution of chordate nervous systems. *Philos. Trans. R. Soc. Lond. B* 370, 20150048.
962. Holland, L.Z. (2016). Tunicates. *Curr. Biol.* 26, R141–R156.
963. Holland, L.Z., Carvalho, J.E., Escriva, H., Laudet, V., Schubert, M., Shimeld, S.M., and Yu, J.-K. (2013). Evolution of bilaterian central nervous systems: a single origin? *EvoDevo* 4, Article 27.
964. Holland, L.Z., and Holland, N.D. (1996). Expression of *AmphiHox-1* and *AmphiPax-1* in amphioxus embryos treated with retinoic acid: insights into evolution and patterning of the chordate nerve cord and pharanx. *Development* 122, 1829–1838.
965. Holland, L.Z., and Holland, N.D. (1999). Chordate origins of the vertebrate central nervous system. *Curr. Opin. Neurobiol.* 9, 596–602.
966. Holland, L.Z., and Holland, N.D. (2007). A revised fate map for amphioxus and the evolution of axial patterning in chordates. *Integr. Comp. Biol.* 47, 360–372.
967. Holland, L.Z., Kene, M., Williams, N.A., and Holland, N.D. (1997). Sequence and embryonic expression of the amphioxus *engrailed* gene (*AmphiEn*): the metameric pattern of transcription resembles that of its segment-polarity homolog in *Drosophila*. *Development* 124, 1723–1732.
968. Holland, N.D. (2003). Early central nervous system evolution: an era of skin brains? *Nat. Rev. Neurosci.* 4, 1–11.
969. Holland, N.D., Holland, L.Z., and Holland, P.W.H. (2015). Scenarios for the making of vertebrates. *Nature* 520, 450–455.

970. Holland, P.W.H. (2001). Beyond the Hox: how widespread is homeobox gene clustering? *J. Anat.* 199, 13–23.

971. Holland, P.W.H. (2013). Evolution of homeobox genes. *WIREs Dev. Biol.* 2, 31–45.

972. Holland, P.W.H., Booth, H.A.F., and Bruford, E.A. (2007). Classification and nomenclature of all human homeobox genes. *BMC Biol.* 5, Article 47.

973. Holley, S.A., Jackson, P.D., Sasai, Y., Lu, B., De Robertis, E.M., Hoffmann, F.M., and Ferguson, E.L. (1995). A conserved system for dorsal-ventral patterning in insects and vertebrates involving *sog* and *chordin. Nature* 376, 249–253.

974. Holt, R.D. (2000). Use it or lose it. *Nature* 407, 689–690.

975. Honda, H., Kodama, R., Takeuchi, T., Yamanaka, H., Watanabe, K., and Eguchi, G. (1984). Cell behaviour in a polygonal cell sheet. *J. Embyrol. Exp. Morph.* 83 (Suppl.), 313–327.

976. Hong, W., and Luo, L. (2014). Genetic control of wiring specificity in the fly olfactory system. *Genetics* 196, 17–29.

977. Honigberg, L., and Kenyon, C. (2000). Establishment of left/right asymmetry in neuroblast migration by UNC-40/DCC, UNC-73/Trio and DPY-19 proteins in *C. elegans. Development* 127, 4655–4668.

978. Hornstein, E.P., O'Carroll, D.C., Anderson, J.C., and Laughlin, S.B. (2000). Sexual dimorphism matches photoreceptor performance to behavioural requirements. *Proc. R. Soc. Lond. B* 267, 2111–2117.

979. Hosaka, Y., Saito, T., Sugita, S., Hikata, T., Kobayashi, H., Fukai, A., Taniguchi, Y., Hirata, M., Akiyama, H., Chung, U.-i., and Kawaguchi, H. (2013). Notch signaling in chondrocytes modulates endochondral ossification and osteoarthritis development. *PNAS* 110, #5, 1875–1880.

980. Houle, M., Sylvestre, J.-R., and Lohnes, D. (2003). Retinoic acid regulates a subset of Cdx1 function in vivo. *Development* 130, 6555–6567.

981. Houweling, A.C., Dildrop, R., Peters, T., Mummenhoff, J., Moorman, A.F.M., Rüther, U., and Christoffels, V.M. (2001). Gene and cluster-specific expression of the *Iroquois* family members during mouse development. *Mech. Dev.* 107, 169–174.

982. Howard, J., Grill, S.W., and Bois, J.S. (2011). Turing's next steps: the mechanochemical basis of morphogenesis. *Nat. Rev. Mol. Cell Biol.* 12, 400–406.

983. Hoy, R.R. (2012). Convergent evolution of hearing. *Science* 338, 894–895.

984. Hoyle, C.H.V. (2011). Evolution of neuronal signalling: transmitters and receptors. *Autonomic Neurosci. Basic Clin.* 165, 28–53.

985. Hozumi, S., Maeda, R., Taniguchi, K., Kanai, M., Shirakabe, S., Sasamura, T., Spéder, P., Noselli, S., Aigaki, T., Murakami, R., and Matsuno, K. (2006). An unconventional myosin in *Drosophila* reverses the default handedness in visceral organs. *Nature* 440, 798–802.

986. Hozumi, S., Maeda, R., Taniguchi-Kanai, M., Okumura, T., Taniguchi, K., Kawakatsu, Y., Nakazawa, N., Hatori, R., and Matsuno, K. (2008). Head region of unconventional Myosin I family members is responsible for the organ-specificity of their roles in left-right polarity in *Drosophila. Dev. Dyn.* 237, 3528–3537.

987. Hsu, Y.-C., Chuang, J.-Z., and Sung, C.-H. (2015). Light regulates the ciliary protein transport and outer segment disc renewal of mammalian photoreceptors. *Dev. Cell* 32, 731–742.

988. Huang, J., Wang, Y., Raghavan, S., Feng, S., Kiesewetter, K., and Wang, J. (2011). Human down syndrome cell adhesion molecules (DSCAMs) are functionally

conserved with *Drosophila* Dscam[TM1] isoforms in controlling neurodevelopment. *Insect Biochem. Mol. Biol.* 41, 778–787.

989. Huang, R., Zhi, Q., Schmidt, C., Wilting, J., Brand-Saberi, B., and Christ, B. (2000). Sclerotomal origin of the ribs. *Development* 127, 527–532.

990. Huberman, A.D. (2009). Mammalian DSCAMs: they won't help you find a partner, but they'll guarantee you some personal space. *Neuron* 64, 441–443.

991. Hudspeth, A.J. (2008). Making an effort to listen: mechanical amplification in the ear. *Neuron* 59, 530–545.

992. Hudspeth, A.J. (2014). Integrating the active process of hair cells with cochlear function. *Nat. Rev. Neurosci.* 15, 600–614.

993. Hueber, S.D., Weiller, G.F., Djordjevic, M.A., and Frickey, T. (2010). Improving Hox protein classification across the major model organisms. *PLoS ONE* 5, #5, e10820.

994. Huelsken, J., Vogel, R., Erdmann, B., Cotsarelis, G., and Birchmeier, W. (2001). β-Catenin controls hair follicle morphogenesis and stem cell differentiation in the skin. *Cell* 105, 533–545.

995. Hughes, C.L., and Kaufman, T.C. (2002). Exploring the myriapod body plan: expression patterns of the ten Hox genes in a centipede. *Development* 129, 1225–1238.

996. Hughes, C.L., and Kaufman, T.C. (2002). Hox genes and the evolution of the arthropod body plan. *Evol. Dev.* 4, 459–499.

997. Hughes, C.L., Liu, P.Z., and Kaufman, T.C. (2004). Expression patterns of the rogue Hox genes *Hox3/zen* and *fushi tarazu* in the apterygote insect *Thermobia domestica*. *Evol. Dev.* 6, 393–401.

998. Hughes, N.C. (2003). Trilobite body patterning and the evolution of arthropod tagmosis. *BioEssays* 25, 386–395.

999. Hui, J.H.L., McDougall, C., Monteiro, A.S., Holland, P.W.H., Arendt, D., Balavoine, G., and Ferrier, D.E.K. (2012). Extensive chordate and annelid macrosynteny reveals ancestral homeobox gene organization. *Mol. Biol. Evol.* 29, 157–165.

1000. Hummel, K.P., and Chapman, D.B. (1959). Visceral inversion and associated anomalies in the mouse. *J. Hered.* 50, 9–13.

1001. Hummel, T., Vasconcelos, M.L., Clemens, J.C., Fishilevich, Y., Vosshall, L.B., and Zipursky, S.L. (2003). Axonal targeting of olfactory receptor neurons in *Drosophila* is controlled by Dscam. *Neuron* 37, 221–231.

1002. Hummer, T.A., and McClintock, M.K. (2009). Putative human pheromone androstadienone attunes the mind specifically to emotional information. *Horm. Behav.* 55, 548–559.

1003. Hunnekuhl, V.S., and Akam, M. (2014). An anterior medial cell population with an apical-organ-like transcriptional profile that pioneers the central nervous system in the centipede *Strigamia maritima*. *Dev. Biol.* 396, 136–149.

1004. Hunt, D.M., Carvalho, L.S., Cowing, J.A., and Davies, W.L. (2009). Evolution and spectral tuning of visual pigments in birds and mammals. *Philos. Trans. R. Soc. Lond. B* 364, 2941–2955.

1005. Hunt, D.M., and Collin, S.P. (2014). The evolution of photoreceptors and visual photopigments in vertebrates. *In* D.M. Hunt, M.W. Hankins, S.P. Collin, and N.J. Marshall (eds.), *Evolution of Visual and Non-Visual Pigments*. Springer, New York, NY, pp. 163–217.

1006. Hunt, D.M., Dulai, K.S., Partridge, J.C., Cottrill, P., and Bowmaker, J.K. (2001). The molecular basis for spectral tuning of rod visual pigments in deep-sea fish. *J. Exp. Biol.* 204, 3333–3344.

1007. Hunt, P., and Krumlauf, R. (1992). Hox codes and positional specification in vertebrate embryonic axes. *Annu. Rev. Cell Biol.* 8, 227–256.

1008. Hunter, T. (2012). Why nature chose phosphates to modify proteins. *Philos. Trans. R. Soc. Lond. B* 367, 2513–2516.

1009. Hyman, S.E. (2005). Neurotransmitters. *Curr. Biol.* 15, R154–R158.

1010. Ide, H. (2012). Bone pattern formation in mouse limbs after amputation at the forearm level. *Dev. Dyn.* 241, 435–441.

1011. Ifkovits, J.L., Addis, R.C., Epstein, J.A., and Gearhart, J.D. (2014). Inhibition of TGFβ signaling increases direct conversion of fibroblasts to induced cardiomyocytes. *PLoS ONE* 9, #2, e89678.

1012. Ikuta, T., Yoshida, N., Satoh, N., and Saiga, H. (2004). *Ciona intestinalis* Hox gene cluster: its dispersed structure and residual colinear expression in development. *PNAS* 101, #42, 15118–15123.

1013. Illergard, K., Ardell, D.H., and Elofsson, A. (2009). Structure is three to ten times more conserved than sequence: a study of structural response in protein cores. *Proteins* 77, 499–508.

1014. Im, S.H., and Galko, M.J. (2012). Pokes, sunburn, and hot sauce: *Drosophila* as an emerging model for the biology of nociception. *Dev. Dyn.* 241, 16–26.

1015. Imai, T. (2012). Positional information in neural map development: lessons from the olfactory system. *Dev. Growth Differ.* 54, 358–365.

1016. Imai, T. (2014). Construction of functional neuronal circuitry in the olfactory bulb. *Semin. Cell Dev. Biol.* 35, 180–188.

1017. Imai, T., Sakano, H., and Vosshall, L.B. (2010). Topographic mapping: the olfactory system. *Cold Spring Harb. Perspect. Biol.* 2, a001776.

1018. Inbal, A., Halachmi, N., Dibner, C., Frank, D., and Salzberg, A. (2001). Genetic evidence for the transcriptional-activating function of Homothorax during adult fly development. *Development* 128, 3405–3413.

1019. Ingber, D.E., and Levin, M. (2007). What lies at the interface of regenerative medicine and developmental biology? *Development* 134, 2541–2547.

1020. Ingham, P.W., and Martinez-Arias, A. (1986). The correct activation of *Antennapedia* and bithorax complex genes requires the *fushi tarazu* gene. *Nature* 324, 592–597.

1021. Innan, H., and Kondrashov, F. (2010). The evolution of gene duplications: classifying and distinguishing between models. *Nat. Rev. Genet.* 11, 97–108.

1022. Irimia, M., Maeso, I., Roy, S.W., and Fraser, H.B. (2013). Ancient *cis*-regulatory constraints and the evolution of genome architecture. *Trends Genet.* 29, 521–528.

1023. Irimia, M., Piñeiro, C., Maeso, I., Gómez-Skarmeta, J.L., Casares, F., and Garcia-Fernàndez, J. (2010). Conserved developmental expression of *Fezf* in chordates and *Drosophila* and the origin of the *Zona Limitans Intrathalamica* (ZLI) brain organizer. *EvoDevo* 1, Article 7.

1024. Irimia, M., Tena, J.J., Alexis, M.S., Fernandez-Miñan, A., Maeso, I., Bogdanovic, O., de la Calle-Mustienes, E., Roy, S.W., Gómez-Skarmeta, J.L., and Fraser, H.B. (2012). Extensive conservation of ancient microsynteny across metazoans due to *cis*-regulatory constraints. *Genome Res.* 22, 2356–2367.

1025. Irish, V.F., Martinez-Arias, A., and Akam, M. (1989). Spatial regulation of the *Antennapedia* and *Ultrabithorax* homeotic genes during *Drosophila* early development. *EMBO J.* 8, 1527–1537.

1026. Irvine, K.D. (1999). Fringe, Notch, and making developmental boundaries. *Curr. Opin. Genet. Dev.* 9, 434–441.

1027. Irvine, K.D., and Rauskolb, C. (2001). Boundaries in development: formation and function. *Annu. Rev. Cell Dev. Biol.* 17, 189–214.

1028. Irvine, K.D., and Wieschaus, E. (1994). *fringe*, a boundary-specific signaling molecule, mediates interactions between dorsal and ventral cells during *Drosophila* wing development. *Cell* 79, 595–606.

1029. Irvine, S.Q., and Martindale, M.Q. (2001). Comparative analysis of Hox gene expression in the polychaete *Chaetopterus*: implications for the evolution of body plan regionalization. *Am. Zool.* 41, 640–651.

1030. Isayama, T., Chen, Y., Kono, M., Fabre, E., Slavsky, M., DeGrip, W.J., Ma, J.-x., Crouch, R.K., and Makino, C.L. (2013). Coexpression of three opsins in cone photoreceptors of the salamander *Ambystoma tigrinum*. *J. Comp. Neurol.* 522, 2249–2265.

1031. Ito, M., Yang, Z., Andl, T., Cui, C., Kim, N., Millar, S.E., and Cotsarelis, G. (2007). Wnt-dependent *de novo* hair follicle regeneration in adult mouse skin after wounding. *Nature* 447, 316–320.

1032. Itskov, P.M., and Ribeiro, C. (2013). The dilemmas of the gourmet fly: the molecular and neuronal mechanisms of feeding and nutrient decision making in *Drosophila*. *Front. Neurosci.* 7, Article 12.

1033. Iwai, M., Yokono, M., and Nakano, A. (2014). Visualizing structural dynamics of thylakoid membranes. *Sci. Rep.* 4, Article 3768.

1034. Iwaki, D.D., and Lengyel, J.A. (2002). A Delta-Notch signaling border regulated by Engrailed/Invected repression specifies boundary cells in the *Drosophila* hindgut. *Mech. Dev.* 114, 71–84.

1035. Iwasaki, Y., Hosoya, T., Takebayashi, H., Ogawa, Y., Hotta, Y., and Ikenaka, K. (2003). The potential to induce glial differentiation is conserved between *Drosophila* and mammalian glial cells missing genes. *Development* 130, 6027–6035.

1036. Izaddoost, S., Nam, S.-C., Bhat, M.A., Bellen, H.J., and Choi, K.-W. (2002). *Drosophila* Crumbs is a positional cue in photoreceptor adherens junctions and rhabdomeres. *Nature* 416, 178–182.

1037. Izpisúa Belmonte, J.C. (1999). How the body tells left from right. *Sci. Am.* 280, #6, 46–51.

1038. Jablonski, N.G. (2010). The naked truth. *Sci. Am.* 302, #2, 42–49.

1039. Jack, T., Regulski, M., and McGinnis, W. (1988). Pair-rule segmentation genes regulate the expression of the homeotic selector gene, *Deformed*. *Genes Dev.* 2, 635–651.

1040. Jackson, D.J., Meyer, N.P., Seaver, E., Pang, K., McDougall, C., Moy, V.N., Gordon, K., Degnan, B.M., Martindale, M.Q., Burke, R.D., and Peterson, K.J. (2010). Developmental expression of *COE* across the Metazoa supports a conserved role in neuronal cell-type specification and mesodermal development. *Dev. Genes Evol.* 220, 221–234.

1041. Jacobo, A., and Hudspeth, A.J. (2014). Reaction–diffusion model of hair-bundle morphogenesis. *PNAS* 111, #43, 15444–15449.

1042. Jacobs, D.K., Hughes, N.C., Fitz-Gibbon, S.T., and Winchell, C.J. (2005). Terminal addition, the Cambrian radiation and the Phanerozoic evolution of bilaterian form. *Evol. Dev.* 7, 498–514.

1043. Jacobs, D.K., Nakanishi, N., Yuan, D., Camara, A., Nichols, S.A., and Hartenstein, V. (2007). Evolution of sensory structures in basal metazoa. *Integr. Comp. Biol.* 47, 712–723.

1044. Jacobs, G.H. (2012). The evolution of vertebrate color vision. *In* C. López-Larrea (ed.), *Sensing in Nature*. Springer, New York, NY, pp. 156–172.

1045. Jacobs, G.H., and Nathans, J. (2009). The evolution of primate color vision. *Sci. Am.* 300, #4, 56–63.

1046. Jacobs, G.H., Williams, G.A., Cahill, H., and Nathans, J. (2007). Emergence of novel color vision in mice engineered to express a human cone photopigment. *Science* 315, 1723–1725.

1047. Jaeger, J. (2011). The gap gene network. *Cell. Mol. Life Sci.* 68, 243–274.

1048. Jaeger, J., Manu, and Reinitz, J. (2012). *Drosophila* blastoderm patterning. *Curr. Opin. Genet. Dev.* 22, 533–541.

1049. Jafari, S., and Alenius, M. (2015). Cis-regulatory mechanisms for robust olfactory sensory neuron class-restricted odorant receptor gene expression in *Drosophila*. *PLoS Genet.* 11, #3, e1005051.

1050. Jafari, S., Alkhori, L., Schleiffer, A., Brochtrup, A., Hummel, T., and Alenius, M. (2012). Combinatorial activation and repression by seven transcription factors specify *Drosophila* odorant receptor expression. *PLoS Biol.* 10, #3, e1001280.

1051. Jahan, I., Pan, N., Elliott, K.L., and Fritzsch, B. (2015). The quest for restoring hearing: understanding ear development more completely. *BioEssays* 37, 1016–1027.

1052. Jane, S.M., Ting, S.B., and Cunningham, J.M. (2005). Epidermal impermeable barriers in mouse and fly. *Curr. Opin. Genet. Dev.* 15, 447–453.

1053. Jankowski, R. (ed.) (2013). *The Evo-Devo Origin of the Nose, Anterior Skull Base and Midface*. Spinger, New York, NY.

1054. Janssen, R., and Budd, G.E. (2013). Deciphering the onychophoran "segmentation gene cascade": gene expression reveals limited involvement of pair rule gene orthologs in segmentation, but a highly conserved segment polarity gene network. *Dev. Biol.* 382, 224–234.

1055. Janssen, R., and Budd, G.E. (2016). Gene expression analysis reveals that Delta/Notch signalling is not involved in onychophoran segmentation. *Dev. Genes Evol.* 226, 69–77.

1056. Janssen, R., and Damen, W.G.M. (2006). The ten *Hox* genes of the millipede *Glomeris marginata*. *Dev. Genes Evol.* 216, 451–465.

1057. Janssen, R., Eriksson, B.J., Budd, G.E., Akam, M., and Prpic, N.-M. (2010). Gene expression patterns in an onychophoran reveal that regionalization predates limb segmentation in pan-arthropods. *Evol. Dev.* 12, 363–372.

1058. Janssen, R., Eriksson, B.J., Tait, N.N., and Budd, G.E. (2014). Onychophoran Hox genes and the evolution of arthropod Hox gene expression. *Front. Zool.* 11, Article 22.

1059. Janssen, R., Jörgensen, M., Prpic, N.-M., and Budd, G.E. (2015). Aspects of dorso-ventral and proximo-distal limb patterning in onychophorans. *Evol. Dev.* 17, 21–33.

1060. Jarman, A.P. (2000). Developmental genetics: vertebrates and insects see eye to eye. *Curr. Biol.* 10, R857–R859.

1061. Jarman, A.P. (2002). Studies of mechanosensation using the fly. *Hum. Mol. Genet.* 11, #10, 1215–1218.

1062. Jarman, A.P., and Groves, A.K. (2013). The role of *Atonal* transcription factors in the development of mechanosensitive cells. *Semin. Cell Dev. Biol.* 24, 438–447.

1063. Jaumouillé, E., Almeida, P.M., Stähli, P., Koch, R., and Nagoshi, E. (2015). Transcriptional regulation via nuclear receptor crosstalk required for the *Drosophila* circadian clock. *Curr. Biol.* 25, 1502–1508.

1064. Jaw, T.J., You, L.-R., Knoepfler, P.S., Yao, L.-C., Pai, C.-Y., Tang, C.-Y., Chang, L.-P., Berthelsen, J., Blasi, F., Kamps, M.P., and Sun, Y.H. (2000). Direct interaction of two homeoproteins, Homothorax and Extradenticle, is essential for EXD nuclear localization and function. *Mech. Dev.* 91, 279–291.

1065. Jaworski, A., Tom, I., Tong, R.K., Gildea, H.K., Koch, A.W., Gonzalez, L.C., and Tessier-Lavigne, M. (2015). Operational redundancy in axon guidance through the multifunctional receptor Robo3 and its ligand NELL2. *Science* 350, 961–965.

1066. Jefferis, G.S.X.E., Potter, C.J., Chan, A.M., Marin, E.C., Rohlfing, T., Maurer, C.R., Jr., and Luo, L. (2007). Comprehensive maps of *Drosophila* higher olfactory centers: spatially segregated fruit and pheromone representation. *Cell* 128, 1187–1203.

1067. Jeffery, G., and Erskine, L. (2005). Variations in the architecture and development of the vertebrate optic chiasm. *Prog. Retin. Eye Res.* 24, 721–753.

1068. Jékely, G. (2013). Global view of the evolution and diversity of metazoan neuropeptide signaling. *PNAS* 110, #21, 8702–8707.

1069. Jékely, G., Paps, J., and Nielsen, C. (2015). The phylogenetic position of ctenophores and the origin(s) of nervous systems. *EvoDevo* 6, Article 1. [*See also* Jager, M., and Manuel, M. (2016). Ctenophores: an evolutionary-developmental perspective. *Curr. Opin. Genet. Dev.* 39, 85–92.]

1070. Jensen, B., Wang, T., Christoffels, V.M., and Moorman, A.F.M. (2013). Evolution and development of the building plan of the vertebrate heart. *Biochim. Biophys. Acta* 1833, 783–794.

1071. Jessell, T.M. (2000). Neuronal specification in the spinal cord: inductive signals and transcription codes. *Nat. Rev. Genet.* 1, 20–29.

1072. Ji, C., Wu, L., Zhao, W., Wang, S., and Lv, J. (2012). Echinoderms have bilateral tendencies. *PLoS ONE* 7, #1, e28978.

1073. Jiang, H., and Edgar, B.A. (2012). Intestinal stem cell function in *Drosophila* and mice. *Curr. Opin. Genet. Dev.* 22, 354–360.

1074. Jiang, X., Shen, S., Cadwell, C.R., Berens, P., Sinz, F., Ecker, A.S., Patel, S., and Tolias, A.S. (2015). Principles of connectivity among morphologically defined cell types in the adult neocortex. *Science* 350, 1055.

1075. Jiao, Y., Lau, T., Hatzikirou, H., Meyer-Hermann, M., Corbo, J.C., and Torquato, S. (2014). Avian photoreceptor patterns represent a disordered hyperuniform solution to a multiscale packing problem. *Phys. Rev. E* 89, Article 022721.

1076. Jimenez-Gurl, E., and Pujades, C. (2011). An ancient mechanism of hindbrain patterning has been conserved in vertebrate evolution. *Evol. Dev.* 13, 38–46.

1077. Jockusch, E.L., Williams, T.A., and Nagy, L. (2004). The evolution of patterning of serially homologous appendages in insects. *Dev. Genes Evol.* 214, 324–338.

1078. Joffe, B., Peichl, L., Hendrickson, A., Leonhardt, H., and Solovei, I. (2014). Diurnality and nocturnality in primates: an analysis from the rod photoreceptor nuclei perspective. *Evol. Biol.* 41, 1–11.

1079. Johansson, J.A., and Headon, D.J. (2014). Regionalisation of the skin. *Semin. Cell Dev. Biol.* 25–26, 3–10.

1080. Johnsen, S. (2012). *The Optics of Life: A Biologist's Guide to Light in Nature*. Princeton University Press, Princeton, NJ.

1081. Johnson, J.-L.F., and Leroux, M.R. (2010). cAMP and cGMP signaling: sensory systems with prokaryotic roots adopted by eukaryotic cilia. *Trends Cell Biol.* 20, 435–444.

1082. Johnston, R.J., Jr., and Desplan, C. (2010). Stochastic mechanisms of cell fate specification that yield random or robust outcomes. *Annu. Rev. Cell Dev. Biol.* 26, 689–719.

1083. Johnston, R.J., Jr., and Hobert, O. (2003). A microRNA controlling left/right neuronal asymmetry in *Caenorhabditis elegans*. *Nature* 426, 845–849.

1084. Johnston, R.J., Jr., Otake, Y., Sood, P., Vogt, N., Behnia, R., Vasiliauskas, D., McDonald, E., Xie, B., Koenig, S., Wolf, R., Cook, T., Gebelein, B., Kussell, E., Nakagoshi, H., and Desplan, C. (2011). Interlocked feedforward loops control cell-type-specific rhodopsin expression in the *Drosophila* eye. *Cell* 145, 956–968.

1085. Joliot, A., and Prochiantz, A. (2008). Homeoproteins as natural Penetratin cargoes with signaling properties. *Adv. Drug Deliv. Rev.* 60, 608–613.

1086. Joly, W., Mugat, B., and Maschat, F. (2007). Engrailed controls the organization of the ventral nerve cord through *frazzled* regulation. *Dev. Biol.* 301, 542–554.

1087. Jonasova, K., and Kozmik, Z. (2008). Eye evolution: lens and cornea as an upgrade of animal visual system. *Semin. Cell Dev. Biol.* 19, 71–81.

1088. Jones, C., Qian, D., Kim, S.M., Li, S., Ren, D., Knapp, L., Sprinzak, D., Avraham, K.B., Matsuzaki, F., Chi, F., and Chen, P. (2014). *Ankrd6* is a mammalian functional homolog of *Drosophila* planar cell polarity gene *diego* and regulates coordinated cellular orientation in the mouse inner ear. *Dev. Biol.* 395, 62–72.

1089. Joseph, R.M., and Carlson, J.R. (2015). *Drosophila* chemoreceptors: a molecular interface between the chemical world and the brain. *Trends Genet.* 31, 683–695.

1090. Jost, M., Fernández-Zapata, J., Polanco, M.C., Ortiz-Guerrero, J.M., Chen, P.Y.-T., Kang, G., Padmanabhan, S., Elías-Arnanz, M., and Drennan, C.L. (2015). Structural basis for gene regulation by a B_{12}-dependent photoreceptor. *Nature* 526, 536–541.

1091. Ju, Y.-E.S., Lucey, B.P., and Holtzman, D.M. (2014). Sleep and Alzheimer disease pathology: a bidirectional relationship. *Nat. Rev. Neurol.* 10, 115–119.

1092. Jung, H., and Dasen, J.S. (2015). Evolution of patterning systems and circuit elements for locomotion. *Dev. Cell* 32, 408–422.

1093. Kageyama, R., Ohtsuka, T., Shimojo, H., and Imayoshi, I. (2009). Dynamic regulation of Notch signaling in neural progenitor cells. *Curr. Opin. Cell Biol.* 21, 733–740.

1094. Kainz, F., Ewen-Campen, B., Akam, M., and Extavour, C.G. (2011). Notch/Delta signalling is not required for segment generation in the basally branching insect *Gryllus bimaculatus*. *Development* 138, 5015–5026.

1095. Kaiser, M. (2015). Neuroanatomy: Connectome connects fly and mammalian brain networks. *Curr. Biol.* 25, R409–R430.

1096. Kaltenbach, J.A., Falzarano, P.R., and Simpson, T.H. (1994). Postnatal development of the hamster cochlea. II. Growth and differentiation of stereocilia bundles. *J. Comp. Neurol.* 350, 187–198.

1097. Kalteziotis, V., Kouroupi, G., Oikonomaki, M., Mantouvalou, E., Stergiopoulos, A., Charonis, A., Rohrer, H., Matsas, R., and Politis, P.K. (2010). Prox1 regulates the Notch1-mediated inhibition of neurogenesis. *PLoS Biol.* 8, #12, e1000565.

1098. Kamikouchi, A. (2013). Auditory neuroscience in fruit flies. *Neurosci. Res.* 76, 113–118.

1099. Kamikouchi, A., Inagaki, H.K., Effertz, T., Hendrich, O., Fiala, A., Göpfert, M.C., and Ito, K. (2009). The neural basis of *Drosophila* gravity-sensing and hearing. *Nature* 458, 165–171.

1100. Kamikouchi, A., Shimada, T., and Ito, K. (2006). Comprehensive classification of the auditory sensory projections in the brain of the fruit fly *Drosophila melanogaster*. *J. Comp. Neurol.* 499, 317–356.

1101. Kammandel, B., Chowdhury, K., Stoykova, A., Aparicio, S., Brenner, S., and Gruss, P. (1999). Distinct *cis*-essential modules direct the time-space pattern of the Pax6 gene activity. *Dev. Biol.* 205, 79–97.

1102. Kammermeier, L., and Reichert, H. (2001). Common developmental genetic mechanisms for patterning invertebrate and vertebrate brains. *Brain Res. Bull.* 55, 675–682.

1103. Kandel, E.R. (2012). The molecular biology of memory: cAMP, PKA, CRE, CREB-1, CREB-2, and CPEB. *Mol. Brain* 5, Article 14.

1104. Kang, J., Hu, J., Karra, R., Dickson, A.L., Tornini, V.A., Nachtrab, G., Gemberling, M., Goldman, J.A., Black, B.L., and Poss, K.D. (2016). Modulation of tissue repair by regeneration enhancer elements. *Nature* 532, 201–206.

1105. Kang, K., Pulver, S.R., Panzano, V.C., Chang, E.C., Griffith, L.C., Theobald, D.L., and Garrity, P.A. (2010). Analysis of *Drosophila* TRPA1 reveals an ancient origin for human chemical nociception. *Nature* 464, 597–600.

1106. Kantor, D.B., and Kolodkin, A.L. (2003). Curbing the excesses of youth: molecular insights into axonal pruning. *Neuron* 38, 849–852.

1107. Kaprielian, Z., Imondi, R., and Runko, E. (2000). Axon guidance at the midline of the developing CNS. *Anat. Rec.* 261, 176–197.

1108. Kaprielian, Z., Runko, E., and Imondi, R. (2001). Axon guidance at the midline choice point. *Dev. Dyn.* 221, 154–181.

1109. Karadge, U.B., Gosto, M., and Nicotra, M.L. (2015). Allorecognition proteins in an invertebrate exhibit homophilic interactions. *Curr. Biol.* 25, 2845–2850.

1110. Karavitaki, K.D., and Corey, D.P. (2010). Sliding adhesion confers coherent motion to hair cell stereocilia and parallel gating to transduction channels. *J. Neurosci.* 30, 9051–9063.

1111. Kardon, G., Heanue, T.A., and Tabin, C.J. (2004). The *Pax/Six/Eya/Dach* network in development and evolution. *In* G. Schlosser and G.P. Wagner (eds.), *Modularity in Development and Evolution.* University of Chicago Press, Chicago, IL, pp. 59–80.

1112. Karten, H.J. (2015). Vertebrate brains and evolutionary connectomics: on the origins of the mammalian "neocortex". *Philos. Trans. R. Soc. Lond. B* 370, 20150060.

1113. Kasahara, M. (2007). The 2R hypothesis: an update. *Curr. Opin. Immunol.* 19, 547–552.

1114. Kashalikar, S.J. (1988). An explanation for the development of decussations in the central nervous system. *Med. Hypotheses* 26, 1–8.

1115. Kato, H.E., Inoue, K., Abe-Yoshizumi, R., Kato, Y., Ono, H., Konno, M., Hososhima, S., Ishizuka, T., Hoque, M.R., Kunitomo, H., Ito, J., Yoshizawa, S., Yamashita, K., Takemoto, M., Nishizawa, T., Taniguchi, R., Kogure, K., Maturana, A.D., Iino, Y., Yawo, H., Ishitani, R., Kandori, H., and Nureki, O. (2015). Structural basis for Na$^+$ transport mechanism by a light-driven Na$^+$ pump. *Nature* 521, 48–53.

1116. Katta, S., Krieg, M., and Goodman, M.B. (2015). Feeling force: physical and physiological principles enabling sensory mechanotransduction. *Annu. Rev. Cell Dev. Biol.* 31, 347–371.

1117. Katz, B., and Minke, B. (2009). *Drosophila* photoreceptors and signaling mechanisms. *Front. Cell. Neurosci.* 3, Article 2.

1118. Katz, M.J., and Grenander, U. (1982). Developmental matching and the numerical matching hypothesis for neuronal cell death. *J. Theor. Biol.* 98, 501–517.

1119. Katz, P.S. (2016). Evolution of central pattern generators and rhythmic behaviours. *Philos. Trans. R. Soc. Lond. B* 371, 20150057.

1120. Katz, P.S., and Harris-Warrick, R.M. (1999). The evolution of neuronal circuits underlying species-specific behavior. *Curr. Opin. Neurobiol.* 9, 628–633.

1121. Kaufman, L. (2005). One fish, two fish, red fish, blue fish: why are coral reefs so colorful? *Natl. Geogr.* 207, #5, 86–109.

1122. Kaufman, T.C., Lewis, R., and Wakimoto, B. (1980). Cytogenetic analysis of chromosome 3 in *Drosophila melanogaster*: the homoeotic gene complex in polytene chromosome interval 84A-B. *Genetics* 94, 115–133.

1123. Kaupp, U.B. (2010). Olfactory signalling in vertebrates and insects: differences and commonalities. *Nat. Rev. Neurosci.* 11, 188–200.

1124. Kavaler, J., Fu, W., Duan, H., Noll, M., and Posakony, J.W. (1999). An essential role for the *Drosophila Pax2* homolog in the differentiation of adult sensory organs. *Development* 126, 2261–2272.

1125. Kavanau, J.L. (2001). Memory failures, dream illusions and mental malfunction. *Neuropsychobiology* 44, 199–211.

1126. Kavlie, R.G., and Albert, J.T. (2013). Chordotonal organs. *Curr. Biol.* 23, R334–R335.

1127. Kavlie, R.G., Fritz, J.L., Nies, F., Göpfert, M.C., Oliver, D., Albert, J.T., and Eberl, D.F. (2015). Prestin is an anion transporter dispensable for mechanical feedback amplification in *Drosophila* hearing. *J. Comp. Physiol. A* 201, 51–60.

1128. Kawamori, A., Shimaji, K., and Yamaguchi, M. (2012). Dynamics of endoreplication during *Drosophila* posterior scutellar macrochaete development. *PLoS ONE* 7, #6, e38714.

1129. Kawasumi, A., Nakamura, T., Iwai, N., Yashiro, K., Saijoh, Y., Belo, J.A., Shiratori, H., and Hamada, H. (2011). Left-right asymmetry in the level of active Nodal protein produced in the node is translated into left-right asymmetry in the lateral plate of mouse embryos. *Dev. Biol.* 353, 321–330.

1130. Kay, J.N., Link, B.A., and Baier, H. (2005). Staggered cell-intrinsic timing of *ath5* expression underlies the wave of ganglion cell neurogenesis in the zebrafish retina. *Development* 132, 2573–2585.

1131. Kayser, M.S., Mainwaring, B., Yue, Z., and Sehgal, A. (2015). Sleep deprivation suppresses aggression in *Drosophila*. *eLife* 4, e07643.

1132. Kazmierczak, P., and Müller, U. (2011). Sensing sound: molecules that orchestrate mechanotransduction by hair cells. *Trends Neurosci.* 35, 220–229.

1133. Kechad, A., Jolicoeur, C., Tufford, A., Mattar, P., Chow, R.W.Y., Harris, W.A., and Cayouette, M. (2012). Numb is required for the production of terminal asymmetric cell divisions in the developing mouse retina. *J. Neurosci.* 32, #48, 17197–17210.

1134. Kefalov, V., Fu, Y., Marsh-Armstrong, N., and Yau, K.-W. (2003). Role of visual pigment properties in rod and cone phototransduction. *Nature* 425, 526–531.

1135. Keil, T.A. (2012). Sensory cilia in arthropods. *Arthropod Struct. Dev.* 41, 515–534.

1136. Kelber, A., and Osorio, D. (2010). From spectral information to animal colour vision: experiments and concepts. *Proc. R. Soc. Lond. B* 277, 1617–1625.

1137. Keller, R. (2012). Physical biology returns to morphogenesis. *Science* 338, 201–203.

1138. Kelly, M., and Chen, P. (2007). Shaping the mammalian auditory sensory organ by the planar cell polarity pathway. *Int. J. Dev. Biol.* 51, 535–547.

1139. Kelly, M.C., and Chen, P. (2009). Development of form and function in the mammalian cochlea. *Curr. Opin. Neurobiol.* 19, 395–401.

1140. Kelly, M.W. (2007). Cellular commitment and differentiation in the organ of Corti. *Int. J. Dev. Biol.* 51, 571–583.

1141. Kenaley, C.P., DeVaney, S.C., and Fjeran, T.T. (2013). The complex evolutionary history of seeing red: molecular phylogeny and the evolution of an adaptive visual system in deep-sea dragonfishes (Stomiiformes: Stomiidae). *Evolution* 68, 996–1013.

1142. Kennedy, B., and Malicki, J. (2009). What drives cell morphogenesis: a look inside the vertebrate photoreceptor. *Dev. Dyn.* 238, 2115–2138.

1143. Kernan, M.J. (2007). Mechanotransduction and auditory transduction in *Drosophila*. *Pflugers Arch.* 454, 703–720.

1144. Kerner, P., Ikmi, A., Coen, D., and Vervoort, M. (2009). Evolutionary history of the *iroquois/Irx* genes in metazoans. *BMC Evol. Biol.* 9, Article 74.

1145. Kerner, P., Simionato, E., Le Gouar, M., and Vervoort, M. (2009). Orthologs of key vertebrate neural genes are expressed during neurogenesis in the annelid *Platynereis dumerilii*. *Evol. Dev.* 11, 513–524.

1146. Kessel, M. (1992). Respecification of vertebral identities by retinoic acid. *Development* 115, 487–501.

1147. Kessel, M., and Gruss, P. (1991). Homeotic transformations of murine vertebrae and concomitant alteration of *Hox* codes induced by retinoic acid. *Cell* 67, 89–104.

1148. Kettle, C., Johnstone, J., Jowett, T., Arthur, H., and Arthur, W. (2003). The pattern of segment formation, as revealed by *engrailed* expression, in a centipede with a variable number of segments. *Evol. Dev.* 5, 198–207.

1149. Kidd, T., Bland, K.S., and Goodman, C.S. (1999). Slit is the midline repellent for the Robo receptor in *Drosophila*. *Cell* 96, 785–794.

1150. Kiecker, C., and Lumsden, A. (2005). Compartments and their boundaries in vertebrate brain development. *Nat. Rev. Neurosci.* 6, 553–564.

1151. Kiecker, C., and Lumsden, A. (2012). The role of organizers in patterning the nervous system. *Annu. Rev. Neurosci.* 35, 347–367.

1152. Kiernan, A.E. (2013). Notch signaling during cell fate determination in the inner ear. *Semin. Cell Dev. Biol.* 24, 470–479.

1153. Kim, H., Kim, K., and Yim, J. (2013). Biosynthesis of drosopterins, the red eye pigments of *Drosophila melanogaster*. *IUBMB Life* 65, 334–340.

1154. Kim, J.H., Jin, P., Duan, R., and Chen, E.H. (2015). Mechanisms of myoblast fusion during muscle development. *Curr. Opin. Genet. Dev.* 32, 162–170.

1155. Kim, S.-G., Ashe, J., Hendrich, K., Ellermann, J.M., Merkle, H., Ugurbil, K., and Georgopoulos, A.P. (1993). Functional magnetic resonance imaging of motor cortex: hemispheric asymmetry and handedness. *Science* 261, 615–617.

1156. Kim, S.Y., Paylor, S.W., Magnuson, T., and Schumacher, A. (2007). Juxtaposed Polycomb complexes co-regulate vertebral identity. *Development* 133, 4957–4968.

1157. Kimelman, D., and Martin, B.L. (2011). Anterior-posterior patterning in early development: three strategies. *WIREs Dev. Biol.* 1, 253–266.

1158. Kimmel, C.B. (1996). Was *Urbilateria* segmented? *Trends Genet.* 12, 329–331.

1159. King, T., and Brown, N.A. (1995). The embryo's one-sided genes. *Curr. Biol.* 5, 1364–1366.

1160. King, T., and Brown, N.A. (1999). Embryonic asymmetry: the left side gets all the best genes. *Curr. Biol.* 9, R18–R22.

1161. Kingsley, M.C.S., and Ramsay, M.A. (1988). The spiral in the tusk of the narwhal. *Arctic* 41, 236–238.

1162. Kinsbourne, M. (2013). Somatic twist: a model for the evolution of decussation. *Neuropsychology* 27, 511–515.

1163. Kipryushina, Y.O., Yakovlev, K.V., and Odintsova, N.A. (2015). Vascular endothelial growth factors: a comparison between invertebrates and vertebrates. *Cytokine Growth Factor Rev.* 26, 687–695.

1164. Kirschner, M.W., and Gerhart, J.C. (2005). *The Plausibility of Life: Resolving Darwin's Dilemma.* Yale University Press, New Haven, CT.

1165. Kiser, P.D., Golczak, M., and Palczewski, K. (2013). Chemistry of the retinoid (visual) cycle. *Chem. Rev.* 114, 194–232.

1166. Klar, A.J.S. (2003). Human handedness and scalp hair-whorl direction develop from a common genetic mechanism. *Genetics* 165, 269–276.

1167. Klar, A.J.S. (2005). A 1927 study supports a current genetic model for inheritance of human scalp hair-whorl orientation and hand-use preference traits. *Genetics* 170, 2027–2030.

1168. Klaus, A., Saga, Y., Taketo, M.M., Tzahor, E., and Birchmeier, W. (2007). Distinct roles of Wnt/β-catenin and Bmp signaling during early cardiogenesis. *PNAS* 104, #47, 18531–18536.

1169. Klein, R., and Kania, A. (2014). Ephrin signalling in the developing nervous system. *Curr. Opin. Neurobiol.* 27, 16–24.

1170. Klein, T., and Martinez Arias, A. (1998). Interactions among Delta, Serrate and Fringe modulate Notch activity during *Drosophila* wing development. *Development* 125, 2951–2962.

1171. Klezovitch, O., and Vasioukhin, V. (2013). Your gut is right to turn left. *Dev. Cell* 26, 553–554.

1172. Kliethermes, C.L. (2015). Conservation of the ethanol-induced locomotor stimulant response among arthropods. *Brain Behav. Evol.* 85, 37–46.

1173. Klomp, J., Athy, D., Kwan, C.W., Bloch, N.I., Sandmann, T., Lemke, S., and Schmidt-Ott, U. (2015). A cysteine-clamp gene drives embryo polarity in the midge *Chironomus*. *Science* 348, 1040–1042.

1174. Klug, W.S., Cummings, M.R., and Spencer, C. (2006). *Concepts in Genetics*, 8th edn. Pearson Prentice Hall, Upper Saddle River, NJ.

1175. Kmita, M., and Duboule, D. (2003). Organizing axes in time and space: 25 years of colinear tinkering. *Science* 301, 331–333.

1176. Knaden, M., and Hansson, B.S. (2014). Mapping odor valence in the brain of flies and mice. *Curr. Opin. Neurobiol.* 24, 34–38.

1177. Kobayashi, D., and Takeda, H. (2012). Ciliary motility: the components and cytoplasmic preassembly mechanisms of the axonemal dyneins. *Differentiation* 83, S23–S29.

1178. Koh, T.-W., Gorur-Shandilya, S., Menuz, K., Larter, N.K., Stewart, S., and Carlson, J.R. (2014). The *Drosophila* IR20a clade of ionotropic receptors are candidate taste and pheromone receptors. *Neuron* 83, 850–865.

1179. Kohl, J., and Jefferis, G.S.X.E. (2011). Neuroanatomy: decoding the fly brain. *Curr. Biol.* 21, R19–R20.

1180. Kohler, R.E. (1994). *Lords of the Fly: Drosophila Genetics and the Experimental Life.* University of Chicago Press, Chicago, IL.

1181. Koirala, S., and Ko, C.-P. (2004). Pruning an axon piece by piece: a new mode of synapse elimination. *Neuron* 44, 578–580.

1182. Kojima, T. (2004). The mechanism of *Drosophila* leg development along the proximodistal axis. *Dev. Growth Differ.* 46, 115–129.

1183. Konopova, B., and Akam, M. (2014). The *Hox* genes *Ultrabithorax* and *abdominal-A* specify three different types of abdominal appendage in the springtail *Orchesella cincta* (Collembola). *EvoDevo* 5, Article 2.

1184. Konstantinides, N., and Averof, M. (2014). A common cellular basis for muscle regeneration in arthropods and vertebrates. *Science* 343, 788–791.

1185. Koop, D., Holland, N.D., Sémon, M., Alvarez, S., Rodriguez de Lera, A., Laudet, V., Holland, L.Z., and Schubert, M. (2010). Retinoic acid signaling targets *Hox* genes during the amphioxus gastrula stage: insights into early anterior-posterior patterning of the chordate body plan. *Dev. Biol.* 338, 98–106.

1186. Kopp, A. (2011). *Drosophila* sex combs as a model of evolutionary innovations. *Evol. Dev.* 13, 504–522.

1187. Kosaki, K., and Casey, B. (1998). Genetics of human left-right axis malformations. *Semin. Cell Dev. Biol.* 9, 89–99.

1188. Koschowitz, M.-C., Fischer, C., and Sander, M. (2014). Beyond the rainbow. *Science* 346, 416–418.

1189. Koshiba-Takeuchi, K., Mori, A.D., Kaynak, B.L., Cebra Thomas, J., Sukonnik, T., Georges, R.O., Latham, S., Beck, L., Henkelman, R.M., Black, B.L., Olson, E.N., Wade, J., Takeuchi, J.K., Nemer, M., Gilbert, S.F., and Bruneau, B.G. (2009). Reptilian heart development and the molecular basis of cardiac chamber evolution. *Nature* 461, 95–98.

1190. Kotkamp, K., Klingler, M., and Schoppmeier, M. (2010). Apparent role of *Tribolium orthodenticle* in anteroposterior blastoderm patterning largely reflects novel functions in dorsoventral axis formation and cell survival. *Development* 137, 1853–1862.

1191. Kottler, B., and van Swinderen, B. (2014). Taking a new look at how flies learn. *eLife* 3, e03978.

1192. Koulakov, A., Gelperin, A., and Rinberg, D. (2007). Olfactory coding with all-or-nothing glomeruli. *J. Neurophysiol.* 98, 3134–3142.

1193. Kowalczyk, A.P., and Moses, K. (2002). Photoreceptor cells in flies and mammals: crumby homology? *Dev. Cell* 2, 253–254.

1194. Koyama, E., Yasuda, T., Minugh-Purvis, N., Kinumatsu, T., Yallowitz, A.R., Wellik, D.M., and Pacifici, M. (2010). *Hox11* genes establish synovial joint organization and phylogenetic characteristics in developing mouse zeugopod skeletal elements. *Development* 137, 3795–3800.

1195. Koyanagi, M., Nagata, T., Katoh, K., Yamashita, S., and Tokunaga, F. (2008). Molecular evolution of arthropod color vision deduced from multiple opsin genes of jumping spiders. *J. Mol. Evol.* 66, 130–137.

1196. Kozlov, A.S., Baumgart, J., Risler, T., Versteegh, C.P.C., and Hudspeth, A.J. (2011). Forces between clustered stereocilia minimize friction in the ear on a subnanometre scale. *Nature* 474, 376–379.

1197. Kozmik, Z. (2005). Pax genes in eye development and evolution. *Curr. Opin. Genet. Dev.* 15, 430–438.

1198. Kozmik, Z. (2008). The role of Pax genes in eye evolution. *Brain Res. Bull.* 75, 335–339.

1199. Kozmikova, I., Smolikova, J., Vlcek, C., and Kozmik, Z. (2011). Conservation and diversification of an ancestral chordate gene regulatory network for dorsoventral patterning. *PLoS ONE* 6, #2, e14650.

1200. Kram, Y.A., Mantey, S., and Corbo, J.C. (2010). Avian cone photoreceptors tile the retina as five independent, self-organizing mosaics. *PLoS ONE* 5, #2, e8992.

1201. Krapp, H.G. (2009). Ocelli. *Curr. Biol.* 19, R435–R437.

1202. Kreahling, J.M., and Graveley, B.R. (2005). The iStem, a long-range RNA secondary structure element required for efficient exon inclusion in the *Drosophila Dscam* pre-mRNA. *Mol. Cell. Biol.* 25, #23, 10251–10260.

1203. Krendel, M., and Mooseker, M.S. (2005). Myosins: tails (and heads) of functional diversity. *Physiology* 20, 239–251.

1204. Krishnan, A., and Schiöth, H.B. (2015). The role of G protein-coupled receptors in the early evolution of neurotransmission and the nervous system. *J. Exp. Biol.* 218, 562–571.

1205. Kronhamn, J., Frei, E., Daube, M., Jiao, R., Shi, Y., Noll, M., and Rasmuson-Lestander, Å. (2002). Headless flies produced by mutations in the paralogous *Pax6* genes *eyeless* and *twin of eyeless*. *Development* 129, 1015–1026.

1206. Krueger, D.D., Tuffy, L.P., Papadopoulos, T., and Brose, N. (2012). The role of neurexins and neuroligins in the formation, maturation, and function of vertebrate synapses. *Curr. Opin. Neurobiol.* 22, 412–422.

1207. Kulakova, M., Bakalenko, N., Novikova, E., Cook, C.E., Eliseeva, E., Steinmetz, P.R.H., Kostyuchenko, R.P., Dondua, A., Arendt, D., Akam, M., and Andreeva, T. (2007). Hox gene expression in larval development of the polychaetes *Nereis virens* and *Platynereis dumerilii* (Annelida, Lophotrochozoa). *Dev. Genes Evol.* 217, 39–54.

1208. Kumar, A., and Shivashankar, G.V. (2012). Mechanical force alters morphogenetic movements and segmental gene expression patterns during *Drosophila* embryogenesis. *PLoS ONE* 7, #3, e33089.

1209. Kumar, J. (2006). The molecular circuitry governing retinal determination. *Biochim. Biophys. Acta* 1789, 306–314.

1210. Kumar, J.P. (2001). Signalling pathways in *Drosophila* and vertebrate retinal development. *Nat. Rev. Genet.* 2, 846–857.

1211. Kumar, J.P. (2012). Building an ommatidium one cell at a time. *Dev. Dyn.* 241, 136–149.

1212. Kumar, J.P., and Ready, D.F. (1995). Rhodopsin plays an essential structural role in *Drosophila* photoreceptor development. *Development* 121, 4359–4370.

1213. Kunst, M., Hughes, M.E., Raccuglia, D., Felix, M., Li, M., Barnett, G., Duah, J., and Nitabach, M.N. (2014). Calcitonin gene-related peptide neurons mediate sleep-specific circadian output in *Drosophila*. *Curr. Biol.* 24, 2652–2664.

1214. Kuo, D.-H., and Weisblat, D.A. (2011). A new molecular logic for BMP-mediated dorsoventral patterning in the leech *Helobdella*. *Curr. Biol.* 21, 1282–1288.

1215. Kuranaga, E., Matsunuma, T., Kanuka, H., Takemoto, K., Koto, A., Kimura, K.-i., and Miura, M. (2011). Apoptosis controls the speed of looping morphogenesis in *Drosophila* male terminalia. *Development* 138, 1493–1499.

1216. Kuroda, J., Nakamura, M., Yoshida, M., Yamamoto, H., Maeda, T., Taniguchi, K., Nakazawa, N., Hatori, R., Ishio, A., Ozaki, A., Shimaoka, S., Ito, T., Iida, H., Okumura, T., Maeda, R., and Matsuno, K. (2012). Canonical Wnt signaling in the visceral muscle is required for left-right asymmetric development of the *Drosophila* midgut. *Mech. Dev.* 128, 625–639.

1217. Kurokawa, M., and Kornbluth, S. (2009). Caspases and kinases in a death grip. *Cell* 138, 838–854.

1218. Kyrchanova, O., Mogila, V., Wolle, D., Magbanua, J.P., White, R., Georgiev, P., and Schedl, P. (2015). The boundary paradox in the Bithorax complex. *Mech. Dev.* 138, 122–132.

1219. Labhart, T., and Nilsson, D.-E. (1995). The dorsal eye of the dragonfly *Sympetrum*: specializations for prey detection against the blue sky. *J. Comp. Physiol. A* 176, 437–453.

1220. Lacalli, T. (2003). Body plans and simple brains. *Nature* 424, 263–264.

1221. Lacalli, T. (2004). Light on ancient photoreceptors. *Nature* 432, 454–455.

1222. Lacalli, T. (2008). Head organization and the head/trunk relationship in protochordates: problems and prospects. *Integr. Comp. Biol.* 48, 620–629.

1223. Lacalli, T. (2010). The emergence of the chordate body plan: some puzzles and problems. *Acta Zool. (Stockholm)* 91, 4–10.

1224. Lacalli, T. (2014). Echinoderm conundrums: Hox genes, heterochrony, and an excess of mouths. *EvoDevo* 5, Article 46.

1225. Lacalli, T.C. (2004). Sensory systems in amphioxus: a window on the ancestral chordate condition. *Brain Behav. Evol.* 64, 148–162.

1226. Lacalli, T.C. (2008). Mucus secretion and transport in amphioxus larvae: organization and ultrastructure of the food trapping system, and implications for head evolution. *Acta Zool. (Stockholm)* 89, 219–230.

1227. Lacin, H., Zhu, Y., Wilson, B.A., and Skeath, J.B. (2014). Transcription factor expression uniquely identifies most postembryonic neuronal lineages in the *Drosophila* thoracic central nervous system. *Development* 141, 1011–1021.

1228. Lacoste, A.M.B., Schoppik, D., Robson, D.N., Haesemeyer, M., Portugues, R., Li, J.M., Randlett, O., Wee, C.L., Engert, F., and Schier, A.F. (2015). A convergent and essential interneuron pathway for Mauthner-cell-mediated escapes. *Curr. Biol.* 25, 1526–1534.

1229. Lai, E.C. (2004). Notch signaling: control of cell communication and cell fate. *Development* 131, 965–973.

1230. Lai, E.C., and Orgogozo, V. (2004). A hidden program in *Drosophila* peripheral neurogenesis revealed: fundamental principles underlying sensory organ diversity. *Dev. Biol.* 269, 1–17.

1231. Lall, S., Ludwig, M.Z., and Patel, N.H. (2003). Nanos plays a conserved role in axial patterning outside of the Diptera. *Curr. Biol.* 13, 224–229.

1232. Lall, S., and Patel, N.H. (2001). Conservation and divergence in molecular mechanisms of axis formation. *Annu. Rev. Genet.* 35, 407–437.

1233. Lamb, T.D. (2011). Evolution of the eye. *Sci. Am.* 305, #1, 64–69.

1234. Lamb, T.D. (2013). Evolution of phototransduction, vertebrate photoreceptors and retina. *Prog. Retin. Eye Res.* 36, 52–119.

1235. Lambrechts, D., and Carmeliet, P. (2004). Sculpting heart valves with NFATc and VEGF. *Cell* 118, 532–534.

1236. LaMendola, N.P., and Bever, T.G. (1997). Peripheral and cerebral asymmetries in the rat. *Science* 278, 483–486.

1237. Land, M.F. (2013). Animal vision: rats watch the sky. *Curr. Biol.* 23, R611–R613.

1238. Land, M.F. (2014). Animal vision: starfish can see at last. *Curr. Biol.* 24, R200–R201.

1239. Land, M.F., and Nilsson, D.-E. (2012). *Animal Eyes*, 2nd edn. Oxford University Press, New York, NY.

1240. Lander, A.D. (2007). Morpheus unbound: reimagining the morphogen gradient. *Cell* 128, 245–256.

1241. Landgraf, M., and Thor, S. (2006). Development of *Drosophila* motoneurons: specification and morphology. *Semin. Cell Dev. Biol.* 17, 3–11.

1242. Lange, A., Nemeschkal, H.L., and Müller, G.B. (2014). Biased polyphenism in polydactylous cats carrying a single point mutation: the Hemingway model for digit novelty. *Evol. Biol.* 41, 262–275.

1243. Langen, M., Agi, E., Altschuler, D.J., Wu, L.F., Altschuler, S.J., and Hiesinger, P.R. (2015). The developmental rules of neural superposition in *Drosophila*. *Cell* 162, 120–133.

1244. Langen, M., Koch, M., Yan, J., De Geest, N., Erfurth, M.-L., Pfeiffer, B.D., Schmucker, D., Moreau, Y., and Hassan, B.A. (2013). Mutual inhibition among post-mitotic neurons regulates robustness of brain wiring in *Drosophila*. *eLife* 2, e00337.

1245. Lanot, R., Zachary, D., Holder, F., and Meister, M. (2001). Postembryonic hemato-poiesis in *Drosophila*. *Dev. Biol.* 230, 247–257.

1246. Lapraz, F., Besnardeau, L., and Lepage, T. (2009). Patterning of the dorsal-ventral axis in echinoderms: insights into the evolution of the BMP-Chordin signaling net-work. *PLoS Biol.* 7, #11, e1000248.

1247. Larroux, C., Fahey, B., Degnan, S.M., Adamski, M., Rokhsar, D.S., and Degnan, B.M. (2007). The NK homeobox gene cluster predates the origin of Hox genes. *Curr. Biol.* 17, 706–710.

1248. Larsen, C., Bardet, P.-L., Vincent, J.-P., and Alexandre, C. (2008). Specification and positioning of parasegment grooves in *Drosophila*. *Dev. Biol.* 321, 310–318.

1249. Larsson, M.C., Domingos, A.I., Jones, W.D., Chiappe, M.E., Amrein, H., and Vosshall, L.B. (2004). Or83b encodes a broadly expressed odorant receptor essential for *Drosophila* olfaction. *Neuron* 43, 703–714.

1250. Larsson, M.L. (2015). Binocular vision, the optic chiasm, and their associations with vertebrate motor behavior. *Front. Ecol. Evol.* 3, Article 89.

1251. Laubichler, M.D., and Maienschein, J. (eds.) (2007). *From Embryology to Evo-Devo: A History of Developmental Evolution*. MIT Press, Cambridge, MA.

1252. Laufer, E., Dahn, R., Orozco, O.E., Yeo, C.-Y., Pisenti, J., Henrique, D., Abbott, U.K., Fallon, J.F., and Tabin, C. (1997). Expression of *Radical fringe* in limb-bud ectoderm regulates apical ectodermal ridge formation. *Nature* 386, 366–373.

1253. Lauri, A., Brunet, T., Handberg-Thorsager, M., Fischer, A.H.L., Simakov, O., Steinmetz, P.R.H., Tomer, R., Keller, P.J., and Arendt, D. (2014). Development of the annelid axochord: insights into notochord evolution. *Science* 345, 1365–1368.

1254. Lavagnino, N.J., Arya, G.H., Korovaichuk, A., and Fanara, J.J. (2013). Genetic archi-tecture of olfactory behavior in *Drosophila melanogaster*: differences and similarities across development. *Behav. Genet.* 43, 348–359.

1255. Lavagnino, N.J., and Fanara, J.J. (2016). Changes across development influence visi-ble and cryptic natural variation of *Drosophila melanogaster* olfactory response. *Evol. Biol.* 43, 96–108.

1256. Lawrence, P.A., and Casal, J. (2013). The mechanisms of planar cell polarity, growth and the Hippo pathway: some known unknowns. *Dev. Biol.* 377, 1–8. [*See also* Devenport, D. (2016). Tissue morphodynamics: translating planar polarity cues into polarized cell behaviors. *Semin. Cell Dev. Biol.* 55, 99–110.]

1257. Layton, W.M., Jr. (1976). Random determination of a developmental process: rever-sal of normal visceral asymmetry in the mouse. *J. Hered.* 67, 336–338.

1258. Le Douarin, N.M., and Dieterlen-Liévre, F. (2013). How studies on the avian embryo have opened new avenues in the understanding of development: a view about the neural and hematopoietic systems. *Dev. Growth Differ.* 55, 1–14.

1259. Leal, W.S. (2013). Odorant reception in insects: roles of receptors, binding proteins, and degrading enzymes. *Annu. Rev. Entomol.* 58, 373–391.

1260. Lebestky, T., Chang, T., Hartenstein, V., and Banerjee, U. (2000). Specification of *Drosophila* hematopoietic lineage by conserved transcription factors. *Science* 288, 146–149.

1261. LeBon, L., Lee, T.V., Sprinzak, D., Jafar-Nejad, H., and Elowitz, M.B. (2014). Fringe proteins modulate Notch-ligand *cis* and *trans* interactions to specify signaling states. *eLife* 3, e02950.

1262. Leclère, L., and Rentzsch, F. (2014). RGM regulates BMP-mediated secondary axis formation in the sea anemone *Nematostella vectensis*. *Cell Rep.* 9, 1921–1930.

1263. Lee, C., Kim, N., Roy, M., and Graveley, B.R. (2010). Massive expansions of *Dscam* splicing diversity via staggered homologous recombination during arthropod evolution. *RNA* 16, 91–105.

1264. Lee, H.-G., Kim, Y.-C., Dunning, J.S., and Han, K.-A. (2008). Recurring ethanol exposure induces disinhibited courtship in *Drosophila*. *PLoS ONE*, #1, e1391.

1265. Lee, M.S.Y., Jago, J.B., García-Bellido, D.C., Edgecombe, G.D., Gehling, J.G., and Paterson, J.R. (2011). Modern optics in exceptionally preserved eyes of early Cambrian arthropods from Australia. *Nature* 474, 631–634.

1266. Lee, Y., and Rio, D.C. (2015). Mechanisms and regulation of alternative pre-mRNA splicing. *Annu. Rev. Biochem.* 84, 291–323.

1267. Lefebvre, J.L., Kostadinov, D., Chen, W.V., Maniatis, T., and Sanes, J.R. (2012). Protocadherins mediate dendritic self-avoidance in the mammalian nervous system. *Nature* 488, 517–521.

1268. Lefebvre, J.L., Sanes, J.R., and Kay, J.N. (2015). Development of dendritic form and function. *Annu. Rev. Cell Dev. Biol.* 31, 741–777.

1269. Leinwand, S.G., and Chalasani, S.H. (2011). Olfactory networks: from sensation to perception. *Curr. Opin. Genet. Dev.* 21, 806–811.

1270. Leite-Castro, J., Beviano, V., Rodrigues, P.N., and Freitas, R. (2016). HoxA genes and the fin-to-limb transition in vertebrates. *J. Dev. Biol.* 4, Article 4010010. [*See also* Gehrke, A.R., and Shubin, N.H. (2016). *Cis*-regulatory programs in the development and evolution of vertebrate paired appendages. *Semin. Cell Dev. Biol.* 57, 31–39.]

1271. Lelli, A., Asai, Y., Forge, A., Holt, J.R., and Géléoc, G.S.G. (2009). Tonotopic gradient in the developmental acquisition of sensory transduction in outer hair cells of the mouse cochlea. *J. Neurophysiol.* 101, 2961–2973.

1272. Lem, J., Krasnoperova, N.V., Calvert, P.D., Kosaras, B., Cameron, D.A., Nicolò, M., Makino, C.L., and Sidman, R.L. (1999). Morphological, physiological, and biochemical changes in rhodopsin knockout mice. *PNAS* 96, 736–741.

1273. LeMasurier, M., and Gillespie, P.G. (2005). Hair-cell mechanotransduction and cochlear amplification. *Neuron* 48, 403–415.

1274. Lemke, S., and Schmidt-Ott, U. (2009). Evidence for a composite anterior determinant in the hover fly *Episyrphus balteatus* (Syrphidae), a cyclorrhaphan fly with an anterodorsal serosa anlage. *Development* 136, 117–127.

1275. Lemke, S., Stauber, M., Shaw, P.J., Rafiqi, A.M., Prell, A., and Schmidt-Ott, U. (2008). *bicoid* occurrence and Bicoid-dependent *hunchback* regulation in lower cyclorrhaphan flies. *Evol. Dev.* 10, 413–420.

1276. Lemon, R.N. (2008). Descending pathways in motor control. *Annu. Rev. Neurosci.* 31, 195–218.

1277. Lemons, D., Fritzenwanker, J.H., Gerhart, J., Lowe, C.J., and McGinnis, W. (2010). Co-option of an anteroposterior head axis patterning system for proximodistal patterning of appendages in early bilaterian evolution. *Dev. Biol.* 344, 358–362.

1278. Lemons, D., and McGinnis, W. (2006). Genomic evolution of Hox gene clusters. *Science* 313, 1918–1922.

1279. Lengyel, J.A., and Iwaki, D.D. (2002). It takes guts: the *Drosophila* hindgut as a model system for organogenesis. *Dev. Biol.* 243, 1–19.

1280. Leondaritis, G., and Eickholt, B.J. (2015). Short lives with long-lasting effects: filopodia protrusions in neuronal branching morphogenesis. *PLoS Biol.* 13, #9, e1002241.

1281. Leroi, A.M. (2003). *Mutants: On Genetic Variety and the Human Body.* Viking Press, New York, NY.

1282. Lesser, M.P., Carleton, K.L., Böttger, S.A., Barry, T.M., and Walker, C.W. (2011). Sea urchin tube feet are photosenory organs that express a rhabdomeric-like opsin and PAX6. *Proc. R. Soc. Lond. B* 278, 3371–3379.

1283. Lettice, L.A., Williamson, I., Devenney, P.S., Kilanowski, F., Dorin, J., and Hill, R.E. (2014). Development of five digits is controlled by a bipartite long-range *cis*-regulator. *Development* 141, 1715–1725.

1284. Letzkus, P., Ribi, W.A., Wood, J.T., Zhu, H., Zhang, S.-W., and Srinivasan, M.V. (2006). Lateralization of olfaction in the honeybee *Apis mellifera. Curr. Biol.* 16, 1471–1476.

1285. Levin, M. (2005). Left-right asymmetry in embryonic development: a comprehensive review. *Mech. Dev.* 122, 3–25.

1286. Levin, M., Johnson, R.L., Stern, C.D., Kuehn, M., and Tabin, C. (1995). A molecular pathway determining left-right asymmetry in chick embryogenesis. *Cell* 82, 803–814.

1287. Levin, M., and Mercola, M. (1998). The compulsion of chirality: toward an understanding of left-right asymmetry. *Genes Dev.* 12, 763–769.

1288. Levin, M., and Palmer, A.R. (2007). Left-right patterning from the inside out: widespread evidence for intracellular control. *BioEssays* 29, 271–287.

1289. Levin, M., Roberts, D.J., Holmes, L.B., and Tabin, C. (1996). Laterality defects in conjoined twins. *Nature* 384, 321.

1290. Levin, M., Thorlin, T., Robinson, K.R., Nogi, T., and Mercola, M. (2002). Asymmetries in H^+/K^+-ATPase and cell membrane potentials comprise a very early step in left-right patterning. *Cell* 111, 77–89.

1291. Levine, J.S., and MacNichol, E.F., Jr. (1982). Color vision in fishes. *Sci. Am.* 246, #2, 140–149.

1292. Levine, M. (2010). Transcriptional enhancers in animal development and evolution. *Curr. Biol.* 20, R754–R763.

1293. Levine, M., Cattoglio, C., and Tijan, R. (2014). Looping back to leap forward: transcription enters a new era. *Cell* 157, 13–25.

1294. Levine, S.S., Weiss, A., Erdjument-Bromage, H., Shao, Z., Tempst, P., and Kingston, R.E. (2002). The core of the Polycomb repressive complex is compositionally and functionally conserved in flies and humans. *Mol. Cell. Biol.* 22, #17, 6070–6078.

1295. Levy, D.L., and Heald, R. (2012). Mechanisms of intracellular scaling. *Annu. Rev. Cell Dev. Biol.* 28, 113–135.

1296. Lewis, E.B. (1978). A gene complex controlling segmentation in *Drosophila. Nature* 276, 565–570.

1297. Lewis, E.B. (1994). Homeosis: the first 100 years. *Trends Genet.* 10, 341–343.

1298. Lewis, E.B., Pfeiffer, B.D., Mathog, D.R., and Celniker, S.E. (2003). Evolution of the homeobox complex in the Diptera. *Curr. Biol.* 13, R587–R588.

1299. Lewis, M.A., and Steel, K.P. (2012). A cornucopia of candidates for deafness. *Cell* 150, 879–881.

1300. Lewitzky, M., Simister, P.C., and Feller, S.M. (2012). Beyond "furballs" and "dumpling soups": towards a molecular architecture of signaling complexes and networks. *FEBS Lett.* 586, 2740–2750.

1301. Li, A., Xue, J., and Peterson, E.H. (2008). Architecture of the mouse utricle: macular organization and hair bundle heights. *J. Neurophysiol.* 99, 718–733.

1302. Li, L., and Ginty, D.D. (2014). The structure and organization of lanceolate mechanosensory complexes at mouse hair follicles. *eLife* 3, e01901.

1303. Li, L., Rutlin, M., Abraira, V.E., Cassidy, C., Kus, L., Gong, S., Jankowski, M.P., Luo, W., Heintz, N., Koerber, R., Woodbury, C.J., and Ginty, D.D. (2011). The functional organization of cutaneous low-threshold mechanosensory neurons. *Cell* 147, 1615–1627.

1304. Li, Q., Barish, S., Okuwa, S., Maciejewski, A., Brandt, A.T., Reinhold, D., Jones, C.D., and Volkan, P.C. (2016). A functionally conserved gene regulatory network module governing olfactory neuron diversity. *PLoS Genet.* 12, #1, e1005780.

1305. Li, Q., and Liberles, S.D. (2015). Aversion and attraction through olfaction. *Curr. Biol.* 25, R120–R129.

1306. Li, S., Sukeena, J.M., Simmons, A.B., Hansen, E.J., Nuhn, R.E., Samuels, I.S., and Fuerst, P.G. (2015). DSCAM promotes refinement in the mouse retina through cell death and restriction of exploring dendrites. *J. Neurosci.* 35, #14, 5640–5654.

1307. Liang, H.-L., Xu, M., Chuang, Y.-C., and Rushlow, C. (2012). Response to the BMP gradient requires highly combinatorial inputs from multiple patterning systems in the *Drosophila* embryo. *Development* 139, 1956–1964.

1308. Liang, X., Holy, T.E., and Taghert, P.H. (2016). Synchronous *Drosophila* circadian pacemakers display nonsynchronous Ca^{2+} rhythms in vivo. *Science* 351, 976–981.

1309. Liang, Z., and Biggin, M.D. (1998). Eve and ftz regulate a wide array of genes in blastoderm embryos: the selector homeoproteins directly or indirectly regulate most genes in *Drosophila*. *Development* 125, 4471–4482.

1310. Liberles, S.D. (2014). Mammalian pheromones. *Annu. Rev. Physiol.* 76, 151–175.

1311. Liberles, S.D., and Buck, L.B. (2006). A second class of chemosensory receptors in the olfactory epithelium. *Nature* 442, 645–650.

1312. Liberman, M.C., Gao, J., He, D.Z.Z., Wu, X., Jia, S., and Zuo, J. (2002). Prestin is required for electromotility of the outer hair cell and for the cochlear amplifier. *Nature* 419, 300–304.

1313. Lichtneckert, R., and Reichert, H. (2005). Insights into the urbilaterian brain: conserved genetic patterning mechanisms in insect and vertebrate brain development. *Heredity* 94, 465–477.

1314. Lienkamp, S., Ganner, A., and Walz, G. (2012). Inversin, Wnt signaling and primary cilia. *Differentiation* 83, S49–S55.

1315. Lilly, B. (2014). We have contact: endothelial cell–smooth muscle cell interactions. *Physiology* 29, 234–241.

1316. Lillywhite, P.G. (1980). The insect's compound eye. *Trends Neurosci.* 3, 169–173.

1317. Lin, C.H., and Rankin, C.H. (2012). Alcohol addiction: chronic ethanol leads to cognitive dependence in *Drosophila*. *Curr. Biol.* 22, R1043–R1044.

1318. Lin, C.Y., Chuang, C.C., Hua, T.E., Chen, C.C., Dickson, B.J., Greenspan, R.J., and Chiang, A.S. (2013). A comprehensive wiring diagram of the protocerebral bridge for visual information processing in the *Drosophila* brain. *Cell Rep.* 3, 1739–1753.

1319. Lin, G., Chen, Y., and Slack, J.M.W. (2013). Imparting regenerative capacity to limbs by progenitor cell transplantation. *Dev. Cell* 24, 41–51.

1320. Lin, L., Rao, Y., and Isacson, O. (2005). Netrin-1 and slit-2 regulate and direct neurite growth of ventral midbrain dopaminergic neurons. *Mol. Cell. Neurosci.* 28, 547–555.

1321. Lin, R., Rittenhouse, D., Sweeney, K., Potluri, P., and Wallace, D.C. (2015). TSPO, a mitochondrial outer membrane protein, controls ethanol-related behaviors in *Drosophila*. *PLoS Genet.* 11, #8, e1005366.

1322. Lin, S.-Y., and Burdine, R.D. (2005). Brain asymmetry: switching from left to right. *Curr. Biol.* 15, R343–R345.

1323. Lin, Y.Y., and Gubb, D. (2009). Molecular dissection of *Drosophila* Prickle isoforms distinguishes their essential and overlapping roles in planar cell polarity. *Dev. Biol.* 325, 386–399.

1324. Lindeman, R.E., Gearhart, M.D., Minkina, A., Krentz, A.D., Bardwell, V.J., and Zarkower, D. (2015). Sexual cell-fate reprogramming in the ovary by DMRT1. *Curr. Biol.* 25, 764–771.

1325. Lindemann, B. (2001). Receptors and transduction in taste. *Nature* 413, 219–225.

1326. Lindsey, D.T., Brown, A.M., Brainard, D.H., and Apicella, C.L. (2015). Hunter-gatherer color naming provides new insight into the evolution of color terms. *Curr. Biol.* 25, 2441–2446.

1327. Lindsley, D.L., and Zimm, G.G. (1992). *The Genome of Drosophila melanogaster.* Academic Press, New York, NY.

1328. Ling, F., Dahanukar, A., Weiss, L.A., Kwon, J.Y., and Carlson, J.R. (2014). The molecular and cellular basis of taste coding in the legs of *Drosophila*. *J. Neurosci.* 34, 7148–7164.

1329. Linnaeus, C. (1758–59). *Systema naturae per regna tria naturae: secundum classes, ordines, genera, species, cum characteribus, differentiis, synonymis, locis.*, 10th edn. Laurentius Salvius, Stockholm.

1330. Linneweber, G.A., Winking, M., and Fischbach, K.-F. (2015). The cell adhesion molecules *Roughest*, *Hibris*, *Kin of Irre* and *Sticks and Stones* are required for long range spacing of the *Drosophila* wing disc sensory sensilla. *PLoS ONE* 10, #6, e0128490.

1331. Liu, G., Li, W., Wang, L., Kar, A., Guan, K.-L., Rao, Y., and Wu, J.Y. (2009). DSCAM functions as a netrin receptor in commissural axon pathfinding. *PNAS* 106, #8, 2951–2956.

1332. Liu, P.Z., and Kaufman, T.C. (2005). Short and long germ segmentation: unanswered questions in the evolution of a developmental mode. *Evol. Dev.* 7, 629–646. [*See also* Mao, Q., and Lecuit, T. (2016). Evo-devo: universal Toll pass for the extension highway? *Curr. Biol.* 26, R680–R683.]

1333. Liu, Q.-X., Hiramoto, M., Ueda, H., Gojobori, T., Hiromi, Y., and Hirose, S. (2009). Midline governs axon pathfinding by coordinating expression of two major guidance systems. *Genes Dev.* 23, 1165–1170.

1334. Liu, Z., Li, G.-H., Huang, J.-F., Murphy, R.W., and Shi, P. (2012). Hearing aid for vertebrates via multiple episodic adaptive events on prestin genes. *Mol. Biol. Evol.* 29, #9, 2187–2198.

1335. Liu, Z., Qi, F.-Y., Zhou, X., Ren, H.-Q., and Shi, P. (2014). Parallel sites implicate functional convergence of the hearing gene *Prestin* among echolocating mammals. *Mol. Biol. Evol.* 31, #9, 2415–2424.

1336. Liu, Z., Yang, C.-P., Sugino, K., Fu, C.-C., Liu, L.-Y., Yao, X., Lee, L.P., and Lee, T. (2015). Opposing intrinsic temporal gradients guide neural stem cell production of varied neuronal fates. *Science* 350, 317–320.

1337. Livesey, F.J., and Cepko, C.L. (2001). Vertebrate neural cell-fate determination: lessons from the retina. *Nat. Rev. Neurosci.* 2, 109–118.

1338. Lo Celso, C., Prowse, D.M., and Watt, F.M. (2004). Transient activation of β-catenin signalling in adult mouse epidermis is sufficient to induce new hair follicles but continuous activation is required to maintain hair follicle tumours. *Development* 131, 1787–1799.

1339. Lodato, S., and Arlotta, P. (2015). Generating neuronal diversity in the mammalian cerebral cortex. *Annu. Rev. Cell Dev. Biol.* 31, 699–720.

1340. Logan, M.A., and Vetter, M.L. (2004). Do-it-yourself tiling: dendritic growth in the absence of homotypic contacts. *Neuron* 43, 439–440.

1341. Lohnes, D. (2003). The Cdx1 homeodomain protein: an integrator of posterior signaling in the mouse. *BioEssays* 25, 971–980.

1342. Löhr, U., and Pick, L. (2005). Cofactor-interaction motifs and the cooption of a homeotic Hox protein into the segmentation pathway of *Drosophila melanogaster*. *Curr. Biol.* 15, 643–649.

1343. Long, H., Sabatier, C., Ma, L., Plump, A., Yuan, W., Ornitz, D.M., Tamada, A., Murakami, F., Goodman, C.S., and Tessier-Lavigne, M. (2004). Conserved roles for Slit and Robo proteins in midline commissural axon guidance. *Neuron* 42, 213–223.

1344. Longley, R.L., Jr., and Ready, D.F. (1995). Integrins and the development of three-dimensional structure in the *Drosophila* compound eye. *Dev. Biol.* 171, 415–433.

1345. Longobardi, L., Li, T., Tagliafierro, L., Temple, J.D., Willcockson, H.H., Ye, P., Esposito, A., Xu, F., and Spagnoli, A. (2015). Synovial joints: from development to homeostasis. *Curr. Osteoporos. Rep.* 13, 41–51.

1346. Looby, P., and Loudon, A.S.I. (2005). Gene duplication and complex circadian clocks in mammals. *Trends Genet.* 12, 46–53.

1347. Lopes, C.S., and Casares, F. (2010). *hth* maintains the pool of eye progenitors and its downregulation by Dpp and Hh couples retinal fate acquisition with cell cycle exit. *Dev. Biol.* 339, 78–88.

1348. Lopez-Rios, J. (2016). The many lives of SHH in limb development and evolution. *Semin. Cell Dev. Biol.* 49, 116–124.

1349. Lorimer, T., Gomez, F., and Stoop, R. (2015). Mammalian cochlea as a physics guided evolution-optimized hearing sensor. *Sci. Rep.* 5, Article 12492.

1350. Losos, J.B. (2011). Convergence, adaptation, and constraint. *Evolution* 65, 1827–1840.

1351. Loudon, A.S.I. (2012). Circadian biology: a 2.5 billion year old clock. *Curr. Biol.* 22, R570–R571.

1352. Lovato, T.L., and Cripps, R.M. (2016). Regulatory networks that direct the development of specialized cell types in the *Drosophila* heart. *J. Cardiovasc. Dev. Dis.* 3, Article 18.

1353. Lovato, T.L., Sensibaugh, C.A., Swingle, K.L., Martinez, M.M., and Cripps, R.M. (2015). The *Drosophila* transcription factors Tinman and Pannier activate and collaborate with Myocyte Enhancer Factor-2 to promote heart cell fate. *PLoS ONE* 10, #7, e0132965.

1354. Lovejoy, N.R. (2000). Reinterpreting recapitulation: systematics of needlefishes and their allies (Teleostei: Beloniformes). *Evolution* 54, 1349–1362.

1355. Lowe, C.J., Clarke, D.N., Medeiros, D.M., Rokhsar, D.S., and Gerhart, J. (2015). The deuterostome context of chordate origins. *Nature* 520, 456–465.

1356. Lowe, C.J., Terasaki, M., Wu, M., Freeman, R.M., Jr., Runft, L., Kwan, K., Haigo, S., Aronowicz, J., Lander, E., Gruber, C., Smith, M., Kirschner, M., and Gerhart, J. (2006). Dorsoventral patterning in hemichordates: insights into early chordate evolution. *PLoS Biol.* 4, #9, e1291.

1357. Lowe, C.J., Wu, M., Salic, A., Evans, L., Lander, E., Stange-Thomann, N., Gruber, C.E., Gerhart, J., and Kirschner, M. (2003). Anteroposterior patterning in hemichordates and the origins of the chordate nervous system. *Cell* 113, 853–865.

1358. Lowe, L.A., Supp, D.M., Sampath, K., Yokoyama, T., Wright, C.V.E., Potter, S.S., Overbeek, P., and Kuehn, M.R. (1996). Conserved left-right asymmetry of nodal expression and alterations in murine *situs inversus. Nature* 381, 158–161.

1359. Lowery, L.A., and Sive, H. (2004). Strategies of vertebrate neurulation and a re-evaluation of teleost neural tube formation. *Mech. Dev.* 121, 1189–1197.

1360. Lowery, L.A., and Sive, H. (2009). Totally tubular: the mystery behind function and origin of the brain ventricular system. *BioEssays* 31, 446–458.

1361. Lu, C.-H., Rincón-Limas, D.E., and Botas, J. (2000). Conserved overlapping and reciprocal expression of *msh/Msx1* and *apterous/Lhx2* in *Drosophila* and mice. *Mech. Dev.* 99, 177–181.

1362. Lu, Q., Senthilan, P.R., Effertz, T., Nadrowski, B., and Göpfert, M.C. (2009). Using *Drosophila* for studying fundamental processes in hearing. *Integr. Comp. Biol.* 49, 674–680.

1363. Lu, X., and Sipe, C.W. (2016). Developmental regulation of planar cell polarity and hair-bundle morphogenesis in auditory hair cells: lessons from human and mouse genetics. *WIREs Dev. Biol.* 5, 85–101.

1364. Lucas, R.J. (2013). Mammalian inner retinal photoreception. *Curr. Biol.* 23, R125–R133.

1365. Luke, G.N., Castro, L.F.C., McLay, K., Bird, C., Coulson, A., and Holland, P.W.H. (2003). Dispersal of NK homeobox gene clusters in amphioxus and humans. *PNAS* 100, #9, 5292–5295.

1366. Lumpkin, E.A., Marshall, K.L., and Nelson, A.M. (2010). The cell biology of touch. *J. Cell Biol.* 191, 237–248.

1367. Lundberg, Y.W., Xu, Y., Thiessen, K.D., and Kramer, K.L. (2015). Mechanisms of otoconia and otolith development. *Dev. Dyn.* 244, 239–253.

1368. Luo, D.-G., and Yau, K.-W. (2008). How vision begins: an odyssey. *PNAS* 105, #29, 9855–9862.

1369. Luo, D.-G., Yue, W.W.S., Ala-Laurila, P., and Yau, K.-W. (2011). Activation of visual pigments by light and heat. *Science* 332, 1307–1312.

1370. Lynch, J., and Desplan, C. (2003). Evolution of development: beyond Bicoid. *Curr. Biol.* 13, R557–R559.

1371. Lynch, J.A., Brent, A.E., Leaf, D.S., Pultz, M.A., and Desplan, C. (2006). Localized maternal *orthodenticle* patterns anterior and posterior in the long germ wasp *Nasonia. Nature* 439, 728–732.

1372. Lynch, J.A., and Roth, S. (2011). The evolution of dorsal-ventral patterning mechanisms in insects. *Genes Dev.* 25, 107–118.

1373. Lynch, M. (2007). *The Origins of Genome Architecture.* Sinauer, Sunderland, MA.

1374. Lyons, D.B., Allen, W.E., Goh, T., Tsai, L., Barnea, G., and Lomvardas, S. (2013). An epigenetic trap stabilizes singular olfactory receptor expression. *Cell* 154, 325–336.

1375. Ma, J.-x., Znoiko, S., Othersen, K.L., Ryan, J.C., Das, J., Isayama, T., Kono, M., Oprian, D.D., Corson, D.W., Cornwall, M.C., Cameron, D.A., Harosi, F.I., Makino, C.L., and Crouch, R.K. (2001). A visual pigment expressed in both rod and cone photoreceptors. *Neuron* 32, 451–461.

1376. Ma, X., Edgecombe, G.D., Hou, X., Goral, T., and Strausfeld, N.J. (2015). Preservation pathways of corresponding brains of a Cambrian euarthropod. *Curr. Biol.* 25, 2969–2975.

1377. Macdonald, P.M., and Struhl, G. (1986). A molecular gradient in early *Drosophila* embryos and its role in specifying the body pattern. *Nature* 324, 537–545.

1378. Mackay, T.F.C., and Anholt, R.R.H. (2006). Of flies and man: *Drosophila* as a model for human complex traits. *Annu. Rev. Genomics Hum. Genet.* 7, 339–367.

1379. Mackin, K.A., Roy, R.A., and Theobald, D.L. (2014). An empirical test of convergent evolution in rhodopsins. *Mol. Biol. Evol.* 31, 85–95.

1380. MacLean, J.A., II, Lorenzetti, D., Hu, Z., Salerno, W.J., Miller, J., and Wilkinson, M.F. (2006). *Rhox* homeobox gene cluster: recent duplication of three family members. *Genesis* 44, 122–129.

1381. MacNeilage, P.F., Rogers, L.J., and Vallortigara, G. (2009). Origins of the left and right brain. *Sci. Am.* 301, #1, 60–67.

1382. Maderspacher, F. (2016). Snail chirality: the unwinding. *Curr. Biol.* 26, R215–R217.

1383. Maeso, I., Irimia, M., Tena, J.J., Casares, F., and Gómez-Skarmeta, J.L. (2013). Deep conservation of *cis*-regulatory elements in metazoans. *Philos. Trans. R. Soc. Lond. B* 368, 20130020.

1384. Maeso, I., Irimia, M., Tena, J.J., González-Pérez, E., Iran, D., Ravi, V., Venkatesh, B., Campuzano, S., Gómez-Skarmeta, J.L., and Garcia-Fernàndez, J. (2012). An ancient genomic regulatory block conserved across bilaterians and its dismantling in tetrapods by retrogene replacement. *Genome Res.* 22, 642–655.

1385. Magariños, M., Contreras, J., Aburto, M.R., and Varela-Nieto, I. (2012). Early development of the vertebrate inner ear. *Anat. Rec.* 295, 1775–1790.

1386. Maier, E.C., Saxena, A., Alsina, B., Bronner, M.E., and Whitfield, T.T. (2014). Sensational placodes: neurogenesis in the otic and olfactory systems. *Dev. Biol.* 389, 50–67.

1387. Maier, J.X., Blankenship, M.L., Li, J.X., and Katz, D.B. (2015). A multisensory network for olfactory processing. *Curr. Biol.* 25, 2642–2650.

1388. Mainguy, G., In der Rieden, P.M.J., Berezikov, E., Woltering, J.M., Plasterk, R.H.A., and Durston, A.J. (2003). A position-dependent organisation of retinoid response elements is conserved in the vertebrate *Hox* clusters. *Trends Genet.* 19, 476–479.

1389. Maizel, A., Bensaude, O., Prochiantz, A., and Joliot, A. (1999). A short region of its homeodomain is necessary for Engrailed nuclear export and secretion. *Development* 126, 3183–3190.

1390. Makino, C.L., Riley, C.K., Looney, J., Crouch, R.K., and Okada, T. (2010). Binding of more than one retinoid to visual opsins. *Biophys. J.* 99, 2366–2373.

1391. Maklad, A., Kamel, S., Wong, E., and Fritzsch, B. (2010). Development and organization of polarity-specific segregation of primary vestibular afferent fibers in mice. *Cell Tissue Res.* 340, 303–321.

1392. Maklad, A., Reed, C., Johnson, N.S., and Fritzsch, B. (2014). Anatomy of the lamprey ear: morphological evidence for occurrence of horizontal semicircular ducts in the labyrinth of *Petromyzon marinus*. *J. Anat.* 224, 432–446.

1393. Malicki, J. (2004). Cell fate decisions and patterning in the vertebrate retina: the importance of timing, asymmetry, polarity and waves. *Curr. Opin. Neurobiol.* 14, 15–21.

1394. Malik, S. (2014). Polydactyly: phenotypes, genetics and classification. *Clin. Genet.* 85, 203–212.

1395. Mallo, M., and Alonso, C.R. (2013). The regulation of Hox gene expression during animal development. *Development* 140, 3951–3963.

1396. Mallo, M., Wellik, D.M., and Deschamps, J. (2010). *Hox* genes and regional patterning of the vertebrate body plan. *Dev. Biol.* 344, 7–15.

1397. Malnic, B., Hirono, J., Sato, T., and Buck, L.B. (1999). Combinatorial receptor codes for odors. *Cell* 96, 713–723.

1398. Mandal, L., Banerjee, U., and Hartenstein, V. (2004). Evidence for a fruit fly hemangioblast and similarities between lymph-gland hematopoiesis in fruit fly and mammal aorta-gonadal-mesonephros mesoderm. *Nat. Genet.* 36, 1019–1023.

1399. Manfroid, I., Caubit, X., Kerridge, S., and Fasano, L. (2003). Three putative murine Teashirt orthologs specify trunk structures in *Drosophila* in the same way as the *Drosophila teashirt* gene. *Development* 131, 1065–1073.

1400. Manjón, C., Sánchez-Herrero, E., and Suzanne, M. (2007). Sharp boundaries of Dpp signalling trigger local cell death required for *Drosophila* leg morphogenesis. *Nature Cell Biol.* 9, 57–63.

1401. Mann, R.S., and Abu-Shaar, M. (1996). Nuclear import of the homeodomain protein Extradenticle in response to Wg and Dpp signalling. *Nature* 383, 630–633.

1402. Mann, R.S., and Carroll, S.B. (2002). Molecular mechanisms of selector gene function and evolution. *Curr. Opin. Genet. Dev.* 12, 592–600.

1403. Männer, J., Wessel, A., and Yelbuz, T.M. (2010). How does the tubular embryonic heart work? Looking for the physical mechanism generating unidirectional blood flow in the valveless embryonic heart tube. *Dev. Dyn.* 239, 1035–1046.

1404. Mannschreck, A., and von Angerer, E. (2011). The scent of roses and beyond: molecular structures, analysis, and practical applications of odorants. *J. Chem. Educ.* 88, 1501–1506.

1405. Mano, H., and Fukada, Y. (2007). A median third eye: pineal gland retraces evolution of vertebrate photoreceptive organs. *Photochem. Photobiol.* 83, 11–18.

1406. Manor, U., Disanza, A., Grati, M.H., Andrade, L., Lin, H., Di Fiore, P.P., Scita, G., and Kachar, B. (2011). Regulation of stereocilia length by Myosin XVa and whirlin depends on the actin-regulatory protein Eps8. *Curr. Biol.* 21, 167–172.

1407. Manoussaki, D., Chadwick, R.S., Ketten, D.R., Arruda, J., Dimitriadis, E.K., and O'Malley, J.T. (2008). The influence of cochlear shape on low-frequency hearing. *PNAS* 105, #16, 6162–6166.

1408. Manoussaki, D., Dimitriadis, E.K., and Chadwick, R.S. (2006). Cochlea's graded curvature effect on low frequency waves. *Phys. Rev. Lett.* 96, Article 088701.

1409. Mansfield, J.H., and McGlinn, E. (2012). Evolution, expression, and developmental function of Hox-embedded miRNAs. *Curr. Top. Dev. Biol.* 99, 31–57.

1410. Manuel, M. (2009). Early evolution of symmetry and polarity in metazoan body plans. *C. R. Biol.* 332, 184–209.

1411. Marcotti, W. (2012). Functional assembly of mammalian cochlear hair cells. *Exp. Physiol.* 97, 438–451.

1412. Marcus, G., Marblestone, A., and Dean, T. (2014). The atoms of neural computation. *Science* 346, 551–552.

1413. Marcus, G.F. (2006). Cognitive architecture and descent with modification. *Cognition* 101, 443–465.

1414. Maricich, S.M., and Zoghbi, H.Y. (2006). Getting back to basics. *Cell* 126, 11–15.

1415. Marie, B., and Blagburn, J.M. (2003). Differential roles of Engrailed paralogs in determining sensory axon guidance and synaptic target recognition. *J. Neurosci.* 23, #21, 7854–7862.

1416. Marlétaz, F., Holland, L.Z., Laudet, V., and Schubert, M. (2006). Retinoic acid signaling and the evolution of chordates. *Int. J. Biol. Sci.* 2, 38–47.

1417. Marlow, H., and Arendt, D. (2014). Ctenophore genomes and the origin of neurons. *Curr. Biol.* 24, R757–R761.

1418. Marlow, H., Tosches, M.A., Tomer, R., Steinmetz, P.R., Lauri, A., Larsson, T., and Arendt, D. (2014). Larval body patterning and apical organs are conserved in animal evolution. *BMC Biol.* 12, Article 7.

1419. Marlow, H.Q., Srivastava, M., Matus, D.Q., Rokhsar, D., and Martindale, M.Q. (2009). Anatomy and development of the nervous system of *Nematostella vectensis*, an anthozoan cnidarian. *Dev. Neurobiol.* 69, 235–254.

1420. Marmor, M.F., Choi, S.S., Zawadzki, R.J., and Werner, J.S. (2008). Visual insignificance of the foveal pit. *Arch. Ophthalmol.* 126, 907–913.

1421. Marquardt, T., Ashery-Padan, R., Andrejewski, N., Scardigli, R., Guillemot, F., and Gruss, P. (2001). Pax6 is required for the multipotent state of retinal progenitor cells. *Cell* 105, 43–55.

1422. Marques-Souza, H., Aranda, M., and Tautz, D. (2008). Delimiting the conserved features of *hunchback* function for the trunk organization of insects. *Development* 135, 881–888.

1423. Marshall, C.R., Raff, E.C., and Raff, R.A. (1994). Dollo's law and the death and resurrection of genes. *PNAS* 91, 12283–12287.

1424. Marshall, J., and Arikawa, K. (2014). Unconventional colour vision. *Curr. Biol.* 24, R1150–R1154.

1425. Marshall, K.L., and Lumpkin, E.A. (2012). The molecular basis of mechanosensory transduction. *In* C. López-Larrea (ed.), *Sensing in Nature.* Landes Bioscience, Austin, TX, pp. 142–155.

1426. Marshall, W.F. (2008). Basal bodies: platforms for building cilia. *Curr. Top. Dev. Biol.* 85, 1–22.

1427. Marshall, W.F. (2015). How cells measure length on subcellular scales. *Trends Cell Biol.* 25, 760–768.

1428. Marshall, W.F., and Nonaka, S. (2006). Cilia: tuning in to the cell's antenna. *Curr. Biol.* 16, R604–R614.

1429. Martin, A., and Orgogozo, V. (2013). The loci of repeated evolution: a catalog of genetic hotspots of phenotypic variation. *Evolution* 67, 1235–1250.

1430. Martin, A., Serano, J.M., Jarvis, E., Bruce, H.S., Wang, J., Ray, S., Barker, C.A., O'Connell, L.C., and Patel, N.H. (2016). CRISPR/Cas9 mutagenesis reveals versatile roles of Hox genes in crustacean limb specification and evolution. *Curr. Biol.* 26, 14–26.

1431. Martin, B.L., and Kimelman, D. (2009). Wnt signaling and the evolution of embryonic posterior development. *Curr. Biol.* 19, R215–R219.

1432. Martin, J.H. (2005). The corticospinal system: from development to motor control. *Neuroscientist* 11, 161–173.

1433. Martin, J.P., Guo, P., Mu, L., Harley, C.M., and Ritzmann, R.E. (2015). Central-complex control of movement in the freely walking cockroach. *Curr. Biol.* 25, 2795–2803.

1434. Martin, J.P., and Hildebrand, J.G. (2010). Innate recognition of pheromone and food odors in moths: a common mechanism in the antennal lobe? *Front. Behav. Neurosci.* 4, Article 159.

1435. Martindale, M.Q., and Hejnol, A. (2009). A developmental perspective: changes in the position of the blastopore during bilaterian evolution. *Dev. Cell* 17, 162–174.

1436. Martinez, S.R., Gay, M.S., and Zhang, L. (2015). Epigenetic mechanisms in heart development and disease. *Drug Discov. Today* 20, 799–811.

1437. Martini, F.H., Ober, W.C., Garrison, C.W., Welch, K., Hutchings, R.T., and Ireland, K. (2004). *Fundamentals of Anatomy and Physiology*, 6th edn. Benjamin Cummings, San Francisco, CA.

1438. Mashanov, V.S., Zueva, O.R., Heinzeller, T., Aschauer, B., and Dolmatov, I.Y. (2007). Developmental origin of the adult nervous system in a holothurian: an attempt to unravel the enigma of neurogenesis in echinoderms. *Evol. Dev.* 9, 244–256.

1439. Mason, A.C., Oshinsky, M.L., and Hoy, R.R. (2011). Hyperacute directional hearing in a microscale auditory system. *Nature* 410, 686–690.

1440. Masse, N.Y., Turner, G.C., and Jefferis, G.S.X.E. (2009). Olfactory information processing in *Drosophila*. *Curr. Biol.* 19, R700–R713.

1441. Matis, M., Russler-Germain, D.A., Hu, Q., Tomlin, C.J., and Axelrod, J. (2014). Microtubules provide directional information for core PCP function. *eLife* 3, e02893.

1442. Matsui, A., Go, Y., and Niimura, Y. (2010). Degeneration of olfactory receptor gene repertoires in primates: no direct link to full trichromatic vision. *Mol. Biol. Evol.* 27, 1192–1200.

1443. Matsunami, M., Sumiyama, K., and Saitou, N. (2010). Evolution of conserved non-coding sequences within the vertebrate Hox clusters through the two-round whole genome duplications revealed by phylogenetic footprinting analysis. *J. Mol. Evol.* 71, 427–436.

1444. Matsuo, E., and Kamikouchi, A. (2013). Neuronal encoding of sound, gravity, and wind in the fruit fly. *J. Comp. Physiol. A* 199, 253–262.

1445. Matthews, B.J., Kim, M.E., Flanagan, J.J., Hattori, D., Clemens, J.C., Zipursky, S.L., and Grueber, W.B. (2007). Dendrite self-avoidance is controlled by *Dscam*. *Cell* 129, 593–604.

1446. Maung, S.M.T.W., and Jenny, A. (2011). Planar cell polarity in *Drosophila*. *Organogenesis* 7, 1–15.

1447. Maurange, C., and Gould, A.P. (2005). Brainy but not too brainy: starting and stopping neuroblast divisions in *Drosophila*. *Trends Neurosci.* 28, 30–36.

1448. May-Simera, H., and Kelley, M.W. (2012). Planar cell polarity in the inner ear. *Curr. Top. Dev. Biol.* 101, 111–140.

1449. Mayer, G. (2006). Structure and development of onychophoran eyes: what is the ancestral visual organ in arthropods? *Arthropod Struct. Dev.* 35, 231–245.

1450. Mayer, G., Kato, C., Quast, B., Chisholm, R.H., Landman, K.A., and Quinn, L.M. (2010). Growth patterns in Onychophora (velvet worms): lack of a localised posterior proliferation zone. *BMC Evol. Biol.* 10, Article 339.

1451. Mayer, G., and Whitington, P.M. (2009). Neural development in Onychophora (velvet worms) suggests a step-wise evolution of segmentation in the nervous system of Panarthropoda. *Dev. Biol.* 335, 263–275.

1452. Maynard Smith, J. (1968). The counting problem. *In* C.H. Waddington (ed.), *Towards a Theoretical Biology. I. Prolegomena*. Aldine, Chicago, IL, pp. 120–124.

1453. Mazzoni, E.O., Celik, A., Wernet, M.F., Vasiliauskas, D., Johnston, R.J., Cook, T.A., Pichaud, F., and Desplan, C. (2008). *Iroquois Complex* genes induce co-expression of rhodopsins in *Drosophila*. *PLoS Biol.* 6, #4, e97.

1454. McAlpine, J.F. (1981). Morphology and terminology: adults. *In* J.F. McAlpine, B.V. Peterson, G.E. Shewell, H.J. Teskey, J.R. Vockeroth, and D.M. Wood (eds.), *Manual of Nearctic Diptera*. Agriculture Canada, Ottawa, pp. 9–63.

1455. McCabe, K.L., Gunther, E.C., and Reh, T.A. (1999). The development of the pattern of retinal ganglion cells in the chick retina: mechanisms that control differentiation. *Development* 126, 5713–5724.

1456. McDevitt, D.S., Brahma, S.K., Jeanny, J.-C., and Hicks, D. (1993). Presence and foveal enrichment of rod opsin in the "all cone" retina of the American chameleon. *Anat. Rec.* 237, 299–307.

1457. McDougall, C., Korchagina, N., Tobin, J.L., and Ferrier, D.E.K. (2011). Annelid Distal-less/Dlx duplications reveal varied post-duplication fates. *BMC Evol. Biol.* 11, Article 241.

1458. McGary, K.L., Park, T.J., Woods, J.O., Cha, H.J., Wallingford, J.B., and Marcotte, E.M. (2010). Systematic discovery of nonobvious human disease models through orthologous phenotypes. *PNAS* 107, #14, 6544–6549.

1459. McGhee, G.R., Jr. (2011). *Convergent Evolution: Limited Forms Most Beautiful.* MIT Press, Cambridge, MA.

1460. McGinnis, W. (1994). A century of homeosis, a decade of homeoboxes. *Genetics* 137, 607–611.

1461. McGinnis, W., and Krumlauf, R. (1992). Homeobox genes and axial patterning. *Cell* 68, 283–302.

1462. McGinnis, W., and Kuziora, M. (1994). The molecular architects of body design. *Sci. Am.* 270, #2, 58–66.

1463. McGinnis, W., Levine, M.S., Hafen, E., Kuroiwa, A., and Gehring, W.J. (1984). A conserved DNA sequence in homoeotic genes of the *Drosophila* Antennapedia and bithorax complexes. *Nature* 308, 428–433.

1464. McGregor, A.P. (2005). How to get ahead: the origin, evolution and function of *bicoid*. *BioEssays* 27, 904–913.

1465. McGregor, A.P. (2006). Wasps, beetles and the beginning of the ends. *BioEssays* 28, 683–686.

1466. McGurk, L., Berson, A., and Bonini, N.M. (2015). *Drosophila* as an *in vivo* model for human neurodegenerative disease. *Genetics* 201, 377–402.

1467. McHenry, M.J. (2005). The morphology, behavior, and biomechanics of swimming in ascidian larvae. *Can. J. Zool.* 83, 62–74.

1468. McInerney, J.O., and O'Connell, M.J. (2014). Ghost locus appears. *Nature* 514, 570–571.

1469. McKinney, R.M., Vernier, C., and Ben-Shahar, Y. (2015). The neural basis for insect pheromonal communication. *Curr. Opin. Insect Sci.* 12, 86–92.

1470. McManus, C. (2005). Reversed bodies, reversed brains, and (some) reversed behaviors: of zebrafish and men. *Dev. Cell* 8, 796–797.

1471. McManus, I.C., Martin, N., Stubbings, G.F., Chung, E.M.K., and Mitchison, H.M. (2004). Handedness and *situs inversus* in primary ciliary dyskinesia. *Proc. R. Soc. Lond. B* 271, 2579–2582.

1472. McMillen, P., and Holley, S.A. (2015). The tissue mechanics of vertebrate body elongation and segmentation. *Curr. Opin. Genet. Dev.* 32, 106–111.

1473. Meadows, R. (2015). Odors help fruit flies escape parasitoid wasps. *PLoS Biol.* 13, #12, e1002317.

1474. Medioni, C., Sénatore, S., Salmand, P.-A., Lalevée, N., Perrin, L., and Sémériva, M. (2009). The fabulous destiny of the *Drosophila* heart. *Curr. Opin. Genet. Dev.* 19, 518–525.

1475. Meganathan, K., Sotiriadou, I., Natarajan, K., Hescheler, J., and Sachinidis, A. (2015). Signaling molecules, transcription growth factors and other regulators revealed from in-vivo and in-vitro models for the regulation of cardiac development. *Int. J. Cardiol.* 183, 117–128.

1476. Mehta, T.K., Ravi, V., Yamasaki, S., Lee, A.P., Lian, M.M., Tay, B.-H., Tohari, S., Yanai, S., Tay, A., Brenner, S., and Venkatesh, B. (2013). Evidence for at least six Hox clusters in the Japanese lamprey (*Lethenteron japonicum*). *PNAS* 110, 16044–16049.

1477. Meier, T., Chabaud, F., and Reichert, H. (1991). Homologous patterns in the embryonic development of the peripheral nervous system in the grasshopper *Schistocerca gregaria* and the fly *Drosophila melanogaster*. *Development* 112, 241–253.

1478. Meijers, R., Puettmann-Holgado, R., Skiniotis, G., Liu, J.-h., Walz, T., Wang, J.-h., and Schmucker, D. (2007). Structural basis of Dscam isoform specificity. *Nature* 449, 487–491.

1479. Meinhardt, H. (2015). Dorsoventral patterning by the Chordin-BMP pathway: a unified model from a pattern-formation perspective for *Drosophila*, vertebrates, sea urchins and *Nematostella*. *Dev. Biol.* 405, 137–148.

1480. Meister, M. (2015). On the dimensionality of odor space. *eLife* 4, e07865.

1481. Melnattur, K., and Shaw, P.J. (2015). Learning and memory: do bees dream? *Curr. Biol.* 25, R1040–R1041.

1482. Menda, G., Shamble, P.S., Nitzany, E.I., Golden, J.R., and Hoy, R.R. (2014). Visual perception in the brain of a jumping spider. *Curr. Biol.* 24, 2580–2585.

1483. Menzel, R., Greggers, U., Smith, A., Berger, S., Brandt, R., Brunke, S., Bundrock, G., Hülse, S., Plümpe, T., Schaupp, F., Schüttler, E., Stach, S., Stindt, J., Stollhoff, N., and Watzl, S. (2005). Honey bees navigate according to a map-like spatial memory. *PNAS* 102, #8, 3040–3045.

1484. Merabet, S., and Mann, R.S. (2016). To be specific or not: the critical relationship between Hox and TALE proteins. *Trends Genet.* 32, 334–347.

1485. Meraldi, P. (2016). Centrosomes in spindle organization and chromosome segregation: a mechanistic view. *Chromosome Res.* 24, 19–34.

1486. Mercader, N., Leonardo, E., Azpiazu, N., Serrano, A., Morata, G., Martínez-A, C., and Torres, M. (1999). Conserved regulation of proximodistal limb axis development by Meis1/Hth. *Nature* 402, 425–429.

1487. Meredith, R.W., Gatesy, J., Emerling, C.A., York, V.M., and Springer, M.S. (2013). Rod monochromacy and the coevolution of cetacean retinal opsins. *PLoS Genet.* 9, #4, e1003432.

1488. Merkel, M., Sagner, A., Gruber, F.S., Etournay, R., Blasse, C., Myers, E., Eaton, S., and Jülicher, F. (2014). The balance of Prickle/Spiny-legs isoforms controls the amount of coupling between core and Fat PCP systems. *Curr. Biol.* 24, 2111–2123.

1489. Merriam, L.C., and Chess, A. (2007). *cis*-regulatory elements within the odorant receptor coding region. *Cell* 131, 844–846.

1490. Meunier, H., Vauclair, J., and Fagard, J. (2012). Human infants and baboons show the same pattern of handedness for a communicative gesture. *PLoS ONE* 7, #3, e33959.

1491. Mhatre, N. (2015). Active amplification in insect ears: mechanics, models and molecules. *J. Comp. Physiol. A* 201, 19–37.

1492. Mieko, C., and Bier, E. (2008). EvoD/Vo: the origins of BMP signalling in the neuroectoderm. *Nat. Rev. Genet.* 9, 663–677.

1493. Mihrshahi, R. (2006). The corpus callosum as an evolutionary innovation. *J. Exp. Zool. B. Mol. Dev. Evol.* 306, 8–17.

1494. Mikeladze-Dvali, T., Desplan, C., and Pistillo, D. (2005). Flipping coins in the fly retina. *Curr. Top. Dev. Biol.* 69, 1–15 (+ color plates).

1495. Mikeladze-Dvali, T., Wernet, M.F., Pistillo, D., Mazzoni, E.O., Teleman, A.A., Chen, Y.-W., Cohen, S., and Desplan, C. (2005). The growth regulators *warts/lats* and *melted* interact in a bistable loop to specify opposite fates in *Drosophila* R8 photoreceptors. *Cell* 122, 775–787.

1496. Milán, M., Weihe, U., Tiong, S., Bender, W., and Cohen, S.M. (2001). *msh* specifies dorsal cell fate in the *Drosophila* wing. *Development* 128, 3263–3268.

1497. Milev, N.B., and Reddy, A.B. (2015). Circadian redox oscillations and metabolism. *Trends Endocr. Metab.* 26, 430–437.

1498. Millar, S.E., Willert, K., Salinas, P.C., Roelink, H., Nusse, R., Sussman, D.J., and Barsh, G.S. (1999). WNT signaling in the control of hair growth and structure. *Dev. Biol.* 207, 133–149.

1499. Millard, S.S., Lu, Z., Zipursky, S.L., and Meinertzhagen, I.A. (2010). *Drosophila* Dscam proteins regulate postsynaptic specificity at multiple-contact synapses. *Neuron* 67, 761–768.

1500. Miller, A. (1941). Position of adult testes in *Drosophila melanogaster* Meigen. *PNAS* 27, 35–41.

1501. Miller, C.J., and Davidson, L.A. (2013). The interplay between cell signalling and mechanics in developmental processes. *Nat. Rev. Genet.* 14, 733–744.

1502. Miller, G. (2009). On the origin of the nervous system. *Science* 325, 24–26.

1503. Minelli, A. (2003). The origin and evolution of appendages. *Int. J. Dev. Biol.* 47, 573–581.

1504. Minelli, A. (2005). A morphologist's perspective on terminal growth and segmentation. *Evol. Dev.* 7, 568–573.

1505. Minelli, A. (2015). Grand challenges in evolutionary developmental biology. *Front. Ecol. Evol.* 2, Article 85. [*See also* Peterson, T., and Müller, G.B. (2016). Phenotypic novelty in EvoDevo: the distinction between continuous and discontinuous variation and its importance in evolutionary theory. *Evol. Biol.* 43, 314–335.]

1506. Minelli, A., and Fusco, G. (2004). Evo-devo perspectives on segmentation: model organisms, and beyond. *Trends Ecol. Evol.* 19, 423–429.

1507. Mirth, C., and Akam, M. (2002). Joint development in the *Drosophila* leg: cell movements and cell populations. *Dev. Biol.* 246, 391–406.

1508. Mishra, M., Oke, A., Lebel, C., McDonald, E.C., Plummer, Z., Cook, T.A., and Zelhof, A.C. (2010). Pph13 and Orthodenticle define a dual regulatory pathway for photoreceptor cell morphogenesis and function. *Development* 137, 2895–2904.

1509. Miskolczi-McCallum, C.M., Scavetta, R.J., Svendsen, P.C., Soanes, K.H., and Brook, W.J. (2005). The *Drosophila melanogaster* T-box genes *midline* and *H15* are conserved regulators of heart development. *Dev. Biol.* 278, 459–472.

1510. Missbach, C., Dweck, H.K.M., Vogel, H., Vilcinskas, A., Stensmyr, M., Hansson, B.S., and Gross-Wilde, E. (2014). Evolution of insect olfactory receptors. *eLife* 3, e02115.

1511. Missler, M., and Südhof, T.C. (1998). Neurexins: three genes and 1001 products. *Trends Genet.* 14, 20–26.

1512. Mitchell, K.J., Doyle, J.L., Serafini, T., Kennedy, T.E., Tessier-Lavigne, M., Goodman, C.S., and Dickson, B. (1996). Genetic analysis of *Netrin* genes in *Drosophila*: netrins guide CNS commissural axons and peripheral motor axons. *Neuron* 17, 203–215.

1513. Mitchell, M.E., Sander, T.L., Klinkner, D.B., and Tomita-Mitchell, A. (2007). The molecular basis of congenital heart disease. *Semin. Thorac. Cardiovasc. Surg.* 19, 228–237.

1514. Mito, T., Nakamura, T., and Noji, S. (2010). Evolution of insect development: to the hemimetabolous paradigm. *Curr. Opin. Genet. Dev.* 20, 355–361.

1515. Mito, T., Shinmyo, Y., Kurita, K., Nakamura, T., Ohuchi, H., and Noji, S. (2011). Ancestral functions of Delta/Notch signaling in the formation of body and leg segments in the cricket *Gryllus bimaculatus*. *Development* 138, 3823–3833.

1516. Mittmann, B., and Scholtz, G. (2001). *Distal-less* expression in embryos of *Limulus polyphemus* (Chelicerata, Xiphosura) and *Lepisma saccharina* (Insecta, Zygentoma) suggests a role in the development of mechanoreceptors, chemoreceptors, and the CNS. *Dev. Genes Evol.* 211, 232–243.

1517. Miura, S.K., Martins, A., Zhang, K.X., Graveley, B.R., and Zipursky, S.L. (2013). Probabilistic splicing of *Dscam1* establishes identity at the level of single neurons. *Cell* 155, 1166–1177.

1518. Miyazono, S., Isayama, T., Delori, F.C., and Makino, C.L. (2011). Vitamin A activates rhodopsin and sensitizes it to ultraviolet light. *Vis. Neurosci.* 28, 485–497.

1519. Mizunami, M. (1994). Processing of contrast signals in the insect ocellar system. *Zool. Sci.* 11, 175–190.

1520. Mizutani, C.M., and Bier, E. (2008). EvoD/Vo: the origins of BMP signalling in the neuroectoderm. *Nat. Rev. Genet.* 9, 663–677.

1521. Mizutani, C.M., Meyer, N., Roelink, H., and Bier, E. (2006). Threshold-dependent BMP-mediated repression: a model for a conserved mechanism that patterns the neuroectoderm. *PLoS Biol.* 4, #10, e1313.

1522. Moczek, A.P., Sears, K.E., Stollewerk, A., Wittkopp, P.J., Diggle, P., Dworkin, I., Ledon-Rettig, C., Matus, D.Q., Roth, S., Abouheif, E., Brown, F.D., Chiu, C.-H., Cohen, S., De Tomaso, A.W., Gilbert, S.F., Hall, B., Love, A.C., Lyons, D.C., Sanger, T.J., Smith, J., Specht, C., Vallejo-Marin, M., and Extavour, C.G. (2015). The significance and scope of evolutionary developmental biology: a vision for the 21st century. *Evol. Dev.* 17, 198–219.

1523. Mogilner, A., and Fogelson, B. (2015). Cytoskeletal chirality: swirling cells tell left from right. *Curr. Biol.* 25, R501–R503.

1524. Möglich, A., Yang, X., Ayers, R.A., and Moffat, K. (2010). Structure and function of plant photoreceptors. *Annu. Rev. Plant Biol.* 61, 21–47.

1525. Mohammad, F., Aryal, S., Ho, J., Stewart, J.C., Norman, N.A., Tan, T.L., Eisaka, A., and Claridge-Chang, A. (2016). Ancient anxiety pathways influence *Drosophila* defense behaviors. *Curr. Biol.* 26, 981–986.

1526. Molina, M.D., de Crozé, N., Haillot, E., and Lepage, T. (2013). Nodal: master and commander of the dorsal-ventral and left-right axes in the sea urchin embryo. *Curr. Opin. Genet. Dev.* 23, 445–453.

1527. Molina, M.D., Neto, A., Maeso, I., Gómez-Skarmeta, J.L., Saló, E., and Cebrià, F. (2011). Noggin and noggin-like genes control dorsoventral axis regeneration in planarians. *Curr. Biol.* 21, 300–305.

1528. Molina, M.D., Saló, E., and Cebrià, F. (2007). The BMP pathway is essential for re-specification and maintenance of the dorsoventral axis in regenerating and intact planarians. *Development* 311, 79–94.

1529. Mombaerts, P. (2004). Genes and ligands for odorant, vomeronasal and taste receptors. *Nat. Rev. Neurosci.* 5, 263–278.

1530. Monahan, K., and Lomvardas, S. (2015). Monoallelic expression of olfactory receptors. *Annu. Rev. Cell Dev. Biol.* 31, 721–740.

1531. Monahan-Earley, R., Dvorak, A.M., and Aird, W.C. (2013). Evolutionary origins of the blood vascular system and endothelium. *J. Thromb. Haemost.* 11 (Suppl. 1), 46–66.

1532. Monier, B., Tevy, F., Perrin, L., Capovilla, M., and Semeriva, M. (2007). Downstream of Hox genes. *Fly* 1, 59–67.

1533. Montcouquiol, M., Sans, N., Huss, D., Kach, J., Dickman, J.D., Forge, A., Rachel, R.A., Copeland, N.G., Jenkins, N.A., Bogani, D., Murdoch, J., Warchol, M.E., Wenthold, R.J., and Kelley, M.W. (2006). Asymmetric localization of Vangl2 and Fz3 indicate novel mechanisms for planar cell polarity in mammals. *J. Neurosci.* 26, #19, 5265–5275.

1534. Montealegre-Z., F., Jonsson, T., Robson-Brown, K.A., Postles, M., and Robert, D. (2012). Convergent evolution between insect and mammalian audition. *Science* 338, 968–971.

1535. Monteiro, A.S., and Ferrier, D.E.K. (2006). Hox genes are not always colinear. *Int. J. Biol. Sci.* 2, 95–103.

1536. Monteiro, A.S., Shierwater, B., Dellaporta, S.L., and Holland, P.W.H. (2006). A low diversity of ANTP class homeobox genes in Placozoa. *Evol. Dev.* 8, 174–182.

1537. Montell, C. (1999). Visual transduction in *Drosophila*. *Annu. Rev. Cell Dev. Biol.* 15, 231–268.

1538. Montell, C. (2009). A taste of the *Drosophila* gustatory receptors. *Curr. Opin. Neurobiol.* 19, 345–353.

1539. Mooi, R., and David, B. (2008). Radial symmetry, the anterior/posterior axis, and echinoderm Hox genes. *Ann. Rev. Ecol. Evol. Syst.* 39, 43–62. [*See also* Arnone, M.I., *et al.* (2016). Echinoderm systems for gene regulatory studies in evolution and development. *Curr. Opin. Genet. Dev.* 39, 129–137.]

1540. Moore, M.S., DeZazzo, J., Luk, A.Y., Tully, T., Singh, C.M., and Heberlein, U. (1998). Ethanol intoxication in *Drosophila*: genetic and pharmacological evidence for regulation by the cAMP signaling pathway. *Cell* 93, 997–1007.

1541. Moorman, A.F.M., Christoffels, V.M., Anderson, R.H., and van den Hoff, M.J.B. (2007). The heart-forming fields: one or mutiple? *Philos. Trans. R. Soc. Lond. B* 362, 1257–1265.

1542. Moran, D.T., Rowley, J.C., III, Jafek, B.W., and Lovell, M.A. (1982). The fine structure of the olfactory mucosa in man. *J. Neurocytol.* 11, 721–746.

1543. Morante, J., and Desplan, C. (2008). The color-vision circuit in the medulla of *Drosophila*. *Curr. Biol.* 18, 553–565.

1544. Morata, G., Macías, A., Urquía, N., and González-Reyes, A. (1990). Homoeotic genes. *Semin. Cell Biol.* 1, 219–227.

1545. Moreira, I.S. (2014). Structural features of the G-protein/GPCR interactions. *Biochim. Biophys. Acta* 1840, 16–33.

1546. Morell, V. (1991). A hand on the bird: and one on the bush. *Science* 254, 33–34.

1547. Moreno, E., De Mulder, K., Salvenmoser, W., Ladurner, P., and Martínez, P. (2010). Inferring the ancestral function of the posterior Hox gene within the bilateria: controlling the maintenance of reproductive structures, the musculature and the nervous system in the acoel flatworm *Isodiametra pulchra*. *Evol. Dev.* 12, 258–266.

1548. Moreno, E., Permanyer, J., and Martinez, P. (2011). The origin of patterning systems in Bilateria: insights from the Hox and ParaHox genes in Acoelomorpha. *Genomics Proteomics Bioinformatics* 9, 65–76.

1549. Morey, C., Da Silva, N.R., Perry, P., and Bickmore, W.A. (2007). Nuclear reorganisation and chromatin decondensation are conserved, but distinct, mechanisms linked to Hox gene activation. *Development* 134, 909–919.

1550. Morgan, D., Goodship, J., Essner, J.J., Vogan, K.J., Turnpenny, L., Yost, H.J., Tabin, C.J., and Strachan, T. (2002). The left-right determinant inversin has highly conserved ankyrin repeat and IQ domains and interacts with calmodulin. *Hum. Genet.* 110, 377–384.

1551. Morgan, D., Turnpenny, L., Goodship, J., Dai, W., Majumder, K., Matthews, L., Gardner, A., Schuster, G., Vein, L., Harrison, W., Elder, F.F.B., Penman-Splitt, M., Overbeek, P., and Strachan, T. (1998). Inversin, a novel gene in the vertebrate left-right axis pathway, is partially deleted in the *inv* mouse. *Nat. Genet.* 20, 149–156.

1552. Morgan, M.J. (1992). On the evolutionary origin of right handedness. *Curr. Biol.* 2, 15–17.

1553. Morgan, R. (2004). Conservation of sequence and function in the *Pax6* regulatory elements. *Trends Genet.* 20, 283–287.

1554. Morgan, R. (2006). *Engrailed*: complexity and economy of a multi-functional transcription factor. *FEBS Lett.* 580, 2531–2533.

1555. Mori, K., Nagao, H., and Yoshihara, Y. (1999). The olfactory bulb: coding and processing of odor molecule information. *Science* 286, 711–715.

1556. Mori, T., Mchaourab, H., and Johnson, C.H. (2015). Circadian clocks: unexpected biochemical cogs. *Curr. Biol.* 25, R827–R844.

1557. Morin, L.P. (1994). The circadian visual system. *Brain Res. Rev.* 67, 102–127.

1558. Morokuma, J., Ueno, M., Kawanishi, H., Saiga, H., and Nishida, H. (2002). *HrNodal*, the ascidian nodal-related gene, is expressed in the left side of the epidermis, and lies upstream of *HrPitx*. *Dev. Genes Evol.* 212, 439–446.

1559. Moroz, L.L. (2009). On the independent origins of complex brains and neurons. *Brain Behav. Evol.* 74, 177–190.

1560. Moroz, L.L., Kocot, K.M., Citarella, M.R., Dosung, S., Norekian, T.P., Povolotskaya, I.S., Grigorenko, A.P., Dailey, C., Berezikov, E., Buckley, K.M., Ptitsyn, A., Reshetov, D., Mukherjee, K., Moroz, T.P., Bobkova, Y., Yu, F., Kapitonov, V.V., Jurka, J., Bobkov, Y.V., Swore, J.J., Girardo, D.O., Fodor, A., Gusev, F., Sanford, R., Bruders, R., Kittler, E., Mills, C.E., Rast, J.P., Derelle, R., Solovyev, V.V., Kondrashov, F.A., Swalla, B.J., Sweedler, J.V., Rogaev, E.I., Halanych, K.M., and Kohn, A.B. (2014). The ctenophore genome and the evolutionary origins of neural systems. *Nature* 510, 109–114.

1561. Moroz, L.L., and Kohn, A.B. (2016). Independent origins of neurons and synapses: insights from ctenophores. *Philos. Trans. R. Soc. Lond. B* 371, 20150041.

1562. Morrissey, M.E., Shelton, S., Brockerhoff, S.E., Hurley, J.B., and Kennedy, B.N. (2011). PRE-1, a *cis* element sufficient to enhance cone- and rod- specific expression in differentiating zebrafish photoreceptors. *BMC Dev. Biol.* 11, Article 3.

1563. Morshedian, A., and Fain, G.L. (2015). Single-photon sensitivity of lamprey rods with cone-like outer segments. *Curr. Biol.* 25, 484–487.

1564. Mosca, T.J., and Luo, L. (2014). Synaptic organization of the *Drosophila* antennal lobe and its regulation by the Teneurins. *eLife* 3, e03726.

1565. Mouchel-Vielh, E., Blin, M., Rigolot, C., and Deutsch, J.S. (2002). Expression of a homologue of the *fushi tarazu* (*ftz*) gene in a ciripede crustacean. *Evol. Dev.* 4, 76–85.

1566. Moussian, B., and Uv, A.E. (2005). An ancient control of epithelial barrier formation and wound healing. *BioEssays* 27, 987–990.

1567. Mu, X., Fu, X., Sun, H., Liang, S., Maeda, H., Firishman, L.J., and Klein, W.H. (2005). Ganglion cells are required for normal progenitor-cell proliferation but not cell-fate determination or patterning in the developing mouse retina. *Curr. Biol.* 15, 525–530.

1568. Muchlinski, M.N. (2010). A comparative analysis of vibrissa count and infraorbital foramen area in primates and other mammals. *J. Hum. Evol.* 58, 447–473.

1569. Muerdter, F., and Stark, A. (2014). Hiding in plain sight. *Nature* 512, 374–375.

1570. Mukckli, L., Naumer, M.J., and Singer, W. (2009). Bilateral visual field maps in a patient with only one hemisphere. *PNAS* 106, #31, 13034–13039.

1571. Mukouyama, Y.-s., Shin, D., Britsch, S., Taniguchi, M., and Anderson, D.J. (2002). Sensory nerves determine the pattern of arterial differentiation and blood vessel branching in the skin. *Cell* 109, 693–705.

1572. Müller, G.B. (2003). Homology: the evolution of morphological organization. *In* G.B. Müller and S.A. Newman (eds.), *Origination of Organismal Form: Beyond the Gene in Developmental and Evolutionary Biology*. MIT Press, Cambridge, MA, pp. 51–69.

1573. Muller, H.J. (1932). Further studies on the nature and causes of gene mutations. *Proc. 6th Int. Congr. Genet.* 1, 213–255.

1574. Müller, P., Rogers, K.W., Yu, S.R., Brand, M., and Schier, A.F. (2013). Morphogen transport. *Development* 140, 1621–1638.

1575. Muneoka, K., Han, M., and Gardiner, D.M. (2008). Regrowing human limbs. *Sci. Am.* 298, #4, 56–63.

1576. Munnamalai, V., and Fekete, D.M. (2013). Wnt signaling during cochlear development. *Semin. Cell Dev. Biol.* 24, 480–489.

1577. Muñoz-Chápuli, R. (2011). Evolution of angiogenesis. *Int. J. Dev. Biol.* 55, 345–351.

1578. Murawala, P., Tanaka, E.M., and Currie, J.D. (2012). Regeneration: the ultimate example of wound healing. *Semin. Cell Dev. Biol.* 23, 954–962.

1579. Murdock, D.J.E., and Donoghue, P.C.J. (2011). Evolutionary origins of animal skeletal biomineralization. *Cells Tissues Organs* 194, 98–102.

1580. Mure, L.S., Cornut, P.-L., Rieux, C., Drouyer, E., Denis, P., Gronfier, C., and Cooper, H.M. (2009). Melanopsin bistability: a fly's eye technology in the human retina. *PLoS ONE* 4, #6, e5991.

1581. Murray, J.D. (1989). *Mathematical Biology*. Springer-Verlag, Berlin.

1582. Musiek, E.S., Xiong, D.D., and Holtzman, D.M. (2014). Sleep, circadian rhythms, and the pathogenesis of Alzheimer Disease. *Exp. Mol. Med.* 47, e148.

1583. Mustafi, D., Engel, A.H., and Palczewski, K. (2009). Structure of cone photoreceptors. *Prog. Retin. Eye Res.* 28, 289–302.

1584. Mustárdy, L., Buttle, K., Steinbach, G., and Garab, G. (2008). The three-dimensional network of thylakoid membranes in plants: quasihelical model of the granum-stroma assembly. *Plant Cell* 20, 2552–2557.

1585. Nacu, E., Gromberg, E., Oliveira, C.R., Drechsel, D., and Tanaka, E.M. (2016). FGF8 and SHH substitute for anterior-posterior tissue interactions to induce limb regeneration. *Nature* 533, 407–410.

1586. Nadeau, J.H., and Taylor, B.A. (1984). Lengths of chromosomal segments conserved since divergence of man and mouse. *PNAS* 81, 814–818.

1587. Nadrowski, B., Albert, J.T., and Göpfert, M.C. (2008). Transducer-based force generation explains active process in *Drosophila* hearing. *Curr. Biol.* 18, 1365–1372.

1588. Naganathan, S.R., Fürthauer, S., Nishikawa, M., Jülicher, F., and Grill, S.W. (2014). Active torque generation by the actomyosin cell cortex drives left–right symmetry breaking. *eLife* 3, e04165.

1589. Nakagawa, T., and Vosshall, L.B. (2009). Controversy and consensus: noncanonical signaling mechanisms in the insect olfactory system. *Curr. Opin. Neurobiol.* 19, 284–292.

1590. Nakamura, T., and Hamada, H. (2012). Left-right patterning: conserved and divergent mechanisms. *Development* 139, 3257–3262.

1591. Nakanishi, K. (1991). Why 11-cis-retinal? *Am. Zool.* 31, 479–489.

1592. Nakano, S., Stillman, B., and Horvitz, H.R. (2011). Replication-coupled chromatin assembly generates a neuronal bilateral asymmetry in *C. elegans*. *Cell* 147, 1525–1536.

1593. Nam, J., and Nei, M. (2005). Evolutionary change of the numbers of homeobox genes in bilaterian animals. *Mol. Biol. Evol.* 22, 2386–2394.

1594. Nam, J.-H., Cotton, J.R., and Grant, W. (2007). A virtual hair cell. I. Addition of gating spring theory into a 3-D bundle mechanical model. *Biophys. J.* 92, 1918–1928.

1595. Namigai, E.K.O., Kenny, N.J., and Shimeld, S.M. (2014). Right across the tree of life: the evolution of left-right asymmetry in the Bilateria. *Genesis* 52, 458–470.

1596. Narayanan, C.H., and Hamburger, V. (1971). Motility in chick embryos with substitution of lumbosacral by brachial and brachial by lumbosacral spinal cord segments. *J. Exp. Zool.* 178, 415–432.

1597. Narayanan, P., Chatterton, P., Ikeda, A., Ikeda, S., Corey, D.P., Ervasti, J.M., and Perrin, B.J. (2015). Length regulation of mechanosensitive stereocilia depends on very slow actin dynamics and filament-severing proteins. *Nat. Commun.* 6, Article 6855.

1598. Narendra, V., Rocha, P.P., An, D., Raviram, R., Skok, J.A., Mazzoni, E.O., and Reinberg, D. (2015). CTCF establishes discrete functional chromatin domains at the *Hox* clusters during differentiation. *Science* 347, 1017–1021.

1599. Närhi, K., Järvinen, E., Birchmeier, W., Taketo, M.M., Mikkola, M.L., and Thesleff, I. (2008). Sustained epithelial β-catenin activity induces precocious hair development but disrupts hair follicle down-growth and hair shaft formation. *Development* 135, 1019–1028.

1600. Narris, P.M. (2001). In a fly's ear. *Nature* 410, 644–645.

1601. Natale, A., Sims, C., Chiusano, M.L., Amoroso, A., D'Aniello, E., Fucci, L., Krumlauf, R., Branno, M., and Locascio, A. (2011). Evolution of anterior *Hox* regulatory elements among chordates. *BMC Evol. Biol.* 11, Article 330.

1602. Nathans, J. (1987). Molecular biology of visual pigments. *Annu. Rev. Neurosci.* 10, 163–194.

1603. Nathans, J. (1999). The evolution and physiology of human color vision: insights from molecular genetic studies of visual pigments. *Neuron* 24, 299–312.

1604. Navis, A., and Bagnat, M. (2015). Developing pressures: fluid forces driving morphogenesis. *Curr. Opin. Genet. Dev.* 32, 24–30.

1605. Nawabi, H., and Castellani, V. (2011). Axonal commissures in the central nervous system: how to cross the midline? *Cell. Mol. Life Sci.* 68, 2539–2553.

1606. Nayak, G.D., Ratnayaka, H.S.K., Goodyear, R.J., and Richardson, G.P. (2007). Development of the hair bundle and mechanotransduction. *Int. J. Dev. Biol.* 51, 597–608.

1607. Negre, B., Casillas, S., Suzanne, M., Sánchez-Herrero, E., Akam, M., Nefedov, M., Barbadilla, A., de Jong, P., and Ruiz, A. (2005). Conservation of regulatory sequences and gene expression patterns in the disintegrating *Drosophila Hox* complex. *Genome Res.* 15, 692–700.

1608. Negre, B., Ranz, J.M., Casals, F., Cáceres, M., and Ruiz, A. (2003). A new split of the Hox gene complex in *Drosophila*: relocation and evolution of the gene *labial*. *Mol. Biol. Evol.* 20, 2042–2054.

1609. Negre, B., and Ruiz, A. (2007). HOM-C evolution in *Drosophila*: is there a need for *Hox* gene clustering? *Trends Genet.* 23, 55–59.

1610. Nei, M., Niimura, Y., and Nozawa, M. (2008). The evolution of animal chemosensory receptor gene repertoires: roles of chance and necessity. *Nat. Rev. Genet.* 9, 951–963.

1611. Nelson, B.R., Gumuscu, B., Hartman, B.H., and Reh, T.A. (2006). Notch activity is downregulated just prior to retinal ganglion cell differentiation. *Dev. Neurosci.* 28, 128–141.

1612. Nelson, C. (2004). Selector genes and the genetic control of developmental modules. *In* G. Schlosser and G.P. Wagner (eds.), *Modularity in Development and Evolution.* University of Chicago Press, Chicago, IL, pp. 17–33.

1613. Nemer, M. (2008). Genetic insights into normal and abnormal heart development. *Cardiovasc. Pathol.* 17, 48–54.

1614. Netter, F.H. (2003). *Atlas of Human Anatomy*, 3rd edn. Icon Learning Systems, Teterboro, NJ.

1615. Neumann, C.J., and Nuesslein-Volhard, C. (2000). Patterning of the zebrafish retina by a wave of Sonic hedgehog activity. *Science* 289, 2137–2139.

1616. Neves, J., Demaria, M., Campisi, J., and Jasper, H. (2015). Of flies, mice, and men: evolutionarily conserved tissue damage responses and aging. *Dev. Cell* 32, 9–18.

1617. Ng, T.H., Chiang, Y.-A., Yeh, Y.-C., and Wang, H.-C. (2014). Review of Dscam-mediated immunity in shrimp and other arthropods. *Dev. Comp. Immunol.* 46, 129–138.

1618. Nichols, C.D. (2006). *Drosophila melanogaster* neurobiology, neuropharmacology, and how the fly can inform central nervous system drug discovery. *Pharmacol. Ther.* 112, 677–700.

1619. Nickle, B., and Robinson, P.R. (2007). The opsins of the vertebrate retina: insights from structural, biochemical, and evolutionary studies. *Cell. Mol. Life Sci.* 64, 2917–2932.

1620. Nicklen, P. (2007). Arctic ivory: hunting the narwhal. *Natl. Geogr.* 212, #2, 110–129.

1621. Nicol, X., and Gaspar, P. (2014). Routes to cAMP: shaping neuronal connectivity with distinct adenylate cyclases. *Eur. J. Neurosci.* 39, 1742–1751.

1622. Nie, J., Mahato, S., Mustill, W., Tipping, C., Bhattacharya, S.S., and Zelhof, A.C. (2012). Cross species analysis of Prominin reveals a conserved cellular role in invertebrate and vertebrate photoreceptor cells. *Dev. Biol.* 371, 312–320.

1623. Niederreither, K., and Dollé, P. (2008). Retinoic acid in development: towards an integrated view. *Nat. Rev. Genet.* 9, 541–553.

1624. Niehrs, C. (2010). On growth and form: a Cartesian coordinate system of Wnt and BMP signaling specifies bilaterian body axes. *Development* 137, 845–857.

1625. Nielsen, C. (2005). Larval and adult brains. *Evol. Dev.* 7, 483–489.

1626. Nielsen, C., and Martinez, P. (2003). Patterns of gene expression: homology or homocracy? *Dev. Genes Evol.* 213, 149–154.

1627. Nilsson, D.-E. (1994). Eyes as optical alarm systems in fan worms and ark clams. *Philos. Trans. R. Soc. Lond. B* 346, 195–212.

1628. Nilsson, D.-E. (1996). Eye ancestry: old genes for new eyes. *Curr. Biol.* 6, 39–42.

1629. Nilsson, D.-E. (2004). Eye evolution: a question of genetic promiscuity. *Curr. Opin. Neurobiol.* 14, 407–414.

1630. Nilsson, D.-E. (2005). Photoreceptor evolution: ancient siblings serve different tasks. *Curr. Biol.* 15, R94–R96.

1631. Nilsson, D.-E. (2009). The evolution of eyes and visually guided behaviour. *Philos. Trans. R. Soc. Lond. B* 364, 2833–2847.

1632. Nilsson, D.-E. (2013). Eye evolution and its functional basis. *Vis. Neurosci.* 30, 5–20.

1633. Nilsson, D.-E., and Arendt, D. (2008). Eye evolution: the blurry beginning. *Curr. Biol.* 18, R1096–R1098.

1634. Nilsson, D.-E., Gislén, L., Coates, M.M., Skogh, C., and Garm, A. (2005). Advanced optics in a jellyfish eye. *Nature* 435, 201–205.

1635. Niswander, L. (2003). Pattern formation: old models out on a limb. *Nat. Rev. Genet.* 4, 133–143.

1636. Nitta, K.R., Jolma, A., Yin, Y., Morgunova, E., Kivioja, T., Akhtar, J., Hens, K., Toivonen, J., Deplancke, B., Furlong, E.E.M., and Taipale, J. (2015). Conservation of transcription factor binding specificities across 600 million years of bilateria evolution. *eLife* 4, e04837.

1637. Niwa, N., Akimoto-Kato, A., Niimi, T., Tojo, K., Machida, R., and Hayashi, S. (2010). Evolutionary origin of the insect wing via integration of two developmental modules. *Evol. Dev.* 12, 168–176.

1638. Niwa, N., Hiromi, Y., and Okabe, M. (2004). A conserved developmental program for sensory organ formation in *Drosophila melanogaster*. *Nat. Genet.* 36, 293–297.

1639. Nödl, M.-T., Fossati, S.M., Domingues, P., Sánchez, F.J., and Zullo, L. (2015). The making of an octopus arm. *EvoDevo* 6, Article 19.

1640. Nolte, C., and Krumlauf, R. (2007). Expression of *Hox* genes in the nervous system of vertebrates. *In* S. Papageorgiou (ed.), *HOX Gene Expression*. Landes Bioscience, Austin, TX, pp. 14–41.

1641. Nomaksteinsky, M., Kassabov, S., Chettouh, Z., Stoeklé, H.-C., Bonnaud, L., Fortin, G., Kandel, E.R., and Brunet, J.-F. (2013). Ancient origin of somatic and visceral neurons. *BMC Biol.* 11, Article 53.

1642. Nomaksteinsky, M., Röttinger, E., Dufour, H.D., Chettouh, Z., Lowe, C.J., Martindale, M.Q., and Brunet, J.-F. (2009). Centralization of the deuterostome nervous system predates chordates. *Curr. Biol.* 19, 1264–1269.

1643. Nonaka, S., Tanaka, Y., Okada, Y., Takeda, S., Harada, A., Kanai, Y., Kido, M., and Hirokawa, N. (1998). Randomization of left-right asymmetry due to loss of nodal cilia generating leftward flow of extraembryonic fluid in mice lacking KIF3B motor protein. *Cell* 95, 829–837 (Erratum: *Cell* 1999, 99, 117).

1644. Noonan, J.P. (2009). Regulatory DNAs and the evolution of human development. *Curr. Opin. Genet. Dev.* 19, 557–564.

1645. Noordermeer, D., and Duboule, D. (2013). Chromatin architectures and *Hox* gene collinearity. *Curr. Top. Dev. Biol.* 104, 113–148.

1646. Noordermeer, D., Leleu, M., Splinter, E., Rougemont, J., De Laat, W., and Duboule, D. (2011). The dynamic architecture of *Hox* gene clusters. *Science* 334, 222–225.

1647. Nordström, K.J.V., Almén, M.S., Edstam, M.M., Fredriksson, R., and Schiöth, H.B. (2011). Independent HHsearch, Needleman-Wunsch-based, and motif analyses reveal the overall hierarchy for most of the G protein-coupled receptor families. *Mol. Biol. Evol.* 28, 2471–2480.

1648. Noro, B., Culi, J., McKay, D.J., Zhang, W., and Mann, R.S. (2006). Distinct functions of homeodomain-containing and homeodomain-less isoforms encoded by *homothorax*. *Genes Dev.* 20, 1636–1650.

1649. Norris, D.P. (2012). Cilia, calcium and the basis of left-right asymmetry. *BMC Biol.* 10, Article 102.

1650. Norris, D.P., and Jackson, P.K. (2016). Calcium contradictions in cilia. *Nature* 531, 582–583.

1651. Northcutt, R.G. (2012). Evolution of centralized nervous systems: two schools of evolutionary thought. *PNAS* 109 (Suppl. 1), 10626–10633.

1652. Nossa, C.W., Havlak, P., Yue, J.-X., Lv, J., Vincent, K.Y., Brockmann, H.J., and Putnam, N.H. (2014). Joint assembly and genetic mapping of the Atlantic horseshoe crab genome reveals ancient whole genome duplication. *GigaScience* 3, Article 9.

1653. Novarino, G., Akizu, N., and Gleeson, J.G. (2011). Modeling human disease in humans: the ciliopathies. *Cell* 147, 70–79.

1654. Nübler-Jung, K., and Arendt, D. (1994). Is ventral in insects dorsal in vertebrates? A history of embryological arguments favouring axis inversion in chordate ancestors. *Roux's Arch. Dev. Biol.* 203, 357–366.

1655. Nübler-Jung, K., and Arendt, D. (1996). Enteropneusts and chordate evolution. *Curr. Biol.* 6, 352–353.

1656. Nürnberger, J., Bacallao, R.L., and Phillips, C.L. (2002). Inversin forms a complex with catenins and N-cadherin in polarized epithelial cells. *Mol. Biol. Cell* 13, 3096–3106.

1657. Nürnberger, J., Kribben, A., Saez, A.O., Heusch, G., Philipp, T., and Phillips, C.L. (2004). The *Invs* gene encodes a microtubule-associated protein. *J. Am. Soc. Nephrol.* 15, 1700–1710.

1658. Nusse, R. (2001). An ancient cluster of Wnt paralogues. *Trends Genet.* 17, 443.

1659. Nüsslein-Volhard, C. (1996). Gradients that organize embryo development. *Sci. Am.* 275, #2, 54–61.

1660. Nweeia, M.T., Eichmiller, F.C., Hauschka, P.V., Donahue, G.A., Orr, J.R., Ferguson, S.H., Watt, C.A., Mead, J.G., Potter, C.W., Dietz, R., Giuseppetti, A.A., Black, S.R., Trachtenberg, A.J., and Kuo, W.P. (2014). Sensory ability in the narwhal tooth organ system. *Anat. Rec.* 297, 599–617.

1661. Nweeia, M.T., Eichmiller, F.C., Hauschka, P.V., Tyler, E., Mead, J.G., Potter, C.W., Angnatsiak, D.P., Richard, P.R., Orr, J.R., and Black, S.R. (2012). Vestigial tooth anatomy and tusk nomenclature for *Monodon monoceros*. *Anat. Rec.* 295, 1006–1016.

1662. O'Connell, L.A. (2013). Evolutionary development of neural systems in vertebrates and beyond. *J. Neurogenet.* 27, 69–85.

1663. O'Grady, G.E., and McIver, S.B. (1987). Fine structure of the compound eye of the black fly *Simulium vittatum* (Diptera: Simuliidae). *Can. J. Zool.* 65, 1454–1469.

1664. O'Leary, D.D.M., and Sahara, S. (2008). Genetic regulation of arealization of the neocortex. *Curr. Opin. Neurobiol.* 18, 90–100.

1665. O'Riordan, M.X.D., Bauler, L.D., Scott, F.L., and Duckett, C.S. (2008). Inhibitor of apoptosis proteins in eukaryotic evolution and development: a model of thematic conservation. *Dev. Cell* 15, 497–508.

1666. Oakley, T.H. (2003). The eye as a replicating and diverging, modular developmental unit. *Trends Ecol. Evol.* 18, 623–627.

1667. Ochoa, C., and Rasskin-Gutman, D. (2015). Evo-devo mechanisms underlying the continuum between homology and homoplasy. *J. Exp. Zool. B. Mol. Dev. Evol.* 324, 91–103.

1668. Ochoa-Espinosa, A., Yu, D., Tsirigos, A., Struffi, P., and Small, S. (2009). Anterior-posterior positional information in the absence of a strong Bicoid gradient. *PNAS* 106, #10, 3823–3828.

1669. Ochoa-Espinosa, A., Yucel, G., Kaplan, L., Pare, A., Pura, N., Oberstein, A., Papatsenko, D., and Small, S. (2005). The role of binding site cluster strength in Bicoid-dependent patterning in *Drosophila*. *PNAS* 102, #14, 4960–4965.

1670. Odden, J.P., Holbrook, S., and Doe, C.Q. (2002). *Drosophila HB9* is expressed in a subset of motoneurons and interneurons, where it regulates gene expression and axon pathfinding. *J. Neurosci.* 22, #21, 9143–9149.

1671. Ogura, A., Ikeo, K., and Gojobori, T. (2004). Comparative analysis of gene expression for convergent evolution of camera eye between octopus and human. *Genome Res.* 14, 1555–1561.

1672. Ogura, A., Yoshida, M.-a., Moritaki, T., Okuda, Y., Sese, J., Shimizu, K.K., Sousounis, K., and Tsonis, P.A. (2013). Loss of the six3/6 controlling pathways might have resulted in pinhole-eye evolution in *Nautilus*. *Sci. Rep.* 3, Article 1432.

1673. Oka, K., Yoshiyama, N., Tojo, K., Machida, R., and Hatakeyama, M. (2010). Characterization of abdominal appendages in the sawfly, *Athalia rosae* (Hymenoptera), by morphological and gene expression analyses. *Dev. Genes Evol.* 220, 53–59.

1674. Okada, T., Sugihara, M., Bondar, A.-N., Elstner, M., Entel, P., and Buss, V. (2004). The retinal conformation and its environment in rhodopsin in light of a new 2.2 Å crystal structure. *J. Mol. Biol.* 342, 571–583.

1675. Okada, Y., Nonaka, S., Tanaka, Y., Saijoh, Y., Hamada, H., and Hirokawa, N. (1999). Abnormal nodal flow precedes situs inversus in *iv* and *inv* mice. *Mol. Cell* 4, 459–468.

1676. Okawa, H., Sampath, A.P., Laughlin, S.B., and Fain, G.L. (2008). ATP consumption by mammalian rod photoreceptors in darkness and light. *Curr. Biol.* 18, 1917–1921.

1677. Okkema, P.G., Ha, E., Haun, C., Chen, W., and Fire, A. (1997). The *Caenorhabditis elegans* NK-2 homeobox gene *ceh-22* activates pharyngeal muscle gene expression in combination with *pha-1* and is required for normal pharyngeal development. *Development* 124, 3965–3973.

1678. Okoruwa, O.E., Weston, M.D., Sanjeevi, D.C., Millemon, A.R., Fritzsch, B., Hallworth, R., and Beisel, K.W. (2008). Evolutionary insights into the unique electromotility motor of mammalian outer hair cells. *Evol. Dev.* 10, 300–315.

1679. Okray, Z., and Hassan, B.A. (2013). Genetic approaches in *Drosophila* for the study neurodevelopmental disorders. *Neuropharmacology* 68, 150–156.

1680. Okumura, T., Fujiwara, H., Taniguchi, K., Kuroda, J., Nakazawa, N., Nakamura, M., Hatori, R., Ishio, A., Maeda, R., and Matsuno, K. (2010). Left-right asymmetric morphogenesis of the anterior midgut depends on the activation of a non-muscle myosin II in *Drosophila*. *Dev. Biol.* 344, 693–706.

1681. Okumura, T., Sasamura, T., Inatomi, M., Hozumi, S., Nakamura, M., Hatori, R., Taniguchi, K., Nakazawa, N., Suzuki, E., Maeda, R., Yamakawa, T., and Matsuno, K. (2015). Class I myosins have overlapping and specialized functions in left-right asymmetric development in *Drosophila*. *Genetics* 199, 1183–1199.

1682. Oland, L.A., and Tolbert, L.P. (2011). Roles of glial cells in neural circuit formation: insights from research in insects. *Glia* 59, 1273–1295.

1683. Olesnicky, E.C., Brent, A.E., Tonnes, L., Walker, M., Pultz, M.A., Leaf, D., and Desplan, C. (2006). A *caudal* mRNA gradient controls posterior development in the wasp *Nasonia*. *Development* 133, 3973–3982.

1684. Oliveira, M.B., Liedholm, S.E., Lopez, J.E., Lochte, A.A., Pazio, M., Martin, J.P., Mörch, P.R., Salakka, S., York, J., Yoshimoto, A., and Janssen, R. (2014). Expression of arthropod distal limb-patterning genes in the onychophoran *Euperipatoides kanangrensis*. *Dev. Genes Evol.* 224, 87–96.

1685. Olivera-Martinez, I., Viallet, J.P., Michon, F., Pearton, D.J., and Dhouailly, D. (2004). The different steps of skin formation in vertebrates. *Int. J. Dev. Biol.* 48, 107–115.

1686. Olson, E.N. (2006). Gene regulatory networks in the evolution and development of the heart. *Science* 313, 1922–1927.

1687. Olson, E.N., and Klein, W.H. (1998). Muscle minus MyoD. *Dev. Biol.* 202, 153–156.

1688. Onai, T., Yu, J.-K., Blitz, I.L., Cho, K.W.Y., and Holland, L.Z. (2010). Opposing Nodal/Vg1 and BMP signals mediate axial patterning in embryos of the basal chordate amphioxus. *Dev. Biol.* 344, 377–389.

1689. Onuma, Y., Takahashi, S., Asashima, M., Kurata, S., and Gehring, W.J. (2002). Conservation of *Pax 6* function and upstream activation by *Notch* signaling in eye development of frogs and flies. *PNAS* 99, #4, 2020–2025.

1690. Oppenheim, R.W. (1991). Cell death during development of the nervous system. *Annu. Rev. Neurosci.* 14, 453–501.

1691. Oppenheimer, J.M. (1974). Asymmetry revisited. *Am. Zool.* 14, 867–879.

1692. Ordan, E., and Volk, T. (2015). A non-signaling role of Robo2 in tendons is essential for Slit processing and muscle patterning. *Development* 142, 3512–3518.

1693. Orii, H., Kato, K., Umesono, Y., Sakurai, T., Agata, K., and Watanabe, K. (1999). The planarian HOM/HOX genes (*Plox*) expressed along the anteroposterior axis. *Dev. Biol.* 210, 456–468.

1694. Orkin, S.H., and Zon, L.I. (2008). Hematopoiesis: an evolving paradigm for stem cell biology. *Cell* 132, 631–644.

1695. Orly, G., Manor, U., and Gov, N.S. (2015). A biophysical model for the staircase geometry of stereocilia. *PLoS ONE* 10, #7, e0127926.

1696. Ortega-Hernández, J. (2015). Lobopodians. *Curr. Biol.* 25, R873–R875.

1697. Ortiz, C.O., Etchberger, J.F., Posy, S.L., Frøkjaer-Jensen, C., Lockery, S., Honig, B., and Hobert, O. (2006). Searching for neuronal left/right asymmetry: genomewide analysis of nematode receptor-type guanylyl cyclases. *Genetics* 173, 131–149.

1698. Osborne, P.W., and Ferrier, D.E.K. (2010). Chordate Hox and ParaHox gene clusters differ dramatically in their repetitive element content. *Mol. Biol. Evol.* 27, 217–220.

1699. Osorio, D. (2007). *Spam* and the evolution of the fly's eye. *BioEssays* 29, 111–115.

1700. Osorio, D., and Bossomaier, T.R.J. (1992). Human cone-pigment spectral sensitivities and the reflectances of natural surfaces. *Biol. Cybern.* 67, 217–222.

1701. Osorio, D., and Vorobyev, M. (2008). A review of the evolution of animal color vision and visual communication signals. *Vision Res.* 48, 2042–2051.

1702. Otsuna, H., Shinomiya, K., and Ito, K. (2014). Parallel neural pathways in higher visual centers of the *Drosophila* brain that mediate wavelength-specific behavior. *Front. Neural Circuits* 8, Article 8. [*See also* Panser, K., *et al.* (2016). Automatic segmentation of *Drosophila* neural compartments using GAL4 expression data reveals novel visual pathways. *Curr. Biol.* 26, 1943–1954.]

1703. Otto, E.A., Schermer, B., Obara, T., O'Toole, J.F., Hiller, K.S., Mueller, A.M., Ruf, R.G., Hoefele, J., Beekmann, F., Landau, D., Foreman, J.W., Goodship, J.A., Strachan, T., Kispert, A., Wolf, M.T., Gagnadoux, M.F., Nivet, H., Antignac, C., Walz, G., Drummond, I.A., Benzing, T., and Hildebrandt, F. (2003). Mutations in *INVS* encoding inversin cause nephronophthisis type 2, linking renal cystic disease to the function of primary cilia and left-right axis determination. *Nat. Genet.* 34, 413–420.

1704. Ou, G., Stuurman, N., D'Ambrosio, M., and Vale, R.D. (2010). Polarized myosin produces unequal-size daughters during asymmetric cell division. *Science* 330, 677–680.

1705. Ovespian, S.V., and Vesselkin, N.P. (2014). Wiring prior to firing: the evolutionary rise of electrical and chemical modes of synaptic transmission. *Rev. Neuroscience* 25, 821–832.

1706. Özbudak, E.M., and Lewis, J. (2008). Notch signalling synchronizes the zebrafish segmentation clock but is not needed to create somite boundaries. *PLoS Genet.* 4, #2, e15.

1707. Özüak, O., Buchta, T., Roth, S., and Lynch, J.A. (2014). Dorsoventral polarity of the *Nasonia* embryo primarily relies on a BMP gradient formed without input from Toll. *Curr. Biol.* 24, 2393–2398.

1708. Pace, R.M., Grbic, M., and Nagy, L.M. (2016). Composition and genomic organization of arthropod Hox clusters. *EvoDevo* 7, Article 11. [*See also* Leite, D.J., and McGregor, A.P. (2016). Arthropod evolution and development: recent insights from chelicerates and myriapods. *Curr. Opin. Genet. Dev.* 39, 93–100.]

1709. Pacifici, M., Koyama, E., and Iwamoto, M. (2005). Mechanisms of synovial joint and articular cartilage formation: recent advances, but many lingering mysteries. *Birth Defects Res. C* 75, 237–248. [*See also* Askary, A., *et al.* (2016). Ancient origin of lubricated joints in bony vertebrates. *eLife* 5, e16415.]

1710. Packard, A. (1972). Cephalopods and fish: the limits of convergence. *Biol. Rev.* 47, 241–307.

1711. Page, D.T. (2002). Inductive patterning of the embryonic brain in *Drosophila*. *Development* 129, 2121–2128.

1712. Paixão-Côrtes, V.R., Salzano, F.M., and Bortolini, M.C. (2015). Origins and evolvability of the *PAX* family. *Semin. Cell Dev. Biol.* 44, 64–74.

1713. Palmer, A.R. (1996). From symmetry to asymmetry: phylogenetic patterns of asymmetry variation in animals and their evolutionary significance. *PNAS* 93, 14279–14286.

1714. Palmer, A.R. (2004). Symmetry breaking and the evolution of development. *Science* 306, 828–833.

1715. Palmer, L.M., Schulz, J.M., Murphy, S.C., Ledergerber, D., Murayama, M., and Larkum, M.E. (2012). The cellular basis of $GABA_B$-mediated interhemispheric inhibition. *Science* 335, 989–993.

1716. Palmer, W.J., and Jiggins, F.M. (2015). Comparative genomics reveals the origins and diversity of arthropod immune systems. *Mol. Biol. Evol.* 32, #8, 2111–2129.

1717. Panda, S., Hogenesch, J.B., and Kay, S.A. (2002). Circadian rhythms from flies to humans. *Nature* 417, 329–335.

1718. Panda, S., Nayak, S.K., Campo, B., Walker, J.R., Hogenesch, J.B., and Jegla, T. (2005). Illumination of the melanopsin signaling pathway. *Science* 307, 600–604.

1719. Pandey, U.B., and Nichols, C.D. (2011). Human disease models in *Drosophila melanogaster* and the role of the fly in therapeutic drug discovery. *Pharmacol. Rev.* 63, 411–436.

1720. Pandur, P. (2005). What does it take to make a heart? *Biol. Cell* 97, 197–210.

1721. Pandur, P., Sirbu, I.O., Kühl, S.J., Philipp, M., and Kühl, M. (2013). Islet1-expressing cardiac progenitor cells: a comparison across species. *Dev. Genes Evol.* 223, 117–129.

1722. Pang, K., and Martindale, M.Q. (2008). Developmental expression of homeobox genes in the ctenophore *Mnemiopsis leidyi*. *Dev. Genes Evol.* 218, 307–319.

1723. Panganiban, G., Irvine, S.M., Lowe, C., Roehl, H., Corley, L.S., Sherbon, B., Grenier, J.K., Fallon, J.F., Kimble, J., Walker, M., Wray, G.A., Swalla, B.J., Martindale, M.Q., and Carroll, S.B. (1997). The origin and evolution of animal appendages. *PNAS* 94, 5162–5166.

1724. Panganiban, G., and Rubenstein, J.L.R. (2002). Developmental functions of the *Distal-less*/Dlx homeobox genes. *Development* 129, 4371–4386.

1725. Pani, A.M., Mullarkey, E.E., Aronowicz, J., Assimacopoulos, S., Grove, E.A., and Lowe, C.J. (2012). Ancient deuterostome origins of vertebrate brain signalling centres. *Nature* 483, 289–294.

1726. Papatsenko, D. (2009). Stripe formation in the early fly embryo: principles, models, and networks. *BioEssays* 31, 1172–1180.

1727. Papatsenko, D., Goltsev, Y., and Levine, M. (2009). Organization of developmental enhancers in the *Drosophila* embryo. *Nucleic Acids Res.* 37, 5665–5677.

1728. Papatsenko, D., Nazina, A., and Desplan, C. (2001). A conserved regulatory element present in all *Drosophila rhodopsin* genes mediates Pax6 functions and participates in the fine-tuning of cell-specific expression. *Mech. Dev.* 101, 143–153.

1729. Papillon, D., Perez, Y., Fasano, L., Le Parco, Y., and Caubit, X. (2005). Restricted expression of a median *Hox* gene in the central nervous system of chaetognaths. *Dev. Genes Evol.* 215, 369–373.

1730. Paris, M., Escriva, H., Schubert, M., Brunet, F., Brtko, J., Ciesielski, F., Roecklin, D., Vivat-Hannah, V., Jamin, E.L., Cravedi, J.-P., Scanlan, T.S., Renaud, J.-P., Holland, N.D., and Laudet, V. (2008). Amphioxus postembryonic development reveals the homology of chordate metamorphosis. *Curr. Biol.* 18, 825–830.

1731. Park, J.H., Scheerer, P., Hofmann, K.P., and Choe, H.-W. (2008). Crystal structure of the ligand-free G-protein-coupled receptor opsin. *Nature* 454, 183–187.

1732. Park, M.S., Nakagawa, E., Schoenberg, M.R., Benbadis, S.R., and Vale, F.L. (2013). Outcome of corpus callosotomy in adults. *Epilepsy Behav.* 28, 181–184.

1733. Parker, H.J., Bronner, M.E., and Krumlauf, R. (2014). A Hox regulatory network of hindbrain segmentation is conserved to the base of vertebrates. *Nature* 514, 490–493.

1734. Parr, B.A., and McMahon, A.P. (1995). Dorsalizing signal *Wnt-7a* required for normal polarity of D-V and A-P axes of mouse limb. *Nature* 374, 350–353.

1735. Parr, B.A., Shea, M.J., Vassileva, G., and McMahon, A.P. (1993). Mouse *Wnt* genes exhibit discrete domains of expression in the early embryonic CNS and limb buds. *Development* 119, 247–261.

1736. Partridge, J.C., and Cuthill, I.C. (2010). Animal behaviour: ultraviolet fish faces. *Curr. Biol.* 20, R318–R320.

1737. Pascual, A., Huang, K.-L., Neveu, J., and Préat, T. (2004). Brain asymmetry and long-term memory. *Nature* 427, 605–606.

1738. Pascual-Anaya, J., Adachi, N., Álvarez, S., Kuratani, S., D'Aniello, S., and Garcia-Fernàndez, J. (2012). Broken colinearity of the amphioxus *Hox* cluster. *EvoDevo* 3, Article 28.

1739. Pascual-Anaya, J., Albuixech-Crespo, B., Somorjai, I.M.L., Carmona, R., Oisi, Y., Álvarez, S., Kuratani, S., Muñoz-Chápuli, R., and Garcia-Fernàndez, J. (2013). The evolutionary origins of chordate hematopoesis and vertebrate endothelia. *Dev. Biol.* 375, 182–192.

1740. Pascual-Anaya, J., D'Aniello, S., Kuratani, S., and Garcia-Fernàndez, J. (2013). Evolution of *Hox* gene clusters in deuterostomes. *BMC Dev. Biol.* 13, Article 26.

1741. Passamaneck, Y.J., Furchheim, N., Hejnol, A., Martindale, M.Q., and Lüter, C. (2011). Ciliary photoreceptors in the cerebral eyes of a protostome larva. *EvoDevo* 2, Article 6.

1742. Patel, A.J., Honoré, E., Lesage, F., Fink, M., Romey, G., and Lazdunski, M. (1999). Inhalational anesthetics activate two-pore-domain background K$^+$ channels. *Nat. Neurosci.* 2, 422–426.

1743. Patel, N.H. (1994). Developmental evolution: insights from studies of insect segmentation. *Science* 266, 581–590.

1744. Patel, N.H., Martin-Blanco, E., Coleman, K.G., Poole, S.J., Ellis, M.C., Kornberg, T.B., and Goodman, C.S. (1989). Expression of *engrailed* proteins in arthropods, annelids, and chordates. *Cell* 58, 955–968.

1745. Patel-Hett, S., and D'Amore, P.A. (2011). Signal transduction in vasculogenesis and developmental angiogenesis. *Int. J. Dev. Biol.* 55, 353–363.

1746. Paterson, J.R., García-Bellido, D.C., Lee, M.S.Y., Brock, G.A., Jago, J.B., and Edgecombe, G.D. (2011). Acute vision in the giant Cambrian predator *Anomalocaris* and the origin of compound eyes. *Nature* 480, 237–240.

1747. Patraquim, P., Warnefors, M., and Alonso, C.R. (2011). Evolution of *Hox* post-transcriptional regulation by alternative polyadenylation and microRNA modulation within 12 *Drosophila* genomes. *Mol. Biol. Evol.* 28, 2453–2460.

1748. Patthey, C., Schlosser, G., and Shimeld, S.M. (2014). The evolutionary history of vertebrate cranial placodes. I. Cell type evolution. *Dev. Biol.* 389, 82–97.

1749. Pause, B.M. (2012). Processing of body odor signals by the human brain. *Chem. Percept.* 5, 55–63.

1750. Pavlou, H.J., and Goodwin, S.F. (2013). Courtship behavior in *Drosophila melanogaster*: towards a "courtship connectome". *Curr. Opin. Neurobiol.* 23, 76–83.

1751. Pearson, J.C., Lemons, D., and McGinnis, W. (2005). Modulating Hox gene functions during animal body patterning. *Nat. Rev. Genet.* 6, 893–904.

1752. Pechmann, M., McGregor, A.P., Schwager, E.E., Feitosa, N.M., and Damen, W.G.M. (2009). Dynamic gene expression is required for anterior regionalization in a spider. *PNAS* 106, #5, 1468–1472.

1753. Pecot, M.Y., Chen, Y., Akin, O., Chen, Z., Tsui, C.Y.K., and Zipursky, S.L. (2014). Sequential axon-derived signals couple target survival and layer specificity in the *Drosophila* visual system. *Neuron* 82, 320–333.

1754. Peel, A. (2004). The evolution of arthropod segmentation mechanisms. *BioEssays* 26, 1108–1116.

1755. Peel, A., and Akam, M. (2003). Evolution of segmentation: rolling back the clock. *Curr. Biol.* 13, R708–R710.

1756. Peel, A.D. (2008). The evolution of developmental gene networks: lessons from comparative studies on holometabolous insects. *Philos. Trans. R. Soc. Lond. B* 363, 1539–1547.

1757. Peel, A.D., Chipman, A.D., and Akam, M. (2005). Arthropod segmentation: beyond the *Drosophila* paradigm. *Nat. Rev. Genet.* 6, 905–916.

1758. Peeters, H., and Devriendt, K. (2006). Human laterality disorders. *Eur. J. Med. Genet.* 49, 349–362.

1759. Peichl, L., Behrmann, G., and Kröger, R.H.H. (2001). For whales and seals the ocean is not blue: a visual pigment loss in marine mammals. *Eur. J. Neurosci.* 13, 1520–1528.

1760. Peirson, S.N., Halford, S., and Foster, R.G. (2009). The evolution of irradiance detection: melanopsin and the non-visual opsins. *Philos. Trans. R. Soc. Lond. B* 364, 2849–2865.

1761. Pelaz, S., Urquía, N., and Morata, G. (1993). Normal and ectopic domains of the homeotic gene *Sex combs reduced* of *Drosophila*. *Development* 117, 917–923.

1762. Pellikka, M., Tanentzapf, G., Pinto, M., Smith, C., McGlade, C.J., Ready, D.F., and Tepass, U. (2002). Crumbs, the *Drosophila* homologue of human CRB1/RP12, is essential for photoreceptor morphogenesis. *Nature* 416, 143–149.

1763. Peng, J., and Charron, F. (2013). Lateralization of motor control in the human nervous system: genetics of mirror movements. *Curr. Opin. Neurobiol.* 23, 109–118.

1764. Peng, Y., and Axelrod, J.D. (2012). Asymmetric protein localization in planar cell polarity: mechanisms, puzzles, and challenges. *Curr. Top. Dev. Biol.* 101, 33–53.

1765. Pennisi, E. (2013). Opsins: not just for eyes. *Science* 339, 754–755.

1766. Pennisi, E. (2015). Of mice and men. *Science* 349, 21–23.

1767. Penzias, A.A., and Wilson, R.W. (1965). A measurement of excess antenna temperature at 4080 Mc/s. *Astrophys. J.* 142, 419–421.

1768. Pérez-Pomares, J.M., González-Rosa, J.M., and Muñoz-Chápuli, R. (2009). Building the vertebrate heart: an evolutionary approach to cardiac development. *Int. J. Dev. Biol.* 53, 1427–1443.

1769. Perry, M.W., Boettiger, A.N., Bothma, J.P., and Levine, M. (2010). Shadow enhancers foster robustness of *Drosophila* gastrulation. *Curr. Biol.* 20, 1562–1567.

1770. Persat, A., and Gitai, Z. (2014). Bacterial evolution: rewiring modules to get in shape. *Curr. Biol.* 24, R522–R524. [*See also* Marshall, W.F. (2016). Cell geometry: how cells count and measure size. *Annu. Rev. Biophys.* 45, 49–64.]

1771. Pertea, M., and Salzberg, S.L. (2010). Between a chicken and a grape: estimating the number of human genes. *Genome Biol.* 11, Article 206.

1772. Perusek, L., and Maeda, T. (2013). Vitamin A derivatives as treatment options for retinal degenerative diseases. *Nutrients* 5, 2646–2666.

1773. Peso, M., Elgar, M.A., and Barron, A.B. (2015). Pheromonal control: reconciling physiological mechanism with signalling theory. *Biol. Rev.* 90, 542–559.

1774. Peterkin, T., Gibson, A., Loose, M., and Patient, R. (2005). The roles of GATA-4, -5 and -6 in vertebrate heart development. *Semin. Cell Dev. Biol.* 16, 83–94.

1775. Petralia, R.S., Mattson, M.P., and Yao, P.J. (2014). Aging and longevity in the simplest animals and the quest for immortality. *Ageing Res. Rev.* 16, 66–82.

1776. Petros, T.J., Rebsam, A., and Mason, C.A. (2008). Retinal axon growth at the optic chiasm: to cross or not to cross. *Annu. Rev. Neurosci.* 31, 295–315.

1777. Petrovic, J., Formosa-Jordan, P., Luna-Escalante, J.C., Abelló, G., Ibañes, M., Neves, J., and Giraldez, F. (2014). Ligand-dependent Notch signaling strength orchestrates lateral induction and lateral inhibition in the developing inner ear. *Development* 141, 2313–2324.

1778. Petrovic, M., and Schmucker, D. (2015). Axonal wiring in neural development: target-independent mechanisms help to establish precision and complexity. *BioEssays* 37, 996–1004.

1779. Petzoldt, A.G., Coutelis, J.-B., Géminard, C., Spéder, P., Suzanne, M., Cerezo, D., and Noselli, S. (2012). DE-Cadherin regulates unconventional Myosin ID and Myosin IC in *Drosophila* left-right asymmetry establishment. *Development* 139, 1874–1884.

1780. Peyer, B. (1947). An early description of *Drosophila*. *J. Hered.* 38, 194–199.

1781. Pham, H., Yu, H., and Laski, F.A. (2008). Cofilin/ADF is required for retinal elongation and morphogenesis of the *Drosophila* rhabdomere. *Dev. Biol.* 318, 82–91.

1782. Piatigorsky, J., and Kozmik, Z. (2004). Cubozoan jellyfish: an Evo/Devo model for eyes and other sensory systems. *Int. J. Dev. Biol.* 48, 719–729.

1783. Pichaud, F. (2014). Transcriptional regulation of tissue organization and cell morphogenesis: the fly retina as a case study. *Dev. Biol.* 385, 168–178.

1784. Pichaud, F., Briscoe, A., and Desplan, C. (1999). Evolution of color vision. *Curr. Opin. Neurobiol.* 9, 622–627.

1785. Pichaud, F., and Casares, F. (2000). *homothorax* and *iroquois-C* genes are required for the establishment of territories within the developing eye disc. *Mech. Dev.* 96, 15–25.

1786. Pichaud, F., and Desplan, C. (2002). Pax genes and eye organogenesis. *Curr. Opin. Genet. Dev.* 12, 430–434.

1787. Pichaud, F., Treisman, J., and Desplan, C. (2001). Reinventing a common strategy for patterning the eye. *Cell* 105, 9–12.

1788. Pick, L., and Heffer, A. (2012). *Hox* gene evolution: multiple mechanisms contributing to evolutionary novelties. *Ann. N. Y. Acad. Sci.* 1256, 15–32.

1789. Pickard, G.E., and Sollars, P.J. (2012). Intrinsically photosensitive retinal ganglion cells. *Rev. Physiol. Biochem. Pharmacol.* 162, 59–90.

1790. Pierce, R.J., Wu, W., Hirai, H., Ivens, A., Murphy, L.D., Noël, C., Johnston, D.A., Artiguenave, F., Adams, M., Cornette, J., Viscogliosi, E., Capron, M., and Balavoine, G. (2005). Evidence for a dispersed *Hox* gene cluster in the platyhelminth parasite *Schistosoma mansoni. Mol. Biol. Evol.* 22, 2491–2503.

1791. Piggins, H.D. (2002). Human clock genes. *Ann. Med.* 34, 394–400.

1792. Piñeiro, C., Lopes, C.S., and Casares, F. (2014). A conserved transcriptional network regulates lamina development in the *Drosophila* visual system. *Development* 141, 2838–2847.

1793. Pinheiro, D., and Bellaïche, Y. (2014). Making the most of the midbody remnant: specification of the dorsal-ventral axis. *Dev. Cell* 28, 219–220.

1794. Pinkoviezky, I., and Gov, N.S. (2014). Traffic jams and shocks of molecular motors inside cellular protrusions. *Phys. Rev. E* 89, Article 052703.

1795. Pinnell, J., Lindeman, P.S., Colavito, S., Lowe, C., and Savage, R.M. (2006). The divergent roles of the segmentation gene *hunchback. Integr. Comp. Biol.* 46, 519–532.

1796. Pirrotta, V., Chan, C.S., McCabe, D., and Qian, S. (1995). Distinct parasegmental and imaginal enhancers and the establishment of the expression pattern of the *Ubx* gene. *Genetics* 141, 1439–1450.

1797. Pitt, J.N., and Kaeberlein, M. (2015). Why is aging conserved and what can we do about it? *PLoS Biol.* 13, #4, e1002131.

1798. Pizzari, T. (2006). Evolution: the paradox of sperm leviathans. *Curr. Biol.* 16, R462–R464.

1799. Plachetzki, D.C., Degnan, B.M., and Oakley, T.H. (2007). The origins of novel protein interactions during animal opsin evolution. *PLoS ONE* 2, #10, e1054.

1800. Plachetzki, D.C., Fong, C.R., and Oakley, T.H. (2010). The evolution of phototransduction from an ancestral cyclic nucleotide gated pathway. *Proc. R. Soc. Lond. B* 277, 1963–1969.

1801. Plachetzki, D.C., and Oakley, T.H. (2007). Key transitions during the evolution of animal phototransduction: novelty, "tree-thinking," co-option, and co-duplication. *Integr. Comp. Biol.* 47, 759–769.

1802. Plachetzki, D.C., Serb, J.M., and Oakley, T.H. (2005). New insights into the evolutionary history of photoreceptor cells. *Trends Ecol. Evol.* 20, 465–467.

1803. Plavicki, J., Mader, S., Pueschel, E., Peebles, P., and Boekhoff-Falk, G. (2012). Homeobox gene *distal-less* is required for neuronal differentiation and neurite outgrowth in the *Drosophila* olfactory system. *PNAS* 109, #5, 1578–1583.

1804. Plavicki, J.S., Squirrell, J.M., Eliceiri, K.W., and Boekhoff-Falk, G. (2016). Expression of the *Drosophila* homeobox gene, *Distal-less*, supports an ancestral role in neural development. *Dev. Dyn.* 245, 87–95.

1805. Pohl, C., and Bao, Z. (2010). Chiral forces organize left-right patterning in *C. elegans* by uncoupling midline and anteroposterior axis. *Dev. Cell* 19, 402–412.

1806. Pollock, J.A., and Benzer, S. (1988). Transcript localization of four opsin genes in the three visual organs of *Drosophila*: RH2 is ocellus specific. *Nature* 333, 779–782.

1807. Ponce, C.R., and Born, R.T. (2008). Stereopsis. *Curr. Biol.* 18, R845–R850.

1808. Ponnambalam, S., and Alberghina, M. (2011). Evolution of the VEGF-regulated vascular network from a neural guidance system. *Mol. Neurobiol.* 43, 192–206.

1809. Poodry, C.A. (1980). Epidermis: morphology and development. *In* M. Ashburner and T.R.F. Wright (eds.), *The Genetics and Biology of Drosophila, Vol. 2d*. Academic Press, New York, NY, pp. 443–497.

1810. Popovici, C., Isnardon, D., Birnbaum, D., and Roubin, R. (2002). *Caenorhabditis elegans* receptors related to mammalian vascular endothelial growth factor receptors are expressed in neural cells. *Neurosci. Lett.* 329, 116–120.

1811. Porcher, A., and Dostatni, N. (2010). The Bicoid morphogen system. *Curr. Biol.* 20, R249–R254.

1812. Porter, M.L., Blasic, J.R., Bok, M.J., Cameron, E.G., Pringle, T., Cronin, T.W., and Robinson, P.R. (2012). Shedding new light on opsin evolution. *Proc. R. Soc. Lond. B* 279, 3–14.

1813. Portugues, R., Severi, K.E., Wyart, C., and Ahrens, M.B. (2013). Optogenetics in a transparent animal: circuit function in the larval zebrafish. *Curr. Opin. Neurobiol.* 23, 119–126.

1814. Poss, K.D. (2010). Advances in understanding tissue regenerative capacity and mechanisms in animals. *Nat. Rev. Genet.* 11, 710–722.

1815. Poulain, F.F., and Yost, H.J. (2015). Heparan sulfate proteoglycans: a sugar code for vertebrate development? *Development* 142, 3456–3467.

1816. Pourquié, O. (2003). Vertebrate somitogenesis: a novel paradigm for animal segmentation? *Int. J. Dev. Biol.* 47, 597–603.

1817. Powell, L.M., and Jarman, A.P. (2008). Context dependence of proneural bHLH proteins. *Curr. Opin. Genet. Dev.* 18, 411–417.

1818. Prasad, B.C., and Reed, R.R. (1999). Chemosensation: molecular mechanisms in worms and mammals. *Trends Genet.* 15, 150–153.

1819. Prasov, L., and Glaser, T. (2012). Pushing the envelope of retinal ganglion cell genesis: context dependent function of *Math5* (*Atoh7*). *Dev. Biol.* 368, 214–230.

1820. Pregitzer, P., Greschista, M., Breer, H., and Krieger, J. (2014). The sensory neurone membrane protein SNMP1 contributes to the sensitivity of a pheromone detection system. *Insect Mol. Biol.* 23, 733–742.

1821. Pribil, M., Labs, M., and Leister, D. (2014). Structure and dynamics of thylakoids in land plants. *J. Exp. Botany* 65, 1955–1972.

1822. Price, D.C., Egizi, A., and Fonseca, D.M. (2015). Characterization of the *doublesex* gene within the *Culex pipiens* complex suggests regulatory plasticity at the base of the mosquito sex determination cascade. *BMC Evol. Biol.* 15, Article 108.

1823. Prieto-Godino, L.L., Diegelmann, S., and Bate, M. (2012). Embryonic origin of olfactory circuitry in *Drosophila*: contact and activity-mediated interactions pattern connectivity in the antennal lobe. *PLoS Biol.* 10, #10, e1001400.

1824. Prince, V.E. (2002). The Hox paradox: more complex(es) than imagined. *Dev. Biol.* 249, 1–15.

1825. Prochiantz, A., and Joliot, A. (2003). Can transcription factors function as cell-cell signalling molecules? *Nat. Rev. Mol. Cell Biol.* 4, 814–819.

1826. Prosser, H.M., Rzadzinska, A.K., Steel, K.P., and Bradley, A. (2008). Mosaic complementation demonstrates a regulatory role for myosin VIIa in actin dynamics of stereocilia. *Mol. Cell. Biol.* 28, #5, 1702–1712.

1827. Provencio, I. (2011). The hidden organ in our eyes. *Sci. Am.* 304, #5, 54–59.

1828. Provencio, I., Rollag, M.D., and Castrucci, A.M. (2002). Photoreceptive net in the mammalian retina. *Nature* 415, 493.

1829. Prpic, N.-M. (2008). Arthropod appendages: a prime example for the evolution of morphological diversity and innovation. *In* A. Minelli and G. Fusco (eds.), *Evolving Pathways: Key Themes in Evolutionary Developmental Biology*. Cambridge University Press, New York, NY, pp. 381–398.

1830. Prpic, N.-M., and Damen, W.G.M. (2009). *Notch*-mediated segmentation of the appendages is a molecular phylotypic trait of the arthropods. *Dev. Biol.* 326, 262–271.

1831. Prud'homme, B., de Rosa, R., Arendt, D., Julien, J.-F., Pajaziti, R., Dorresteijn, A.W.C., Adoutte, A., Wittbrodt, J., and Balavoine, G. (2003). Arthropod-like expression patterns of *engrailed* and *wingless* in the annelid *Platynereis dumerilii* suggest a role in segment formation. *Curr. Biol.* 13, 1876–1881.

1832. Pueyo, J.I., and Couso, J.P. (2005). Parallels between the proximal-distal development of vertebrate and arthropod appendages: homology without an ancestor? *Curr. Opin. Genet. Dev.* 15, 439–446.

1833. Pugh, E.N. (2001). Rods are rods and cones cones, and (never) the twain shall meet. *Neuron* 32, 375–380.

1834. Pujol, R., Pickett, S.B., Nguyen, T.B., and Stone, J.S. (2014). Large basolateral processes on type II hair cells are novel processing units in mammalian vestibular organs. *J. Comp. Neurol.* 522, 3141–3159.

1835. Punzo, C., Kurata, S., and Gehring, W.J. (2001). The *eyeless* homeodomain is dispensable for eye development in *Drosophila*. *Genes Dev.* 15, 1716–1723.

1836. Purcell, P., Oliver, G., Mardon, G., Donner, A.L., and Maas, R.L. (2005). Pax6-dependence of Six3, Eya1 and Dach1 expression during lens and nasal placode induction. *Gene Expr. Patterns* 6, 110–118.

1837. Purschke, G., Arendt, D., Hausen, H., and Müller, M.C.M. (2006). Photoreceptor cells and eyes in Annelida. *Arthropod Struct. Dev.* 35, 211–230.

1838. Putnam, N.H., Butts, T., Ferrier, D.E.K., Furlong, R.F., Hellsten, U., Kawashima, T., Robinson-Rechavi, M., Shoguchi, E., Terry, A., Yu, J.-K., Benito-Gutiérrez, È., Dubchak, I., Garcia-Fernàndez, J., Gibson-Brown, J.J., Grigoriev, I.V., Horton, A.C., de Jong, P.J., Jurka, J., Kapitonov, V.V., Kohara, Y., Kuroki, Y., Lindquist, E., Lucas, S., Osoegawa, K., Pennacchio, L.A., Salamov, A.A., Satou, Y., Sauka-Spengler, T., Schmutz, J., Shin-I, T., Toyoda, A., Bronner-Fraser, M., Fujiyama, A., Holland, L.Z., Holland, P.W.H., Satoh, N., and Rokhsar, D.S. (2008). The amphioxus genome and the evolution of the chordate karyotype. *Nature* 453, 1064–1071.

1839. Pyrpassopoulos, S., Feeser, E.A., Mazerik, J.N., Tyska, M.J., and Ostap, E.M. (2012). Membrane-bound Myo1c powers asymmetric motility of actin filaments. *Curr. Biol.* 22, 1688–1692.

1840. Qian, L., Liu, J., and Bodmer, R. (2005). *Neuromancer* Tbx20-related genes (*H15/midline*) promote cell fate specification and morphogenesis of the *Drosophila* heart. *Dev. Biol.* 279, 509–524.

1841. Qian, L., Wythe, J.D., Liu, J., Cartry, J., Vogler, G., Mohapatra, B., Otway, R.T., Huang, Y., King, I.N., Maillet, M., Zheng, Y., Crawley, T., Taghli-Lamallem, O.,

Semsarian, C., Dunwoodie, S., Winlaw, D., Harvey, R.P., Fatkin, D., Towbin, J.A., Molkentin, J.D., Srivastava, D., Ocorr, K., Bruneau, B.G., and Bodmer, R. (2011). Tinman/Nkx2-5 acts via miR-1 and upstream of Cdc42 to regulate heart function across species. *J. Cell Biol.* 193, 1181–1196.

1842. Quan, X.-J., and Hassan, B.A. (2005). From skin to nerve: flies, vertebrates and the first helix. *Cell. Mol. Life Sci.* 62, 2036–2049.

1843. Quiquand, M., Yanze, N., Schmich, J., Schmid, V., Galliot, B., and Piraino, S. (2009). More constraint on *ParaHox* than *Hox* gene families in early metazoan evolution. *Dev. Biol.* 328, 173–187.

1844. Quiring, R., Walldorf, U., Kloter, U., and Gehring, W.J. (1994). Homology of the *eyeless* gene of *Drosophila* to the *Small eye* gene in mice and *Aniridia* in humans. *Science* 265, 785–789.

1845. Rabe, N., Gezelius, H., Vallstedt, A., Memic, F., and Kullander, K. (2009). Netrin-1-dependent spinal interneuron subtypes are required for the formation of left-right alternating locomotor circuitry. *J. Neurosci.* 29, #50, 15642–15649.

1846. Rader, A.J., Anderson, G., Isin, B., Khorana, H.G., Bahar, I., and Klein-Seetharaman, J. (2004). Identification of core amino acids stabilizing rhodopsin. *PNAS* 101, #19, 7246–7251.

1847. Raff, R.A. (1996). *The Shape of Life: Genes, Development, and the Evolution of Animal Form.* University of Chicago Press, Chicago, IL.

1848. Raff, R.A., and Kaufman, T.C. (1983). *Embryos, Genes, and Evolution: The Developmental-Genetic Basis of Evolutionary Change.* Macmillan, New York, NY.

1849. Raft, S., and Groves, A.K. (2015). Segregating neural and mechanosensory fates in the developing ear: patterning, signaling, and transcriptional control. *Cell Tissue Res.* 359, 315–332.

1850. Raines, A.M., Magella, B., Adam, M., and Potter, S.S. (2015). Key pathways regulated by *HoxA9,10,11/HoxD9,10,11* during limb development. *BMC Dev. Biol.* 15, Article 28.

1851. Raj, B., and Blencowe, B.J. (2015). Alternative splicing in the mammalian nervous system: recent insights into mechanisms and functional roles. *Neuron* 87, 14–27.

1852. Rajagopalan, S., Vivancos, V., Nicolas, E., and Dickson, B.J. (2000). Selecting a longitudinal pathway: Robo receptors specify the lateral position of axons in the *Drosophila* CNS. *Cell* 103, 1033–1045.

1853. Rakic, P. (2009). Evolution of the neocortex: a perspective from developmental biology. *Nat. Rev. Neurosci.* 10, 724–735.

1854. Ramaekers, A., Magnenat, E., Marin, E.C., Gendre, N., Jefferis, G.S.X.E., Luo, L., and Stocker, R.F. (2005). Glomerular maps without cellular redundancy at successive levels of the *Drosophila* larval olfactory circuit. *Curr. Biol.* 15, 982–992.

1855. Ramanathan, D.S., Gulati, T., and Ganguly, K. (2015). Sleep-dependent reactivation of ensembles in motor cortex promotes skill consolidation. *PLoS Biol.* 13, #9, e1002263.

1856. Ramdya, P., and Benton, R. (2010). Evolving olfactory systems on the fly. *Trends Genet.* 26, 307–316.

1857. Ramel, M.-C., and Hill, C.S. (2013). The ventral to dorsal BMP activity gradient in the early zebrafish embryo is determined by graded expression of BMP ligands. *Dev. Biol.* 378, 170–182.

1858. Ramirez, M.D., and Oakley, T.H. (2015). Eye-independent, light-activated chromatophore expansion (LACE) and expression of phototransduction genes in the skin of *Octopus bimaculoides*. *J. Exp. Biol.* 218, 1513–1520.

1859. Ramos, C., and Robert, B. (2005). *msh/Msx* gene family in neural development. *Trends Genet.* 21, 624–632.

1860. Ramos, O.M., Barker, D., and Ferrier, D.E.K. (2012). Ghost loci imply Hox and ParaHox existence in the last common ancestor of animals. *Curr. Biol.* 22, 1951–1956.

1861. Ranade, S.S., Yang-Zhou, D., Kong, S.W., McDonald, E.C., Cook, T.A., and Pignoni, F. (2008). Analysis of the Otd-dependent transcriptome supports the evolutionary conservation of CRX/OTX/OTD functions in flies and vertebrates. *Dev. Biol.* 315, 521–534.

1862. Ranganayakulu, G., Elliott, D.A., Harvey, R.P., and Olson, E.N. (1998). Divergent roles for *NK-2* class homeobox genes in cardiogenesis in flies and mice. *Development* 125, 3037–3048.

1863. Rao-Mirotznik, R., Harkins, A.B., Buchsbaum, G., and Sterling, P. (1995). Mammalian rod terminal: architecture of a binary synapse. *Neuron* 14, 561–569.

1864. Rasband, K., Hardy, M., and Chien, C.-B. (2003). Generating X: formation of the optic chiasm. *Neuron* 39, 885–888.

1865. Rauskolb, C. (2001). The establishment of segmentation in the *Drosophila* leg. *Development* 128, 4511–4521.

1866. Ravni, A., Qu, Y., Goffinet, A.M., and Tissir, F. (2009). Planar cell polarity cadherin Celsr1 regulates skin hair patterning in the mouse. *J. Invest. Dermatol.* 129, 2507–2509.

1867. Ray, A., van der Goes van Naters, W., Shiraiwa, T., and Carlson, J.R. (2007). Mechanisms of odor receptor gene choice in *Drosophila*. *Neuron* 53, 353–369.

1868. Raya, Á., and Izpisúa Belmonte, J.C. (2006). Left-right asymmetry in the vertebrate embryo: from early information to higher-level integration. *Nat. Rev. Genet.* 7, 283–293.

1869. Raybaud, C. (2010). The corpus callosum, the other great forebrain commissures, and the septum pellucidum: anatomy, development, and malformation. *Neuroradiology* 52, 447–477.

1870. Raymond, P.A., and Barthel, L.K. (2004). A moving wave patterns the cone photoreceptor mosaic array in the zebrafish retina. *Int. J. Dev. Biol.* 48, 935–945.

1871. Raymond, P.A., Colvin, S.M., Jabeen, Z., Nagashima, M., Barthel, L.K., Hadidjojo, J., Popova, L., Pejaver, V.R., and Lubensky, D.K. (2014). Patterning the cone mosaic array in zebrafish retina requires specification of ultraviolet-sensitive cones. *PLoS ONE* 9, #1, e85325.

1872. Ready, D.F., Hanson, T.E., and Benzer, S. (1976). Development of the *Drosophila* retina, a neurocrystalline lattice. *Dev. Biol.* 53, 217–240.

1873. Rebay, I., Silver, S.J., and Tootle, T.L. (2005). New vision from Eyes absent: transcription factors as enzymes. *Trends Genet.* 21, 163–171.

1874. Rebeiz, M., Castro, B., Liu, F., Yue, F., and Posakony, J.W. (2012). Ancestral and conserved *cis*-regulatory architectures in developmental control genes. *Dev. Biol.* 362, 282–294.

1875. Rebeiz, M., Stone, T., and Posakony, J.W. (2005). An ancient transcriptional regulatory linkage. *Dev. Biol.* 281, 299–308.

1876. Reddien, P.W., Bermange, A.L., Kicza, A.M., and Sánchez Alvarado, A. (2007). BMP signaling regulates the dorsal planarian midline and is needed for asymmetric regeneration. *Development* 134, 4043–4051.

1877. Reddy, P.C., Unni, M.K., Gungi, A., Agarwal, P., and Galande, S. (2015). Evolution of Hox-like genes in Cnidaria: study of Hydra Hox repertoire reveals tailor-made Hox-code for Cnidarians. *Mech. Dev.* 138, 87–96.

1878. Reed, R.R. (2004). After the Holy Grail: establishing a molecular basis for mammalian olfaction. *Cell* 116, 329–336.

1879. Reese, B.E. (2011). Development of the retina and optic pathway. *Vision Res.* 51, 613–632.

1880. Reese, B.E., and Keeley, P.W. (2015). Design principles and developmental mechanisms underlying retinal mosaics. *Biol. Rev.* 90, 854–876.

1881. Reese, B.E., and Tan, S.-S. (1998). Clonal boundary analysis in the developing retina using X-inactivation transgenic mosaic mice. *Semin. Cell Dev. Biol.* 9, 285–292.

1882. Reeves, R.R., and Mitchell, E. (1981). The whale behind the tusk. *Nat. Hist.* 90, #8, 50–57.

1883. Rehorn, K.-P., Thelen, H., Michelson, A.M., and Reuter, R. (1996). A molecular aspect of hematopoiesis and endoderm development common to vertebrates and *Drosophila*. *Development* 122, 4023–4031.

1884. Reichert, H. (2009). Evolutionary conservation of mechanisms for neural regionalization, proliferation and interconnection in brain development. *Biol. Lett.* 5, 112–116.

1885. Reilly, S.K., Yin, J., Ayoub, A.E., Emera, D., Long, J., Cotney, J., Sarro, R., Rakic, P., and Noonan, J.P. (2015). Evolutionary changes in promoter and enhancer activity during human corticogenesis. *Science* 347, 1155–1159.

1886. Reim, I., and Frasch, M. (2010). Genetic and genomic dissection of cardiogenesis in the *Drosophila* model. *Pediatr. Cardiol.* 31, 325–334.

1887. Reim, I., Mohler, J.P., and Frasch, M. (2005). *Tbx20*-related genes, *mid* and *H15*, are required for *tinman* expression, proper patterning, and normal differentiation of cardioblasts in *Drosophila*. *Mech. Dev.* 122, 1056–1069.

1888. Reingruber, J., Holcman, D., and Fain, G.L. (2015). How rods respond to single photons: key adaptations of a G-protein cascade that enable vision at the physical limit of perception. *BioEssays* 37, 1243–1252.

1889. Reiter, L.T., Potocki, L., Chien, S., Gribskov, M., and Bier, E. (2001). A systematic analysis of human disease-associated gene sequences in *Drosophila melanogaster*. *Genome Res.* 11, 1114–1125.

1890. Repérant, J., Lemire, M., Miceli, D., and Peyrichoux, J. (1976). A radioautographic study of the visual system in fresh water teleosts following intraocular injection of tritiated fucose and proline. *Brain Res.* 118, 123–131.

1891. Reppert, S.M., and Weaver, D.R. (2002). Coordination of circadian timing in mammals. *Nature* 418, 935–941.

1892. Reuter, M., Mäntylä, K., and Gustafsson, M.K.S. (1998). Organization of the orthogon: main and minor nerve cords. *Hydrobiologia* 383, 175–182.

1893. Rezával, C., Fabre, C.C.G., and Goodwin, S.F. (2011). Invertebrate neuroethology: food play and sex. *Curr. Biol.* 21, R960–R962.

1894. Rezsohazy, R., Saurin, A.J., Maurel-Zaffran, C., and Graba, Y. (2015). Cellular and molecular insights into Hox protein action. *Development* 142, 1212–1227.

1895. Rhinn, M., and Dollé, P. (2012). Retinoic acid signalling during development. *Development* 139, 843–858.

1896. Riabinina, O., Dai, M., Duke, T., and Albert, J.T. (2011). Active process mediates species-specific tuning of *Drosophila* ears. *Curr. Biol.* 21, 658–664.

1897. Riccomagno, M.M., and Kolodkin, A.L. (2015). Sculpting neural circuits by axon and dendrite pruning. *Annu. Rev. Cell Dev. Biol.* 31, 779–805.

1898. Richards, T.A., and Gomes, S.L. (2015). How to build a microbial eye. *Nature* 523, 166–167.

1899. Richardson, M.K. (2009). The *Hox* complex: an interview with Denis Duboule. *Int. J. Dev. Biol.* 53, 717–723.

1900. Richmond, D.L., and Oates, A.C. (2012). The segmentation clock: inherited trait or universal design principle? *Curr. Opin. Genet. Dev.* 22, 600–606.

1901. Rida, P.C.G., and Chen, P. (2009). Line up and listen: planar cell polarity regulation in the mammalian inner ear. *Semin. Cell Dev. Biol.* 20, 978–985.

1902. Riddle, R.D., Ensini, M., Nelson, C., Tsuchida, T., Jessell, T.M., and Tabin, C. (1995). Induction of the LIM homeobox gene Lmx1 by WNT7a establishes dorsoventral pattern in the vertebrate limb. *Cell* 83, 631–640.

1903. Ridley, M. (2004). *Evolution*, 3rd edn. Blackwell, Malden, MA.

1904. Rieckhof, G.E., Casares, F., Ryoo, H.D., Abu-Shaar, M., and Mann, R.S. (1997). Nuclear translocation of Extradenticle requires *homothorax*, which encodes an Extradenticle-related homeodomain protein. *Cell* 91, 171–183.

1905. Rieder, L.E., and Larschan, E.N. (2014). Wisdom from the fly. *Trends Genet.* 30, 479–481.

1906. Rieke, F., and Rudd, M.E. (2009). The challenges natural images pose for visual adaptation. *Neuron* 64, 605–616.

1907. Riley, B.B., Chiang, M.-Y., Farmer, L., and Heck, R. (1999). The *deltaA* gene of zebrafish mediates lateral inhibition of hair cells in the inner ear and is regulated by *pax2.1*. *Development* 126, 5669–5678.

1908. Rincón-Limas, D.E., Lu, C.-H., Canal, I., Calleja, M., Rodríguez-Esteban, C., Izpisúa-Belmonte, J.C., and Botas, J. (1999). Conservation of the expression and function of *apterous* orthologs in *Drosophila* and mammals. *PNAS* 96, 2165–2170.

1909. Rister, J., and Desplan, C. (2011). The retinal mosaics of opsin expression in invertebrates and vertebrates. *Dev. Neurobiol.* 71, 1212–1226. [*See also* Perry, M., *et al.* (2016). Molecular logic behind the three-way stochastic choices that expand butterfly colour vision. *Nature* 535, 280–284.]

1910. Rister, J., Desplan, C., and Vasiliauskas, D. (2013). Establishing and maintaining gene expression patterns: insights from sensory receptor patterning. *Development* 140, 493–503.

1911. Rister, J., Razzaq, A., Boodram, P., Desai, N., Tsanis, C., Chen, H., Jukam, D., and Desplan, C. (2015). Single-base pair differences in a shared motif determine differential *Rhodopsin* expression. *Science* 350, 1258–1261.

1912. Rivera, A.S., and Weisblat, D.A. (2009). And Lophotrochozoa makes three: Notch/Hes signaling in annelid segmentation. *Dev. Genes Evol.* 219, 37–43.

1913. Robert, D., and Hoy, R.R. (2007). Auditory systems in insects. *In* G. North and R.J. Greenspan (eds.), *Invertebrate Neurobiology*. Cold Spring Harbor Laboratory Press, Cold Spring Harbor, NY, pp. 155–184.

1914. Robert, J.S. (2001). Interpreting the homeobox: metaphors of gene action and activation in development and evolution. *Evol. Dev.* 3, 287–295.

1915. Robinson, B.G., Khurana, S., Kuperman, A., and Atkinson, N.S. (2012). Neural adaptation leads to cognitive ethanol dependence. *Curr. Biol.* 22, 2338–2341.

1916. Robinson, B.G., Khurana, S., Pohl, J.B., Li, W.-k., Ghezzi, A., Cady, A.M., Najjar, K., Hatch, M.M., Shah, R.R., Bhat, A., Hariri, O., Haroun, K.B., Young, M.C., Fife, K., Hooten, J., Tran, T., Goan, D., Desai, F., Husain, F., Godinez, R.M., Sun, J.C., Corpuz, J., Moran, J., Zhong, A.C., Chen, W.Y., and Atkinson, N.S. (2011). A low concentration of ethanol impairs learning but not motor and sensory behavior in *Drosophila* larvae. *PLoS ONE* 7, #5, e37394.

1917. Robledo, R.F., Rajan, L., Li, X., and Lufkin, T. (2002). The *Dlx5* and *Dlx6* homeobox genes are essential for craniofacial, axial, and appendicular skeletal development. *Genes Dev.* 16, 1089–1101.

1918. Rochlin, K., Yu, S., Roy, S., and Baylies, M.K. (2010). Myoblast fusion: when it takes more to make one. *Dev. Biol.* 341, 66–83.

1919. Rodriguez, I., Greer, C.A., Mok, M.Y., and Mombaerts, P. (2000). A putative pheromone receptor gene expressed in human olfactory mucosa. *Nat. Genet.* 26, 18–19.

1920. Rodriguez-Estaban, C., Schwabe, J.W.R., de la Peña, J., Foys, B., Eshelman, B., and Izpisua Belmonte, J.C. (1997). *Radical fringe* positions the apical ectodermal ridge at the dorsoventral boundary of the vertebrate limb. *Nature* 386, 360–366.

1921. Rodríguez-Trelles, F., Tarrío, R., and Ayala, F.J. (2005). Is ectopic expression caused by deregulatory mutations or due to gene-regulation leaks with evolutionary potential? *BioEssays* 27, 592–601.

1922. Rogers, G.E. (2004). Hair follicle differentiation and regulation. *Int. J. Dev. Biol.* 48, 163–170.

1923. Rogers, K.W., and Schier, A.F. (2011). Morphogen gradients: from generation to interpretation. *Annu. Rev. Cell Dev. Biol.* 27, 377–407.

1924. Rogers, L.J. (2014). Asymmetry of brain and behavior in animals: its development, function, and human relevance. *Genesis* 52, 555–571.

1925. Rogers, L.J., Vallortigara, G., and Andrew, R.J. (2013). *Divided Brains: The Biology and Behaviour of Brain Asymmetries.* Cambridge University Press, Cambridge.

1926. Rogulja-Ortmann, A., Lüer, K., Seibert, J., Rickert, C., and Technau, G.M. (2007). Programmed cell death in the embryonic central nervous system of *Drosophila melanogaster. Development* 134, 105–116.

1927. Rokas, A. (2008). The origins of multicellularity and the early history of the genetic toolkit for animal development. *Annu. Rev. Genet.* 42, 235–251.

1928. Root, C.M., Denny, C.A., Hen, R., and Axel, R. (2014). The participation of cortical amygdala in innate odour-driven behaviour. *Nature* 515, 269–273.

1929. Rosato, E., and Kyriacou, C.P. (2011). The role of natural selection in circadian behaviour: a molecular-genetic approach. *Essays Biochem.* 49, 71–85.

1930. Rosbash, M. (2009). The implications of multiple circadian clock origins. *PLoS Biol.* 7, #3, e1000062.

1931. Rosenbaum, D.M., Rasmussen, S.G.F., and Kobilka, B.K. (2009). The structure and function of G-protein-coupled receptors. *Nature* 459, 356–363.

1932. Rosenberg, M.I., Lynch, J.A., and Desplan, C. (2009). Heads and tails: evolution of antero-posterior patterning in insects. *Biochim. Biophys. Acta* 1789, 333–342.

1933. Rosenstein, J.M., Krum, J.M., and Ruhrberg, C. (2010). VEGF in the nervous system. *Organogenesis* 6, 107–114.

1934. Rosenwasser, A.M. (2009). Functional neuroanatomy of sleep and circadian rhythms. *Brain Res. Rev.* 61, 281–306.

1935. Rosselló, R.A., Chen, C.-C., Dai, R., Howard, J.T., Hochgeschwender, U., and Jarvis, E.D. (2013). Mammalian genes induce partially reprogrammed pluripotent stem cells in non-mammalian vertebrate and invertebrate species. *eLife* 2, e00036.

1936. Roth, G. (2015). Convergent evolution of complex brains and high intelligence. *Philos. Trans. R. Soc. Lond. B* 370, 20150049.

1937. Roth, S., and Lynch, J. (2012). Does the Bicoid gradient matter? *Cell* 149, 511–512.

1938. Roth, S., and Panfilio, K.A. (2012). Making waves for segments. *Science* 336, 306–307.

1939. Roth, S., Stein, D., and Nüsslein-Volhard, C. (1989). A gradient of nuclear localization of the *dorsal* protein determines dorsoventral pattern in the *Drosophila* embryo. *Cell* 59, 1189–1202.

1940. Röttinger, E., and Lowe, C.J. (2012). Evolutionary crossroads in developmental biology: hemichordates. *Development* 139, 2463–2475.

1941. Roush, S., and Slack, F.J. (2008). The *let-7* family of microRNAs. *Trends Cell Biol.* 18, 505–516.

1942. Rovira, M., Saavedra, P., Casal, J., and Lawrence, P.A. (2015). Regions within a single epidermal cell of *Drosophila* can be planar polarised independently. *eLife* 4, e06303.

1943. Roy, S.W. (2006). Intron-rich ancestors. *Trends Genet.* 22, 468–471.

1944. Royo, J.L., Maeso, I., Irimia, M., Gao, F., Peter, I.S., Lopes, C.S., D'Aniello, S., Casares, F., Davidson, E.H., Garcia-Fernàndez, J., and Gómez-Skarmeta, J.L. (2011). Transphyletic conservation of developmental regulatory state in animal evolution. *PNAS* 108, #34, 14186–14191.

1945. Ru, H., Chambers, M.G., Fu, T.-M., Tong, A.B., Liao, M., and Wu, H. (2015). Molecular mechanism of V(D)J recombination from synaptic RAG1-RAG2 complex structures. *Cell* 163, 1138–1152.

1946. Rubin, G.M. (2001). The draft sequences: comparing species. *Nature* 409, 820–821.

1947. Rubinstein, R., Thu, C.A., Goodman, K.M., Wolcott, H.N., Bahna, F., Mannepalli, S., Ahlsen, G., Chevee, M., Halim, A., Clausen, H., Maniatis, T., Shapiro, L., and Honig, B. (2105). Molecular logic of neuronal self-recognition through protocadherin domain interactions. *Cell* 163, 629–642.

1948. Rudrapatna, V.A., Cagan, R.L., and Das, T.K. (2012). *Drosophila* cancer models. *Dev. Dyn.* 241, 107–118.

1949. Rushlow, C.A., Han, K., Manley, J.L., and Levine, M. (1989). The graded distribution of the *dorsal* morphogen is initiated by selective nuclear transport in *Drosophila*. *Cell* 59, 1165–1177.

1950. Rusten, T.E., Cantera, R., Kafatos, F.C., and Barrio, R. (2002). The role of TGFβ signaling in the formation of the dorsal nervous system is conserved between *Drosophila* and chordates. *Development* 129, 3575–3584.

1951. Ruzickova, J., Piatigorsky, J., and Kozmik, Z. (2009). Eye-specific expression of an ancestral jellyfish *PaxB* gene interferes with *Pax6* function despite its conserved Pax6/Pax2 characteristics. *Int. J. Dev. Biol.* 53, 469–482.

1952. Ryan, J.F., and Baxevanis, A.D. (2007). Hox, Wnt, and the evolution of the primary body axis: insights from the early-divergent phyla. *Biol. Direct* 2, Article 37.

1953. Ryan, J.F., and Chiodin, M. (2015). Where is my mind? How sponges and placozoans may have lost neural cell types. *Philos. Trans. R. Soc. Lond. B* 370, 20150059.

1954. Ryan, J.F., Mazza, M.E., Pang, K., Matus, D.Q., Baxevanis, A.D., Martindale, M.Q., and Finnerty, J.R. (2007). Pre-bilaterian origins of the *Hox* cluster and the *Hox* code: evidence from the sea anemone, *Nematostella vectensis*. *PLoS ONE* 2, #1, e153.

1955. Ryan, T.J., and Grant, S.G.N. (2009). The origin and evolution of synapses. *Nat. Rev. Neurosci.* 10, 701–712.

1956. Ryoo, H.D., Marty, T., Casares, F., Affolter, M., and Mann, R.S. (1999). Regulation of Hox target genes by a DNA bound Homothorax/Hox/Extradenticle complex. *Development* 126, 5137–5148.

1957. Rytz, R., Croset, V., and Benton, R. (2013). Ionotropic receptors (IRs): chemosensory ionotropic glutamate receptors in *Drosophila* and beyond. *Insect Biochem. Mol. Biol.* 43, 888–897.

1958. Ryu, J.-R., Najand, N., and Brook, W.J. (2011). Tinman is a direct activator of *midline* in the *Drosophila* dorsal vessel. *Dev. Dyn.* 240, 86–95.

1959. Saavedra, P., Vincent, J.-P., Palacios, I.M., Lawrence, P.A., and Casal, J. (2014). Plasticity of both planar cell polarity and cell identity during the development of *Drosophila*. *eLife* 3, e01569.

1960. Saburi, S., Hester, I., Goodrich, L., and McNeill, H. (2012). Functional interactions between Fat family cadherins in tissue morphogenesis and planar polarity. *Development* 139, 1806–1820.

1961. Sadaf, S., Reddy, O.V., Sane, S.P., and Hasan, G. (2015). Neural control of wing coordination in flies. *Curr. Biol.* 25, 80–86.

1962. Saga, Y. (2012). The mechanism of somite formation in mice. *Curr. Opin. Genet. Dev.* 22, 331–338.

1963. Saha, D., and Raman, B. (2015). Relating early olfactory processing with behavior: a perspective. *Curr. Opin. Insect Sci.* 12, 54–63.

1964. Sakamaki, K., Imai, K., Tomii, K., and Miller, D.J. (2015). Evolutionary analyses of caspase-8 and its paralogs: deep origins of the apoptotic signaling pathways. *BioEssays* 37, 767–776.

1965. Sakmar, T.P. (2012). Redder than red. *Science* 338, 1299–1300.

1966. Sakurai, T., Namiki, S., and Kanzaki, R. (2014). Molecular and neural mechanisms of sex pheromone reception and processing in the silkmoth *Bombyx mori*. *Front. Physiol.* 5, Article 125.

1967. Salcedo, E., Farrell, D.M., Zheng, L., Phistry, M., Bagg, E.E., and Britt, S.G. (2009). The green-absorbing *Drosophila* Rh6 visual pigment contains a blue-shifting amino acid substitution that is conserved in vertebrates. *J. Biol. Chem.* 284, 5717–5722.

1968. Salcedo, E., Huber, A., Henrich, S., Chadwell, L.V., Chou, W.-H., Paulsen, R., and Britt, S.G. (1999). Blue- and green-absorbing visual pigments of *Drosophila*: ectopic expression and physiological characterization of the R8 photoreceptor cell-specific Rh5 and Rh6 rhodopsins. *J. Neurosci.* 19, #24, 10716–10726.

1969. Sallé, J., Campbell, S.D., Gho, M., and Audibert, A. (2012). CycA is involved in the control of endoreplication dynamics in the *Drosophila* bristle lineage. *Development* 139, 547–557.

1970. Salzberg, A., and Bellen, H.J. (1996). Invertebrate versus vertebrate neurogenesis: variations on the same theme? *Dev. Genet.* 18, 1–10.

1971. Samadi, L., and Steiner, G. (2010). Expression of *Hox* genes during the larval development of the snail, *Gibbula varia* (L.): further evidence of non-colinearity in molluscs. *Dev. Genes Evol.* 220, 161–172.

1972. Sánchez, L., and Guerrero, I. (2001). The development of the *Drosophila* genital disc. *BioEssays* 23, 698–707.

1973. Sánchez-Gracia, A., Vieira, F.G., and Rozas, J. (2009). Molecular evolution of the major chemosensory gene families in insects. *Heredity* 103, 208–216.

1974. Sánchez-Herrero, E., Vernós, I., Marco, R., and Morata, G. (1985). Genetic organization of *Drosophila* bithorax complex. *Nature* 313, 108–113.

1975. Sander, M., Neubüser, A., Ee, H.C., Martin, G.R., and German, M.S. (1997). Genetic analysis reveals that PAX6 is required for normal transcription of pancreatic hormone genes and islet development. *Genes Dev.* 11, 1662–1673.

1976. Sanes, J.R., and Zipursky, S.L. (2010). Design principles of insect and vertebrate visual systems. *Neuron* 66, 15–36.

1977. Santiago, C., Labrador, J.-P., and Bashaw, G.J. (2014). The homeodomain transcription factor Hb9 controls axon guidance in *Drosophila* through the regulation of Robo receptors. *Cell Rep.* 7, 153–165.

1978. Sarin, S., O'Meara, M., Flowers, E.B., Antonio, C., Poole, R.J., Didiano, D., Johnston, R.J., Jr., Chang, S., Narula, S., and Hobert, O. (2007). Genetic screens for *Caenorhabditis elegans* mutants defective in left/right asymmetric neuronal fate specification. *Genetics* 176, 2109–2130.

1979. Sarnat, H.B., and Netsky, M.G. (1981). *Evolution of the Nervous System*. Oxford University Press, New York, NY.

1980. Sarnat, H.B., and Netsky, M.G. (2002). When does a ganglion become a brain? Evolutionary origin of the central nervous system. *Semin. Pediatr. Neurol.* 9, 240–253.

1981. Sarrazin, A.F., Peel, A.D., and Averof, M. (2012). A segmentation clock with two-segment periodicity in insects. *Science* 336, 338–341.

1982. Sasai, Y. (2001). Roles of Sox factors in neural determination: conserved signaling in evolution? *Int. J. Dev. Biol.* 45, 321–326.

1983. Sassi, N., Laadhar, L., Driss, M., Kallel-Sellami, M., Sellami, S., and Makni, S. (2011). The role of the Notch pathway in healthy and osteoarthritic articular cartilage: from experimental models to *ex vivo* studies. *Arthritis Res. Ther.* 13, Article 208.

1984. Sato, K., Pellegrino, M., Nakagawa, T., Nakagawa, T., Vosshall, L.B., and Touhara, K. (2008). Insect olfactory receptors are heteromeric ligand-gated ion channels. *Nature* 452, 1002–1006.

1985. Sato, M., Suzuki, T., and Nakai, Y. (2013). Waves of differentiation in the fly visual system. *Dev. Biol.* 380, 1–11.

1986. Satoh, A.K., Li, B.X., Xia, H., and Ready, D.F. (2008). Calcium-activated Myosin V closes the *Drosophila* pupil. *Curr. Biol.* 18, 951–955.

1987. Satoh, N., Tagawa, K., and Takahashi, H. (2012). How was the notochord born? *Evol. Dev.* 14, 56–75.

1988. Satou, Y., and Satoh, N. (2006). Gene regulatory networks for the development and evolution of the chordate heart. *Genes Dev.* 20, 2634–2638.

1989. Saudemont, A., Dray, N., Hudry, B., Le Gouar, M., Vervoort, M., and Balavoine, G. (2008). Complementary striped expression patterns of *NK* homeobox genes during segment formation in the annelid *Platynereis*. *Dev. Biol.* 317, 430–443.

1990. Savory, J.G.A., Pilon, N., Grainger, S., Sylvestre, J.-R., Béland, M., Houle, M., Oh, K., and Lohnes, D. (2009). Cdx1 and Cdx2 are functionally equivalent in vertebral patterning. *Dev. Biol.* 330, 114–122.

1991. Sawaya, M.R., Wojtowicz, W.M., Andre, I., Qian, B., Wu, W., Baker, D., Eisenberg, D., and Zipursky, S.L. (2008). A double S shape provides the structural basis for the extraordinary binding specificity of Dscam isoforms. *Cell* 134, 1007–1018.

1992. Scheerer, P., Park, J.H., Hildebrand, P.W., Kim, Y.J., Krauss, N., Choe, H.-W., Hofmann, K.P., and Ernst, O.P. (2008). Crystal structure of opsin in its G-protein-interacting conformation. *Nature* 455, 497–502.

1993. Schetelig, M.F., Schmid, B.G.M., Zimowska, G., and Wimmer, E.A. (2008). Plasticity in mRNA expression and localization of *orthodenticle* within higher Diptera. *Evol. Dev.* 10, 700–704.

1994. Schier, A.F. (2009). Nodal morphogens. *Cold Spring Harb. Perspect. Biol.* 1, a003459.

1995. Schilling, T.F., and Knight, R.D. (2001). Origins of anteroposterior patterning and *Hox* gene regulation during chordate evolution. *Philos. Trans. R. Soc. Lond. B* 356, 1599–1613.

1996. Schilling, T.F., Nie, Q., and Lander, A.D. (2012). Dynamics and precision in retinoic acid morphogen gradients. *Curr. Opin. Genet. Dev.* 22, 562–569.

1997. Schinko, J.B., Kreuzer, N., Offen, N., Posnien, N., Wimmer, E.A., and Bucher, G. (2008). Divergent functions of *orthodenticle*, *empty spiracles* and *buttonhead* in early head patterning of the beetle *Tribolium castaneum* (Coleoptera). *Dev. Biol.* 317, 600–613.

1998. Schippers, K.J., and Nichols, S.A. (2014). Deep, dark secrets of melatonin in animal evolution. *Cell* 159, 9–10.

1999. Schlosser, G., Patthey, C., and Shimeld, S.M. (2014). The evolutionary history of vertebrate cranial placodes. II. Evolution of ectodermal patterning. *Dev. Biol.* 389, 98–119.

2000. Schmidt, J., Francois, V., Bier, E., and Kimelman, D (1995). *Drosophila short gastrulation* induces an ectopic axis in *Xenopus*: evidence for conserved mechanisms of dorsal-ventral patterning. *Development* 121, 4319–4328.

2001. Schmidt, M.H. (2014). The energy allocation function of sleep: a unifying theory of sleep, torpor, and continuous wakefulness. *Neurosci. Biobehav. Rev.* 47, 122–153.

2002. Schmidt, T.M., Chen, S.-K., and Hattar, S. (2011). Intrinsically photosensitive retinal ganglion cells: many subtypes, diverse functions. *Trends Neurosci.* 34, 572–580.

2003. Schmidt-Rhaesa, A. (2007). *The Evolution of Organ Systems.* Oxford University Press, New York, NY.

2004. Schmidt-Ullrich, R., and Paus, R. (2005). Molecular principles of hair follicle induction and morphogenesis. *BioEssays* 27, 247–261.

2005. Schmucker, D., and Chen, B. (2009). Dscam and DSCAM: complex genes in simple animals, complex animals yet simple genes. *Genes Dev.* 23, 147–156.

2006. Schmucker, D., Clemens, J.C., Shu, H., Worby, C.A., Xiao, J., Muda, M., Dixon, J.E., and Zipursky, S.L. (2000). *Drosophila* Dscam is an axon guidance receptor exhibiting extraordinary molecular diversity. *Cell* 101, 671–684.

2007. Schnaitmann, C., Garbers, C., Wachtler, T., and Tanimoto, H. (2013). Color discrimination with broadband photoreceptors. *Curr. Biol.* 23, 2375–2382.

2008. Schneider, M.D. (2016). Heartbreak hotel: a convergence in cardiac regeneration. *Development* 143, 1435–1441.

2009. Schneider, M.R., Schmidt-Ullrich, R., and Paus, R. (2009). The hair follicle as a dynamic miniorgan. *Curr. Biol.* 19, R132–R142.

2010. Schoenwolf, G.C., Bleyl, S.B., Brauer, P.R., and Francis-West, P.H. (2009). *Larsen's Human Embryology*, 4th edn. Churchill Livingstone, Philadelphia, PA.

2011. Scholtz, G. (2002). The Articulata hypothesis: or what is a segment? *Org. Divers. Evol.* 2, 197–215.

2012. Schön, P., Tsuchiya, K., Lenoir, D., Mochizuki, T., Guichard, C., Takai, S., Maiti, A.K., Nihei, H., Weil, J., Yokoyama, T., and Bouvagnet, P. (2002). Identification, genomic organization, chromosomal mapping and mutation analysis of the human INV gene, the ortholog of a murine gene implicated in left-right axis development and biliary atresia. *Hum. Genet.* 110, 157–165.

2013. Schonegg, S., Hyman, A.A., and Wood, W.B. (2014). Timing and mechanism of the initial cue establishing handed left-right asymmetry in *Caenorhabditis elegans* embryos. *Genesis* 52, 572–580.

2014. Schoppmeier, M., Fischer, S., Schmitt-Engel, C., Löhr, U., and Klingler, M. (2009). An ancient anterior patterning system promotes *caudal* repression and head formation in Ecdysozoa. *Curr. Biol.* 19, 1811–1815.

2015. Schreiner, D., Nguyen, T.-M., Russo, G., Heber, S., Patrignani, A., Ahrné, E., and Scheiffele, P. (2014). Targeted combinatorial alternative splicing generates brain region-specific repertoires of neurexins. *Neuron* 84, 386–398.

2016. Schröder, R. (2003). The genes *orthodenticle* and *hunchback* substitute for *bicoid* in the beetle *Tribolium*. *Nature* 422, 621–625.

2017. Schroeder, J.A., Jackson, L.F., Lee, D.C., and Camenisch, T.D. (2003). Form and function of developing heart valves: coordination by extracellular matrix and growth factor signaling. *J. Mol. Med.* 81, 392–403.

2018. Schroeder, M.D., Greer, C., and Gaul, U. (2011). How to make stripes: deciphering the transition from non-periodic to periodic patterns in *Drosophila* segmentation. *Development* 138, 3067–3078.

2019. Schubert, M., Escriva, H., Xavier-Neto, J., and Laudet, V. (2006). Amphioxus and tunicates as evolutionary model systems. *Trends Ecol. Evol.* 21, 269–277.

2020. Schubert, M., Yu, J.-K., Holland, N.D., Escriva, H., Laudet, V., and Holland, L.Z. (2005). Retinoic acid signaling acts via Hox1 to establish the posterior limit of the pharynx in the chordate amphioxus. *Development* 132, 61–73.

2021. Schughart, K., Kappen, C., and Ruddle, F.H. (1988). Mammalian homeobox-containing genes: genome organization, structure, expression and evolution. *Br. J. Cancer* 58 (Suppl. IX), 9–13.

2022. Schulte, D., and Bumsted-O'Brien, K.M. (2008). Molecular mechanisms of vertebrate retina development: implications for ganglion cell and photoreceptor patterning. *Brain Res.* 1192, 151–164.

2023. Schulze, J., and Schierenberg, E. (2009). Embryogenesis of *Romanomermis culicivorax*: an alternative way to construct a nematode. *Dev. Biol.* 334, 10–21.

2024. Schuster, S., Strauss, R., and Götz, K.G. (2002). Virtual-reality techniques resolve the visual cues used by fruit flies to evaluate object distances. *Curr. Biol.* 12, 1591–1594.

2025. Schwab, I.R. (2012). *Evolution's Witness: How Eyes Evolved*. Oxford University Press, New York, NY.

2026. Schwabe, T., Borycz, J.A., Meinertzhagen, I.A., and Clandinin, T.R. (2014). Differential adhesion determines the organization of synaptic fascicles in the *Drosophila* visual system. *Curr. Biol.* 24, 1304–1313.

2027. Schwander, M., Kachar, B., and Müller, U. (2010). The cell biology of hearing. *J. Cell Biol.* 190, 9–20.

2028. Schwartz, T.W., and Hubbell, W.L. (2008). A moving story of receptors. *Nature* 455, 473–474.

2029. Schwartz, W.J. (2004). Sunrise and sunset in fly brains. *Nature* 431, 751–752.

2030. Schwarzer, W., and Spitz, F. (2014). The architecture of gene expression: integrating dispersed *cis*-regulatory modules into coherent regulatory domains. *Curr. Opin. Genet. Dev.* 27, 74–82.

2031. Schweickert, A., Walentek, P., Thumberger, T., and Danilchik, M. (2012). Linking early determinants and cilia-driven leftward flow in left-right axis specificiation of *Xenopus laevis*: a theoretical approach. *Differentiation* 83, S67–S77.

2032. Schweitzer, R., Zelzer, E., and Volk, T. (2010). Connecting muscles to tendons: tendons and musculoskeletal development in flies and vertebrates. *Development* 137, 2807–2817.

2033. Scotland, R.W. (2011). What is parallelism? *Evol. Dev.* 13, 214–227.

2034. Scott, C.A., and Dahanukar, A. (2014). Sensory coding of olfaction and taste. *In* J. Dubnau (ed.), *Behavioral Genetics of the Fly (Drosophila melanogaster)*. Cambridge University Press, New York, NY, pp. 49–65.

2035. Scott, M.P. (1994). Intimations of a creature. *Cell* 79, 1121–1124.

2036. Scott, M.P. (2000). Development: the natural history of genes. *Cell* 100, 27–40.

2037. Scott, M.P. (2016). Homeodomains, hedgehogs, and happiness. *Curr. Top. Dev. Biol.* 117, 331–337.

2038. Scott, M.P., and Carroll, S.B. (1987). The segmentation and homeotic gene network in early *Drosophila* development. *Cell* 51, 689–698.

2039. Scott, M.P., and Weiner, A.J. (1984). Structural relationships among genes that control development: sequence homology between the Antennapedia, Ultrabithorax, and fushi tarazu loci of *Drosophila*. *PNAS* 81, 4115–4119

2040. Seaver, E.C. (2003). Segmentation: mono- or polyphyletic? *Int. J. Dev. Biol.* 47, 583–595.

2041. Sebé-Pedrós, A., Burkhardt, P., Sánchez-Pons, N., Fairclough, S.R., Lang, B.F., King, N., and Ruiz-Trillo, I. (2013). Insights into the origin of metazoan filopodia and microvilli. *Mol. Biol. Evol.* 30, 2013–2023.

2042. Seeds, A.M., Ravbar, P., Chung, P., Hampel, S., Midgley, F.M., Jr., Mensh, B.D., and Simpson, J.H. (2014). A suppression hierarchy among competing motor programs drives sequential grooming in *Drosophila*. *eLife* 3, e02951.

2043. Seeger, M., Tear, G., Ferres-Marco, D., and Goodman, C.S. (1993). Mutations affecting growth cone guidance in *Drosophila*: genes necessary for guidance toward or away from the midline. *Neuron* 10, 409–426.

2044. Seelig, J.D., and Jayaraman, V. (2015). Neural dynamics for landmark orientation and angular path integration. *Nature* 521, 186–191.

2045. Sehadova, H., Glaser, F.T., Gentile, C., Simoni, A., Giesecke, A., and Albert, J.T. (2009). Temperature entrainment of *Drosophila*'s circadian clock involves the gene *nocte* and signaling from peripheral sensory tissues to the brain. *Neuron* 64, 251–266.

2046. Seibert, J., Volland, D., and Urbach, R. (2009). Ems and Nkx6 are central regulators in dorsoventral patterning of the *Drosophila* brain. *Development* 136, 3937–3947.

2047. Seifert, A.W., Kiama, S.G., Seifert, M.G., Goheen, J.R., Palmer, T.M., and Maden, M. (2012). Skin shedding and tissue regeneration in African spiny mice (*Acomys*). *Nature* 489, 561–565.

2048. Seifert, A.W., Monaghan, J.R., Voss, S.R., and Maden, M. (2012). Skin regeneration in adult axolotls: a blueprint for scar-free healing in vertebrates. *PLoS ONE* 7, #4, e32875.

2049. Seifert, J.R.K., and Mlodzik, M. (2007). Frizzled/PCP signalling: a conserved mechanism regulating cell polarity and directed motility. *Nat. Rev. Genet.* 8, 126–138.

2050. Seipel, K., Eberhardt, M., Müller, P., Pescia, E., Yanze, N., and Schmid, V. (2004). Homologs of vascular endothelial growth factor and receptor, VEGF and VEGFR, in the jellyfish *Podocoryne carnea*. *Dev. Dyn.* 231, 303–312.

2051. Sekharan, S., and Morokuma, K. (2011). Why 11-*cis*-retinal? Why not 7-*cis*-, 9-*cis*-, or 13-*cis*-retinal in the eye? *J. Am. Chem. Soc.* 133, 19052–19055.

2052. Selverston, A.I. (2010). Invertebrate central pattern generator circuits. *Philos. Trans. R. Soc. Lond. B* 365, 2329–2345.

2053. Semmelhack, J.L., and Wang, J.W. (2009). Select *Drosophila* glomeruli mediate innate olfactory attraction and aversion. *Nature* 459, 218–223.

2054. Sen, S., Reichert, H., and VijayRaghavan, K. (2013). Conserved roles of *ems/Emx* and *otd/Otx* genes in olfactory and visual system development in *Drosophila* and mouse. *Open Biol.* 3, Article 120177.

2055. Senoo, H., Sesaki, H., and Iijima, M. (2016). A GPCR handles bacterial sensing in chemotaxis and phagocytosis. *Dev. Cell* 36, 354–356.

2056. Senthilan, P.R., Piepenbrock, D., Ovezmyradov, G., Nadrowski, B., Bechstedt, S., Pauls, S., Winkler, M., Möbius, W., Howard, J., and Göpfert, M.C. (2012). *Drosophila* auditory organ genes and genetic hearing defects. *Cell* 150, 1042–1054.

2057. Seo, H.-C., Edvardsen, R.B., Maeland, A.D., Bjordal, M., Jensen, M.F., Hansen, A., Flaat, M., Weissenbach, J., Lehrach, H., Wincker, P., Reinhardt, R., and Chourrout, D. (2004). *Hox* cluster disintegration with persistent anteroposterior order of expression in *Oikopleura dioica*. *Nature* 431, 67–71.

2058. Serizawa, S., Miyamichi, K., Nakatani, H., Suzuki, M., Saito, M., Yoshihara, Y., and Sakano, H. (2003). Negative feedback regulation ensures the one receptor-one olfactory neuron rule in mouse. *Science* 302, 2088–2094.

2059. Serizawa, S., Miyamichi, K., Takeuchi, H., Yamagishi, Y., Suzuki, M., and Sakano, H. (2006). A neuronal identity code for the odorant receptor-specific and activity-dependent axon sorting. *Cell* 127, 1057–1069.

2060. Shapiro, L. (2007). Self-recognition at the atomic level: understanding the astonishing molecular diversity of homophilic Dscams. *Neuron* 56, 10–13.

2061. Sharma, P.P., Santiago, M.A., González-Santillán, E., Monod, L., and Wheeler, W.C. (2015). Evidence of duplicated Hox genes in the most recent common ancestor of extant scorpions. *Evol. Dev.* 17, 347–355.

2062. Shaw, P.J., and Franken, P. (2003). Perchance to dream: solving the mystery of sleep through genetic analysis. *J. Neurobiol.* 54, 179–202.

2063. Sheeba, C.J., Andrade, R.P., and Palmeirim, I. (2016). Getting a handle on embryo limb development: molecular interactions driving limb outgrowth and patterning. *Semin. Cell Dev. Biol.* 49, 92–101.

2064. Shen, H.H. (2014). Inner workings: discovering the split mind. *PNAS* 111, #51, 18097.

2065. Shen, M.M. (2007). Nodal signaling: developmental roles and regulation. *Development* 134, 1023–1034.

2066. Shen, W.L., Kwon, Y., Adegbola, A.A., Luo, J., Chess, A., and Montell, C. (2011). Function of rhodopsin in temperature discrimination in *Drosophila*. *Science* 331, 1333–1336.

2067. Shepherd, G.M. (2006). Smell images and the flavour system in the human brain. *Nature* 444, 316–321.

2068. Sheth, R., Grégoire, D., Dumouchel, A., Scotti, M., Pham, J.M.T., Nemec, S., Bastida, M.F., Ros, M.A., and Kmita, M. (2013). Decoupling the function of Hox and Shh in developing limb reveals multiple inputs of Hox genes on limb growth. *Development* 140, 2130–2138.

2069. Sheth, R., Marcon, L., Bastida, M.F., Junco, M., Quintana, L., Dahn, R., Kmita, M., Sharpe, J., and Ros, M.A. (2012). *Hox* genes regulate digit patterning by controlling the wavelength of a Turing-type mechanism. *Science* 338, 1476–1480.

2070. Shibazaki, Y., Shimizu, M., and Kuroda, R. (2004). Body handedness is directed by genetically determined cytoskeletal dynamics in the early embryo. *Curr. Biol.* 14, 1462–1467.

2071. Shichida, Y., and Matsuyama, T. (2009). Evolution of opsins and phototransduction. *Philos. Trans. R. Soc. Lond. B* 364, 2881–2895.

2072. Shimeld, S.M., and Levin, M. (2006). Evidence for the regulation of left-right asymmetry in *Ciona intestinalis* by ion flux. *Dev. Dyn.* 235, 1543–1553.

2073. Shimizu, H., and Fujisawa, T. (2003). Peduncle of *Hydra* and the heart of higher organisms share a common ancestral origin. *Genesis* 36, 182–186.

2074. Shimomura, Y., Agalliu, D., Vonica, A., Luria, V., Wajid, M., Baumer, A., Belli, S., Petukhova, L., Schinzel, A., Brivanlou, A.H., Barres, B.A., and Christiano, A.M. (2010). APCDD1 is a novel Wnt inhibitor mutated in hereditary hypotrichosis simplex. *Nature* 464, 1043–1047.

2075. Shimozono, S., Iimura, T., Kitaguchi, T., Higashijima, S.-I., and Miyawaki, A. (2013). Visualization of an endogenous retinoic acid gradient across embryonic development. *Nature* 496, 363–366.

2076. Shinbrot, T., and Young, W. (2008). Why decussate? Topological constraints on 3D wiring. *Anat. Rec.* 291, 1278–1292.

2077. Shinohara, K., Chen, D., Nishida, T., Misaki, K., Yonemura, S., and Hamada, H. (2015). Absence of radial spokes in mouse node cilia is required for rotational movement but confers ultrastructural instability as a trade-off. *Dev. Cell* 35, 236–246.

2078. Shinomiya, K., Takemura, S.-Y., Rivlin, P.K., Plaza, S.M., Scheffer, L.K., and Meinertzhagen, I.A. (2015). A common evolutionary origin for the ON- and OFF-edge motion detection pathways of the *Drosophila* visual system. *Front. Neural Circuits* 9, Article 33.

2079. Shippy, T.D., Ronshaugen, M., Cande, J., He, J.P., Beeman, R.W., Levine, M., Brown, S.J., and Denell, R. (2008). Analysis of the *Tribolium* homeotic complex: insights into mechanisms constraining insect Hox clusters. *Dev. Genes Evol.* 218, 127–139.

2080. Shirai, T., Yorimitsu, T., Kiritooshi, N., Matsuzaki, F., and Nakagoshi, H. (2007). Notch signaling relieves the joint-suppressive activity of Defective proventriculus in the *Drosophila* leg. *Dev. Biol.* 312, 147–156.

2081. Shiraiwa, T. (2008). Multimodal chemosensory integration through the maxillary palp in *Drosophila*. *PLoS ONE* 3, #5, e2191.

2082. Shirasaki, R., and Pfaff, S.L. (2002). Transcriptional codes and the control of neuronal identity. *Annu. Rev. Neurosci.* 25, 251–281.

2083. Shlyueva, D., Stampfel, G., and Stark, A. (2014). Transcriptional enhancers: from properties to genome-wide predictions. *Nat. Rev. Genet.* 15, 272–286.

2084. Shohat-Ophir, G., Kaun, K.R., Azanchi, R., and Heberlein, U. (2012). Sexual deprivation increases ethanol intake in *Drosophila*. *Science* 335, 1351–1355.

2085. Sholtis, S.J., and Noonan, J.P. (2010). Gene regulation and the origins of human biological uniqueness. *Trends Genet.* 26, 110–118.

2086. Shubin, N., Tabin, C., and Carroll, S. (1997). Fossils, genes and the evolution of animal limbs. *Nature* 388, 639–648.

2087. Shubin, N., Tabin, C., and Carroll, S. (2009). Deep homology and the origins of evolutionary novelty. *Nature* 457, 818–823.

2088. Sick, S., Reinker, S., Timmer, J., and Schlake, T. (2006). WNT and DKK determine hair follicle spacing through a reaction-diffusion mechanism. *Science* 314, 1447–1450.

2089. Siegert, S., Cabuy, E., Scherf, B.G., Kohler, H., Panda, S., Le, Y.-Z., Fehling, H.J., Gaidatzis, D., Stadler, M.B., and Roska, B. (2012). Transcriptional code and disease map for adult retinal cell types. *Nat. Neurosci.* 15, 487–495.

2090. Sienknecht, U.J. (2015). Current concepts of hair cell differentiation and planar cell polarity in inner ear sensory organs. *Cell Tissue Res.* 361, 25–32.

2091. Sienknecht, U.J., Anderson, B.K., Parodi, R.M., Fantetti, K.N., and Fekete, D.M. (2011). Non-cell-autonomous planar cell polarity propagation in the auditory sensory epithelium of vertebrates. *Dev. Biol.* 352, 27–39.

2092. Sillar, K.T. (2009). Escape behaviour: reciprocal inhibition ensures effective escape trajectory. *Curr. Biol.* 19, R697–R699.

2093. Silver, S.J., and Rebay, I. (2005). Signaling circuitries in development: insights from the retinal determination gene network. *Development* 132, 3–13.

2094. Silverman, H.B., and Dunbar, M.J. (1980). Aggressive tusk use by the narwhal (*Monodon monoceros* L.). *Nature* 284, 57–58.

2095. Simakov, O., Marletaz, F., Cho, S.-J., Edsinger-Gonzales, E., Havlak, P., Hellsten, U., Kuo, D.-H., Larsson, T., Lv, J., Arendt, D., Savage, R., Osoegawa, K., de Jong, P., Grimwood, J., Chapman, J.A., Shapiro, H., Aerts, A., Otillar, R.P., Terry, A.Y., Boore, J.L., Grigoriev, I.V., Lindberg, D.R., Seaver, E.C., Weisblat, D.A., Putnam, N.H., and Rokhsar, D.S. (2013). Insights into bilaterian evolution from three spiralian genomes. *Nature* 493, 526–531.

2096. Simeone, A., Puelles, E., and Acampora, D. (2002). The Otx family. *Curr. Opin. Genet. Dev.* 12, 409–415.

2097. Simionato, E., Kerner, P., Dray, N., Le Gouar, M., Ledent, V., Arendt, D., and Vervoort, M. (2008). *atonal-* and *achaete-scute*-related genes in the annelid *Platynereis dumerilii*: insights into the evolution of neural basic-Helix-Loop-Helix genes. *BMC Evol. Biol.* 8, Article 170.

2098. Simmonds, A.J., Brook, W.J., Cohen, S.M., and Bell, J.B. (1995). Distinguishable functions for *engrailed* and *invected* in anterior-posterior patterning in the *Drosophila* wing. *Nature* 376, 424–427.

2099. Simões-Costa, M.S., Vasconcelos, M., Sampaio, A.C., Cravo, R.M., Linhares, V.L., Hochgreb, T., Yan, C.Y.I., Davidson, B., and Xavier-Neto, J. (2005). The evolutionary origin of cardiac chambers. *Dev. Biol.* 277, 1–15.

2100. Simon, A., and Tanaka, E.M. (2013). Limb regeneration. *WIREs Dev. Biol.* 2, 291–300.

2101. Simonnet, F., Célérier, M.-L., and Quéinnec, E. (2006). *Orthodenticle* and *empty spiracles* genes are expressed in a segmental pattern in chelicerates. *Dev. Genes Evol.* 216, 467–480.

2102. Simpson, J.H., Bland, K.S., Fetter, R.D., and Goodman, C.S. (2000). Short-range and long-range guidance by Slit and its Robo receptors: a combinatorial code of Robo receptors controls lateral position. *Cell* 103, 1019–1032.

2103. Simpson, P. (2007). The stars and stripes of animal bodies: evolution of regulatory elements mediating pigment and bristle patterns in *Drosophila*. *Trends Genet.* 23, 350–358.

2104. Singh, A., Tare, M., Puli, O.R., and Kango-Singh, M. (2011). A glimpse into dorso-ventral patterning of the *Drosophila* eye. *Dev. Dyn.* 241, 69–84.

2105. Singh, D., and Pohl, C. (2014). Coupling of rotational cortical flow, asymmetric midbody positioning, and spindle rotation mediates dorsoventral axis formaiton in *C. elegans*. *Dev. Cell* 28, 253–267.

2106. Singh, N.P., and Mishra, R.K. (2008). A double-edged sword to force posterior dominance of Hox genes. *BioEssays* 30, 1058–1061.

2107. Singh, S., Stellrecht, C.M., Tang, H.K., and Saunders, G.F. (2000). Modulation of PAX6 homeodomain function by the paired domain. *J. Biol. Chem.* 275, #23, 17306–17313.

2108. Singla, V., and Reiter, J.F. (2006). The primary cilium as the cell's antenna: signaling at a sensory organelle. *Science* 313, 629–633.

2109. Siniscalchi, M., Lusito, R., Vallortigara, G., and Quaranta, A. (2013). Seeing left- or right-asymmetric tail wagging produces different emotional responses in dogs. *Curr. Biol.* 23, 2279–2282.

2110. Sitaraman, D., Aso, Y., Jin, X., Chen, N., Felix, M., Rubin, G.M., and Nitabach, M.N. (2015). Propagation of homeostatic sleep signals by segregated synaptic microcircuits of the *Drosophila* mushroom body. *Curr. Biol.* 25, 2915–2927.

2111. Sivanantharajah, L., and Percival-Smith, A. (2015). Differential pleiotropy and HOX functional organization. *Dev. Biol.* 398, 1–10.

2112. Skeldon, K.D., Reid, L.M., McInally, V., Dougan, B., and Fulton, C. (1998). Physics of the Theremin. *Am. J. Phys.* 66, 945–955.

2113. Skoyles, J.R. (2006). Human balance, the evolution of bipedalism and dysequilibrium syndrome. *Med. Hypotheses* 66, 1060–1068.

2114. Slack, J.M.W., Holland, P.W.H., and Graham, C.F. (1993). The zootype and the phylotypic stage. *Nature* 361, 490–492.

2115. Smallwood, P.M., Ölveczky, B.P., Williams, G.L., Jacobs, G.H., Reese, B.E., Meister, M., and Nathans, J. (2003). Genetically engineered mice with an additional class of cone receptors: implications for the evolution of color vision. *PNAS* 100, #20, 11706–11711.

2116. Smallwood, P.M., Wang, Y., and Nathans, J. (2002). Role of a locus control region in the mutually exclusive expression of human red and green cone pigment genes. *PNAS* 99, #2, 1008–1011.

2117. Smear, M., Resulaj, A., Zhang, J., Bozza, T., and Rinberg, D. (2013). Multiple perceptible signals from a single olfactory glomerulus. *Nat. Neurosci.* 16, 1687–1691.

2118. Smetacek, V. (1992). Mirror-script and left-handedness. *Nature* 355, 118–119.

2119. Smith, A.B. (2008). Deuterostomes in a twist: the origins of a radical new body plan. *Evol. Dev.* 10, 493–503.

2120. Smith, A.T. (2015). Binocular vision: joining up the eyes. *Curr. Biol.* 25, R661–R663.

2121. Smith, F.W., Boothby, T.C., Giovannini, I., Rebecchi, L., Jockusch, E.L., and Goldstein, B. (2016). The compact body plan of tardigrades evolved by the loss of a large body region. *Curr. Biol.* 26, 224–229.

2122. Smyth, V.A., Di Lorenzo, D., and Kennedy, B.N. (2008). A novel, evolutionarily conserved enhancer of cone photoreceptor-specific expression. *J. Biol. Chem.* 283, #16, 10881–10891.

2123. Soba, P., Zhu, S., Emoto, K., Younger, S., Yang, S.-J., Yu, H.-H., Lee, T., Jan, L.Y., and Jan, Y.-N. (2007). *Drosophila* sensory neurons require Dscam for dendritic self-avoidance and proper dendritic field organization. *Neuron* 54, 403–416.

2124. Sohaskey, M.L., Yu, J., Diaz, M.A., Plaas, A.H., and Harland, R.M. (2008). JAWS coordinates chondrogenesis and synovial joint positioning. *Development* 135, 2215–2220.

2125. Soler, C., Daczewska, M., Da Ponte, J.P., Dastugue, B., and Jagla, K. (2004). Coordinated development of muscles and tendons of the *Drosophila* leg. *Development* 131, 6041–6051.

2126. Solnica-Krezel, L. (2005). Conserved patterns of cell movements during vertebrate gastrulation. *Curr. Biol.* 15, R213–R228.

2127. Solomon, S.G., and Lennie, P. (2007). The machinery of colour vision. *Nat. Rev. Neuro.* 8, 276–286.

2128. Solovei, I., Kreysing, M., Lanctôt, C., Kösem, S., Peichi, L., Cremer, T., Guck, J., and Joffe, B. (2009). Nuclear architecture of rod photoreceptor cells adapts to vision in mammalian evolution. *Cell* 137, 356–368.

2129. Somel, M., Liu, X., and Khaitovich, P. (2013). Human brain evolution: transcripts, metabolites and their regulators. *Nat. Rev. Neurosci.* 14, 112–127.

2130. Sommer, A., and Vyas, K.S. (2012). A global clinical view on vitamin A and carotenoids. *Am. J. Clin. Nutr.* 96 (Suppl.), 1204S–1206S.

2131. Sommer, R., and Tautz, D. (1991). Segmentation gene expression in the housefly *Musca domestica*. *Development* 113, 419–430.

2132. Song, E., de Bivort, B., Dan, C., and Kunes, S. (2012). Determinants of the *Drosophila* odorant receptor pattern. *Dev. Cell* 22, 363–376.

2133. Song, H., Hu, J., Chen, W., Elliott, G., Andre, P., Gao, B., and Yang, Y. (2010). Planar cell polarity breaks bilateral symmetry by controlling ciliary positioning. *Nature* 466, 378–382.

2134. Song, K., Nam, Y.-J., Luo, X., Qi, X., Tan, W., Huang, G.N., Acharya, A., Smith, C.L., Tallquist, M.D., Neilson, E.G., Hill, J.A., Bassel-Duby, R., and Olson, E.N. (2012). Heart repair by reprogramming non-myocytes with cardiac transcription factors. *Nature* 485, 599–604.

2135. Song, Z., Postma, M., Billings, S.A., Coca, D., Hardie, R.C., and Juusola, M. (2012). Stochastic, adaptive sampling of information by microvilli in fly photoreceptors. *Curr. Biol.* 22, 1371–1380.

2136. Sopko, R., Lin, Y.B., Makhijani, K., Alexander, B., Perrimon, N., and Brückner, K. (2015). A systems-level interrogation identifies regulators of *Drosophila* blood cell number and survival. *PLoS Genet.* 11, #3, e1005056.

2137. Sopko, R., and McNeill, H. (2009). The skinny on Fat: an enormous cadherin that regulates cell adhesion, tissue growth, and planar cell polarity. *Curr. Opin. Cell Biol.* 21, 717–723.

2138. Sorrentino, R.P., Gajewski, K.M., and Schulz, R.A. (2005). GATA factors in *Drosophila* heart and blood cell development. *Semin. Cell Dev. Biol.* 16, 107–116.

2139. Soshnikova, N., Dewaele, R., Janvier, P., Krumlauf, R., and Duboule, D. (2013). Duplications of hox gene clusters and the emergence of vertebrates. *Dev. Biol.* 378, 194–199.

2140. Sotomayor, M., Weihofen, W.A., Gaudet, R., and Corey, D.P. (2012). Structure of a force-conveying cadherin bond essential for inner-ear mechanotransduction. *Nature* 492, 128–132.

2141. Soukup, V., and Kozmik, Z. (2016). Zoology: a new mouth for amphioxus. *Curr. Biol.* 26, R367–R368.

2142. Soukup, V., Yong, L.W., Lu, T.-M., Huang, S.-W., Kozmik, Z., and Yu, J.-K. (2015). The Nodal signaling pathway controls left-right asymmetric development in amphioxus. *EvoDevo* 6, Article 5.

2143. Soustelle, L., Trousse, F., Jacques, C., Ceron, J., Cochard, P., Soula, C., and Giangrande, A. (2007). Neurogenic role of Gcm transcription factors is conserved in chicken spinal cord. *Development* 134, 625–634.

2144. Southwell, D.G., Paredes, M.F., Galvao, R.P., Jones, D.L., Froemke, R.C., Sebe, J.Y., Alfaro-Cervello, C., Tang, Y., Garcia-Verdugo, J.M., Rubenstein, J.L., Baraban, S.C., and Alvarez-Buylla, A. (2012). Intrinsically determined cell death of developing cortical interneurons. *Nature* 491, 109–113.

2145. Spanagel, R. (2009). Alcoholism: a systems approach from molecular physiology to addictive behavior. *Physiol. Rev.* 89, 649–705.

2146. Spéder, P., Ádám, G., and Noselli, S. (2006). Type ID unconventional myosin controls left-right asymmetry in *Drosophila*. *Nature* 440, 803–807.

2147. Spéder, P., and Noselli, S. (2007). Left-right asymmetry: class I myosins show the direction. *Curr. Opin. Cell Biol.* 19, 82–87.

2148. Spehr, M., and Munger, S.D. (2009). Olfactory receptors: G protein-coupled receptors and beyond. *J. Neurochem.* 109, 1570–1583.

2149. Spence, C. (2013). Multisensory flavour perception. *Curr. Biol.* 23, R365–R370.

2150. Sperry, R.W. (1963). Chemoaffinity in the orderly growth of nerve fiber patterns and connections. *PNAS* 50, 703–710.

2151. Spitz, F., Herkenne, C., Morris, M.A., and Duboule, D. (2005). Inversion-induced disruption of the *Hoxd* cluster leads to the partition of regulatory landscapes. *Nature Genet.* 37, 889–893.

2152. Spitzweck, B., Brankatschk, M., and Dickson, B.J. (2010). Distinct protein domains and expresion patterns confer divergent axon guidance functions for *Drosophila* Robo receptors. *Cell* 140, 409–420.

2153. Spletter, M.L., and Luo, L. (2009). A new family of odorant receptors in *Drosophila*. *Cell* 136, 23–25.

2154. Spoon, C., Moravec, W.J., Rowe, M.H., Grant, J.W., and Peterson, E.H. (2011). Steady-state stiffness of utricular hair cells depends on macular location and hair bundle structure. *J. Neurophysiol.* 106, 2950–2963.

2155. Spoon, J.M. (2001). Situs inversus totalis. *Neonatal Netw.* 20, 59–63.

2156. Sprecher, S.G., and Desplan, C. (2008). Switch of *rhodopsin* expression in terminally differentiated *Drosophila* sensory neurons. *Nature* 454, 533–537.

2157. Sprecher, S.G., and Reichert, H. (2003). The urbilaterian brain: developmental insights into the evolutionary origin of the brain in insects and vertebrates. *Arthropod Struct. Dev.* 32, 141–156.

2158. Sprecher, S.G., Reichert, H., and Hartenstein, V. (2007). Gene expression patterns in primary neuronal clusters of the *Drosophila* embryonic brain. *Gene Expr. Patterns* 7, 584–595.

2159. Sproul, D., Gilbert, N., and Bickmore, W.A. (2005). The role of chromatin structure in regulating the expression of clustered genes. *Nat. Rev. Genet.* 6, 775–781.

2160. Spudich, J.L., Sineshchekov, O.A., and Govorunova, E.G. (2014). Mechanism divergence in microbial rhodopsins. *Biochim. Biophys. Acta* 1837, 546–552.

2161. Spudich, J.L., Yang, C.-S., Jung, K.-H., and Spudich, E.N. (2000). Retinylidene proteins: structures and functions from archaea to humans. *Annu. Rev. Cell Dev. Biol.* 16, 365–392.

2162. Srinivasan, M., Zhang, S., and Reinhard, J. (2006). Small brains, smart minds: vision, perception, navigation, and "cognition" in insects. *In* E. Warrant and D.-E. Nilsson (eds.), *Invertebrate Vision*. Cambridge University Press, New York, NY, pp. 462–493.

2163. Srivastava, D. (2006). Making or breaking the heart: from lineage determination to morphogenesis. *Cell* 126, 1037–1048.

2164. Srour, M., Rivière, J.-B., Pham, J.M.T., Dubé, M.-P., Girard, S., Morin, S., Dion, P.A., Asselin, G., Rochefort, D., Hince, P., Diab, S., Sharafaddinzadeh, N., Chouinard, S., Théoret, H., Charron, F., and Rouleau, G.A. (2010). Mutations in *DCC* cause congenital mirror movements. *Science* 328, 592–593.

2165. Stanewsky, R. (2003). Genetic analysis of the circadian system in *Drosophila melanogaster* and mammals. *J. Neurobiol.* 54, 111–147.

2166. Stark, W.S., Wagner, R.H., and Gillespie, C.M. (1994). Ultraviolet sensitivity of three cone types in the aphakic observer determined by chromatic adaptation. *Vision Res.* 34, 1457–1459.

2167. Stauber, M., Jäckle, H., and Schmidt-Ott, U. (1999). The anterior determinant *bicoid* of *Drosophila* is a derived *Hox* class 3 gene. *PNAS* 96, 3786–3789.

2168. Stauber, M., Prell, A., and Schmidt-Ott, U. (2002). A single *Hox3* gene with composite *bicoid* and *zerknüllt* expression characteristics in non-Cyclorrhaphan flies. *PNAS* 99, #1, 274–279.

2169. Stavenga, D.G. (2002). Colour in the eyes of insects. *J. Comp. Physiol.* A 188, 337–348.

2170. Stavenga, D.G., and Hardie, R.C. (2011). Metarhodopsin control by arrestin, light-filtering screening pigments, and visual pigment turnover in invertebrate microvillar photoreceptors. *J. Comp. Physiol.* A 197, 227–241.

2171. Steele, R.E., David, C.N., and Technau, U. (2010). A genomic view of 500 million years of cnidarian evolution. *Trends Genet.* 27, 7–13.

2172. Stein, E., and Tessier-Lavigne, M. (2001). Hierarchical organization of guidance receptors: silencing of Netrin attraction by Slit through a Robo/DCC receptor complex. *Science* 291, 1928–1938.

2173. Steinmetz, P.R.H., Kostyuchenko, R.P., Fischer, A., and Arendt, D. (2011). The segmental pattern of *otx*, *gbx*, and *Hox* genes in the annelid *Platynereis dumerilii*. *Evol. Dev.* 13, 72–79.

2174. Steinmetz, P.R.H., Urbach, R., Posnien, N., Eriksson, J., Kostyuchenko, R.P., Brena, C., Guy, K., Akam, M., Bucher, G., and Arendt, D. (2010). *Six3* demarcates the anterior-most developing brain region in bilaterian animals. *EvoDevo* 1, Article 14.

2175. Stengl, M., and Funk, N.W. (2013). The role of the coreceptor Orco in insect olfactory transduction. *J. Comp. Physiol.* A 199, 897–909.

2176. Stennard, F.A., and Harvey, R.P. (2005). T-box transcription factors and their roles in regulatory hierarchies in the developing heart. *Development* 132, 4897–4910.

2177. Stephen, L.A., Johnson, E.J., Davis, G.M., McTeir, L., Pinkham, J., Jaberi, N., and Davey, M.G. (2014). The chicken left right organizer has nonmotile cilia which are lost in a stage-dependent manner in the *talpid³* ciliopathy. *Genesis* 52, 600–613.

2178. Stephenson-Jones, M., Samuelsson, E., Ericsson, J., Robertson, B., and Grillner, S. (2011). Evolutionary conservation of the basal ganglia as a common vertebrate mechanism for action selection. *Curr. Biol.* 21, 1081–1091.

2179. Stergiopoulos, A., Elkouris, M., and Politis, P.K. (2015). Prospero-related homeobox 1 (Prox1) at the crossroads of diverse pathways during adult neural fate specification. *Front. Cell. Neurosci.* 8, Article 454.

2180. Stern, C. (1941). The growth of testes in *Drosophila*. I. The relation between vas deferens and testis within various species. *J. Exp. Zool.* 87, 113–158.

2181. Stern, C., and Tokunaga, C. (1968). Autonomous pleiotropy in *Drosophila*. *PNAS* 60, 1252–1259.

2182. Stern, C.D. (1990). Two distinct mechanisms for segmentation? *Semin. Dev. Biol.* 1, 109–116.

2183. Stern, C.D., and Piatkowska, A.M. (2015). Multiple roles of timing in somite formation. *Semin. Cell Dev. Biol.* 42, 134–139.

2184. Stern, D.L. (2013). The genetic causes of convergent evolution. *Nat. Rev. Genet.* 14, 751–764.

2185. Stern, K., and McClintock, M.K. (1998). Regulation of ovulation by human pheromones. *Nature* 392, 177–179.

2186. Steward, R. (1989). Relocalization of the *dorsal* protein from the cytoplasm to the nucleus correlates with its function. *Cell* 59, 1179–1188.

2187. Stocker, R.F. (2001). *Drosophila* as a focus in olfactory research: mapping of olfactory sensilla by fine structure, odor specificity, odorant receptor expression, and central connectivity. *Microsc. Res. Tech.* 55, 284–296.

2188. Stocum, D.L., and Cameron, J.A. (2011). Looking proximally and distally: 100 years of limb regeneration and beyond. *Dev. Dyn.* 240, 943–968.

2189. Stoick-Cooper, C.L., Moon, R.T., and Weidinger, G. (2007). Advances in signaling in vertebrate regeneration as a prelude to regenerative medicine. *Genes Dev.* 21, 1292–1315.

2190. Stokes, M.D., and Holland, N. (1998). The lancelet. *Am. Sci.* 86, #6, 552–560.

2191. Stollewerk, A. (2008). Evolution of neurogenesis in arthropods. *In* A. Minelli and G. Fusco (eds.), *Evolving Pathways: Key Themes in Evolutionary Developmental Biology.* Cambridge University Press, New York, NY, pp. 359–380.

2192. Stollewerk, A., and Simpson, P. (2005). Evolution of early development of the nervous system: a comparison between arthropods. *BioEssays* 27, 874–883.

2193. Störtkuhl, K.F., and Fiala, A. (2011). The smell of blue light: a new approach toward understanding an olfactory neuronal network. *Front. Neurosci.* 5, Article 72.

2194. Strausfeld, N.J. (2005). The evolution of crustacean and insect optic lobes and the origins of chiasmata. *Arthropod Struct. Dev.* 34, 235–256.

2195. Strausfeld, N.J. (2009). Brain organization and the origin of insects: an assessment. *Proc. R. Soc. Lond. B* 276, 1929–1937.

2196. Strausfeld, N.J., and Hildebrand, J.G. (1999). Olfactory systems: common design, uncommon origins? *Curr. Opin. Neurobiol.* 9, 634–639.

2197. Strausfeld, N.J., and Hirth, F. (2013). Deep homology of arthropod central complex and vertebrate basal ganglia. *Science* 340, 157–161. [*See also* Turner-Evans, D.B., and Jayaraman, V. (2016). The insect central complex. *Curr. Biol.* 26, R453–R457.]

2198. Strausfeld, N.J., and Hirth, F. (2013). Homology versus convergence in resolving transphyletic correspondences of brain organization. *Brain Behav. Evol.* 82, 215–219.

2199. Streelman, J.T. (2014). Advancing evolutionary developmental biology. In *Advances in Evolutionary Developmental Biology.* Wiley-Blackwell, New York, NY, pp. 203–217.

2200. Strigini, M. (2005). Mechanisms of morphogen transport. *J. Neurobiol.* 64, 324–333.

2201. Strilic, B., Kucera, T., and Lammert, E. (2010). Formation of cardiovascular tubes in invertebrates and vertebrates. *Cell. Mol. Life Sci.* 67, 3209–3218.

2202. Struhl, G., Struhl, K., and Macdonald, P.M. (1989). The gradient morphogen *bicoid* is a concentration-dependent transcriptional activator. *Cell* 57, 1259–1273.

2203. Strutt, H., Warrington, S.J., and Strutt, D. (2011). Dynamics of core planar polarity protein turnover and stable assembly into discrete membrane subdomains. *Dev. Cell* 20, 511–525.

2204. Sturtevant, A.H. (1965). *A History of Genetics.* Harper & Row, New York, NY.

2205. Sturtevant, A.H. (1970). Studies on the bristle pattern of *Drosophila. Dev. Biol.* 21, 48–61.

2206. Su, C.-Y., Menuz, K., and Carlson, J.R. (2009). Olfactory perception: receptors, cells, and circuits. *Cell* 139, 45–59.

2207. Suárez, R., García-González, D., and de Castro, F. (2012). Mutual influences between the main olfactory and vomeronasal systems in development and evolution. *Front. Neuroanat.* 6, Article 50.

2208. Suga, H., Schmid, V., and Gehring, W.J. (2008). Evolution and functional diversity of jellyfish opsins. *Curr. Biol.* 18, 51–55.

2209. Sun, T., Patoine, C., Abu-Khalil, A., Visvader, J., Sum, E., Cherry, T.J., Orkin, S.H., Geschwind, D.H., and Walsh, C.A. (2005). Early asymmetry of gene transcription in embryonic human left and right cerebral cortex. *Science* 308, 1794–1798.

2210. Sun, T., and Walsh, C.A. (2006). Molecular approaches to brain asymmetry and handedness. *Nat. Rev. Neurosci.* 7, 655–662.

2211. Sun, Y., Kanekar, S.L., Vetter, M.L., Gorski, S., Jan, Y.-N., Glaser, T., and Brown, N.L. (2003). Conserved and divergent functions of *Drosophila atonal*, amphibian, and mammalian *Ath5* genes. *Evol. Dev.* 5, 532–541.

2212. Sung, C.-H., and Chuang, J.-Z. (2010). The cell biology of vision. *J. Cell Biol.* 190, 953–963.

2213. Supp, D.M., Witte, D.P., Potter, S.S., and Brueckner, M. (1997). Mutation of an axonemal dynein affects left-right asymmetry in *inversus viscerum* mice. *Nature* 389, 963–966.

2214. Sutcliffe, B., Forero, M.G., Zhu, B., Robinson, I.M., and Hidalgo, A. (2013). Neuron-type specific functions of DNT1, DNT2 and Spz at the *Drosophila* neuromuscular junction. *PLoS ONE* 8, #10, e75902.

2215. Suzanne, M., Petzoldt, A.G., Spéder, P., Coutelis, J.-B., Steller, H., and Noselli, S. (2010). Coupling of apoptosis and L/R patterning controls stepwise organ looping. *Curr. Biol.* 20, 1773–1778.

2216. Suzuki, D.G., Murakami, Y., Escriva, H., and Wada, H. (2015). A comparative examination of neural circuit and brain patterning between the lamprey and amphioxus reveals the evolutionary origin of the vertebrate visual center. *J. Comp. Neurol.* 523, 251–261.

2217. Suzuki, T. (2013). How is digit identity determined during limb development? *Dev. Growth Differ.* 55, 130–138.

2218. Suzuki, Y., and Palopoli, M.F. (2001). Evolution of insect abdominal appendages: are prolegs homologous or convergent traits? *Dev. Genes Evol.* 211, 486–492.

2219. Sweeney, E., Fryer, A., Mountford, R., Green, A., and McIntosh, I. (2003). Nail patella syndrome: a review of the phenotype aided by developmental biology. *J. Med. Genet.* 40, 153–162.

2220. Tabin, C., and Wolpert, L. (2007). Rethinking the proximodistal axis of the vertebrate limb in the molecular era. *Genes Dev.* 21, 1433–1442.

2221. Tabin, C.J., Carroll, S.B., and Panganiban, G. (1999). Out on a limb: parallels in vertebrate and invertebrate limb patterning and the origin of appendages. *Am. Zool.* 39, 650–663.

2222. Tabin, C.J., and Vogan, K.J. (2003). A two-cilia model for vertebrate left-right axis specification. *Genes Dev.* 17, 1–6.

2223. Tabuchi, K., and Südhof, T.C. (2002). Structure and evolution of neurexin genes: insight into the mechanism of alternative splicing. *Genomics* 79, 849–859.

2224. Tahayato, A., Sonneville, R., Pichaud, F., Wernet, M.F., Papatsenko, D., Beaufils, P., Cook, T., and Desplan, C. (2003). Otd/Crx, a dual regulator for the specification of ommatidia subtypes in the *Drosophila* retina. *Dev. Cell* 5, 391–402.

2225. Tajiri, R., Misaki, K., Yonemura, S., and Hayashi, S. (2010). Dynamic shape changes of ECM-producing cells drive morphogenesis of ball-and-socket joints in the fly leg. *Development* 137, 2055–2063.

2226. Tajiri, R., Misaki, K., Yonemura, S., and Hayashi, S. (2011). Joint morphology in the insect leg: evolutionary history inferred from *Notch* loss-of-function phenotypes in *Drosophila*. *Development* 138, 4621–4626.

2227. Takahashi, M., and Osumi, N. (2002). Pax6 regulates specification of ventral neurone subtypes in the hindbrain by establishing progenitor domains. *Development* 129, 1327–1338.

2228. Takashima, S., Gold, D., and Hartenstein, V. (2013). Stem cells and lineages of the intestine: a developmental and evolutionary perspective. *Dev. Genes Evol.* 223, 85–102.

2229. Takashima, S., Yoshimori, H., Yamasaki, N., Matsuno, K., and Murakami, R. (2002). Cell-fate choice and boundary formation by combined action of *Notch* and *engrailed* in the *Drosophila* hindgut. *Dev. Genes Evol.* 212, 534–541.

2230. Takeda, S., and Narita, K. (2012). Structure and function of vertebrate cilia, towards a new taxonomy. *Differentiation* 83, S4–S11.

2231. Talpalar, A.E., Bouvier, J., Borgius, L., Fortin, G., Pierani, A., and Kiehn, O. (2013). Dual-mode operation of neuronal networks involved in left-right alternation. *Nature* 500, 85–88.

2232. Tam, P.P.L., Loebel, D.A.F., and Tanaka, S.S. (2006). Building the mouse gastrula: signals, asymmetry and lineages. *Curr. Opin. Genet. Dev.* 16, 419–425.

2233. Tamakoshi, T., Itakura, T., Chandra, A., Uezato, T., Yang, Z., Xue, X.-D., Wang, B., Hackett, B.P., Yokoyama, T., and Miura, N. (2006). Roles of the Foxj1 and Inv genes in the left-right determination of internal organs in mice. *Biochem. Biophys. Res. Comm.* 339, 932–938.

2234. Tan, X., Pecka, J.L., Tang, J., Okoruwa, O.E., Zhang, Q., Beisel, K.W., and He, D.Z.Z. (2011). From zebrafish to mammal: functional evolution of prestin, the motor protein of cochlear outer hair cells. *J. Neurophysiol.* 105, 36–44.

2235. Tanaka, E.M. (2003). Regeneration: if they can do it, why can't we? *Cell* 113, 559–562. [*See also* Tanaka, E.M. (2016). The molecular and cellular choreography of appendage regeneration. *Cell* 165, 1598–1608.]

2236. Tanaka, E.M. (2012). Skin, heal thyself. *Nature* 489, 508–510.

2237. Tanaka, E.M., and Reddien, P.W. (2011). The cellular basis for animal regeneration. *Dev. Cell* 21, 172–185.

2238. Tanaka, M. (2016). Fins into limbs: autopod acquisition and anterior elements reduction by modifying gene networks involving 5'*Hox*, *Gli3*, and *Shh*. *Dev. Biol.* 413, 1–7.

2239. Tanaka, M., Kasahara, H., Bartunkova, S., Schinke, M., Komuro, I., Inagaki, H., Lee, Y., Lyons, G.E., and Izumo, S. (1998). Vertebrate homologs of *tinman* and *bagpipe*: roles of the homeobox genes in cardiovascular development. *Dev. Genet.* 22, 239–249.

2240. Tanaka, Y., Okada, Y., and Hirokawa, N. (2005). FGF-induced vesicular release of Sonic hedgehog and retinoic acid in leftward nodal flow is critical for left-right determination. *Nature* 435, 172–177.

2241. Tang, M., Yuan, W., Bodmer, R., Wu, X., and Ocorr, K. (2013). The role of *pygopus* in the differentiation of intracardiac valves in *Drosophila*. *Genesis* 52, 19–28.

2242. Tang, W.J., Fernandez, J.G., Sohn, J.J., and Amemiya, C.T. (2015). Chitin is endogenously produced in vertebrates. *Curr. Biol.* 25, 897–900.

2243. Taniguchi, K., Hozumi, S., Maeda, R., Ooike, M., Sasamura, T., Aigaki, T., and Matsuno, K. (2007). D-JNK signaling in visceral muscle cells controls the laterality of the *Drosophila* gut. *Dev. Biol.* 311, 251–263.

2244. Tao, Y., and Schulz, R.A. (2007). Heart development in *Drosophila. Semin. Cell Dev. Biol.* 18, 3–15.

2245. Tarazona, O.A., Slota, L.A., Lopez, D.H., Zhang, G., and Cohn, M.J. (2016). The genetic program for cartilage development has deep homology within Bilateria. *Nature* 533, 86–89. [*See also* Brunet, T., and Arendt, D. (2016). Animal evolution: the hard problem of cartilage origins. *Curr. Biol.* 26, R685–R688.]

2246. Tarchini, B., Jolicoeur, C., and Cayouette, M. (2013). A molecular blueprint at the apical surface establishes planar asymmetry in cochlear hair cells. *Dev. Cell* 27, 88–102.

2247. Taschner, M., Bhogaraju, S., and Lorentzen, E. (2012). Architecture and function of IFT complex proteins in ciliogenesis. *Differentiation* 83, S12–S22.

2248. Tateya, T., Sakamoto, S., Imayoshi, I., and Kageyama, R. (2015). In vivo overactivation of the Notch signaling pathway in the developing cochlear epithelium. *Hear. Res.* 327, 209–217.

2249. Tautz, D. (2004). Segmentation. *Dev. Cell* 7, 301–312.

2250. Tautz, D., and Sommer, R.J. (1995). Evolution of segmentation genes in insects. *Trends Genet.* 11, 23–27.

2251. Taverna, E., Götz, M., and Huttner, W.B. (2015). The cell biology of neurogenesis: toward an understanding of the development and evolution of the neocortex. *Annu. Rev. Cell Dev. Biol.* 30, 465–502.

2252. Taylor, J.S., and Raes, J. (2004). Duplication and divergence: the evolution of new genes and old ideas. *Annu. Rev. Genet.* 38, 615–643.

2253. Technau, G.M., Berger, C., and Urbach, R. (2006). Generation of cell diversity and segmental pattern in the embryonic central nervous system of *Drosophila. Dev. Dyn.* 235, 861–869.

2254. Tee, Y.H., Shemesh, T., Thiagarajan, V., Hariadi, R.F., Anderson, K.L., Page, C., Volkmann, N., Hanein, D., Sivaramakrishnan, S., Kozlov, M.M., and Bershadsky, A.D. (2015). Cellular chirality arising from the self-organization of the actin cytoskeleton. *Nat. Cell Biol.* 17, #4, 445–457.

2255. Teixeira, C.S.S., Cerqueira, N.M.F.S.A., and Ferreira, A.C.S. (2016). Unravelling the olfactory sense: from the gene to odor perception. *Chem. Senses* 41, 105–121.

2256. Telford, M.J. (2000). Evidence for the derivation of the *Drosophila fushi tarazu* gene from a Hox gene orthologous to lophotrochozoan *Lox5. Curr. Biol.* 10, 349–352.

2257. Telford, M.J. (2007). A single origin of the central nervous system? *Cell* 129, 237–239.

2258. Telford, M.J., and Budd, G.E. (2003). The place of phylogeny and cladistics in Evo-Devo research. *Int. J. Dev. Biol.* 47, 479–490.

2259. Telford, M.J., Budd, G.E., and Philippe, H. (2015). Phylogenomic insights into animal evolution. *Curr. Biol.* 25, R876–R887.

2260. Temple, S.E. (2011). Why different regions of the retina have different spectral sensitivities: a review of mechanisms and functional significance of intraretinal variability in spectral sensitivity in vertebrates. *Vis. Neurosci.* 28, 281–293.

2261. Terakita, A. (2005). The opsins. *Genome Biol.* 6, Article 213.

2262. Terrell, D., Xie, B., Workman, M., Mahato, S., Zelhof, A.C., Gebelein, B., and Cook, T. (2012). OTX2 and CRX rescue overlapping and photoreceptor-specific functions in the *Drosophila* eye. *Dev. Dyn.* 241, 215–228.

2263. Tessmar-Raible, K., Raible, F., Christodoulou, F., Guy, K., Rembold, M., Hausen, H., and Arendt, D. (2007). Conserved sensory-neurosecretory cell types in annelid and fish forebrain: insights into hypothalamus evolution. *Cell* 129, 1389–1400.

2264. Tettamanti, G., Cattaneo, A.G., Gornati, R., de Eguileor, M., Bernardini, G., and Binelli, G. (2010). Phylogenesis of brain-derived neurotrophic factor (*BDNF*) in vertebrates. *Gene* 450, 85–93.

2265. Tettamanti, G., Grimaldi, A., Valvassori, R., Rinaldi, R., and de Eguileor, M. (2003). Vascular endothelial growth factor is involved in neoangiogenesis in *Hirudo medicinalis* (Annelida, Hirudinea). *Cytokine* 22, 168–179.

2266. Thanawala, S.U., Rister, J., Goldberg, G.W., Zuskov, A., Olesnicky, E.C., Flowers, J.M., Jukam, D., Purugganan, M.D., Gavis, E.R., Desplan, C., and Johnston, R.J., Jr. (2013). Regional modulation of a stochastically expressed factor determines photoreceptor subtypes in the *Drosophila* retina. *Dev. Cell* 25, 93–105.

2267. Tharadra, S.K., Medina, A., and Ray, A. (2013). Advantage of the highly restricted odorant receptor expression pattern in chemosensory neurons of *Drosophila*. *PLoS ONE* 8, #6, e66173.

2268. Theveneau, E., and Mayor, R. (2012). Neural crest delamination and migration: from epithelium-to-mesenchyme transition to collective cell migration. *Dev. Biol.* 366, 34–54.

2269. Thoen, H.H., How, M.J., Chiou, T.-H., and Marshall, J. (2014). A different form of color vision in mantis shrimp. *Science* 343, 411–413.

2270. Thomas, A.L., Davis, S.M., and Dierick, H.A. (2015). Of fighting flies, mice, and men: are some of the molecular and neuronal mechanisms of aggression universal in the animal kingdom? *PLoS Genet.* 11, #8, e1005416.

2271. Thompson, B.J. (2013). Cell polarity: models and mechanisms from yeast, worms, and flies. *Development* 140, 13–21.

2272. Thompson, D., Regev, A., and Roy, S. (2015). Comparative analysis of gene regulatory networks: from network reconstruction to evolution. *Annu. Rev. Cell Dev. Biol.* 31, 399–428.

2273. Thompson, H., Shaw, M.K., Dawe, H.R., and Shimeld, S.M. (2012). The formation and positioning of cilia in *Ciona intestinalis* embryos in relation to the generation and evolution of chordate left-right asymmetry. *Dev. Biol.* 364, 214–223.

2274. Thor, S. (2013). Stem cells in multiple time zones. *Nature* 498, 441–443.

2275. Thor, S., and Thomas, J.B. (2002). Motor neuron specification in worms, flies and mice: conserved and "lost" mechanisms. *Curr. Opin. Genet. Dev.* 12, 558–564.

2276. Tickle, C., and Barker, H. (2013). The Sonic hedgehog gradient in the developing limb. *WIREs Dev. Biol.* 2, 275–290.

2277. Tilney, L.G., Connelly, P., Smith, S., and Guild, G.M. (1996). F-Actin bundles in *Drosophila* bristles are assembled from modules composed of short filaments. *J. Cell Biol.* 135, 1291–1308.

2278. Tilney, L.G., Cotanche, D.A., and Tilney, M.S. (1992). Actin filaments, stereocilia and hair cells of the bird cochlea. VI. How the number and arrangement of stereocilia are determined. *Development* 116, 213–226.

2279. Tilney, L.G., and DeRosier, D.J. (2005). How to make a curved *Drosophila* bristle using straight actin bundles. *PNAS* 102, #52, 18785–18792.

2280. Tilney, L.G., Tilney, M.S., and Cotanche, D.A. (1988). Actin filaments, stereocilia, and hair cells of the bird cochlea. V. How the staircase pattern of stereociliary lengths is generated. *J. Cell Biol.* 106, 355–365.

2281. Tilney, L.G., Tilney, M.S., and DeRosier, D.J. (1992). Actin filaments, stereocilia, and hair cells: how cells count and measure. *Annu. Rev. Cell Biol.* 8, 257–274.

2282. Tilney, L.G., Tilney, M.S., and Guild, G.M. (1995). F-Actin bundles in *Drosophila* bristles. I. Two filament cross-links are involved in bundling. *J. Cell Biol.* 130, 629–638.

2283. Tingler, M., Ott, T., Tözser, J., Kurz, S., Getwan, M., Tisler, M., Schweickert, A., and Blum, M. (2014). Symmetry breakage in the frog *Xenopus*: role of Rab11 and the ventral-right blastomere. *Genesis* 52, 588–599.

2284. Tiozzo, S., Christiaen, L., Deyts, C., Manni, L., Joly, J.-S., and Burighel, P. (2005). Embryonic versus blastogenetic development in the compound ascidian *Botryllus schlosseri*: insights from *Pitx* expression patterns. *Dev. Dyn.* 232, 468–478.

2285. Tissir, F., and Goffinet, A.M. (2010). Planar cell polarity signaling in neural development. *Curr. Opin. Neurobiol.* 20, 572–577.

2286. Tittel, J.N., and Steller, H. (2000). A comparison of programmed cell death between species. *Genome Biol.* 1, #3, 1–6.

2287. Todi, S.V., Franke, J.D., Kiehart, D.P., and Eberl, D.F. (2005). Myosin VIIA defects, which underlie the Usher 1B syndrome in humans, lead to deafness in *Drosophila*. *Curr. Biol.* 15, 862–868. [*See also* Li, T., *et al.* (2016). The E3 ligase Ubr3 regulates Usher syndrome and *MYH9* disorder proteins in the auditory organs of *Drosophila* and mammals. *eLife* 5, e15258.]

2288. Todi, S.V., Sivan-Loukianova, E., Jacobs, J.S., Kiehart, D.P., and Eberl, D.F. (2008). Myosin VIIA, important for human auditory function, is necessary for *Drosophila* auditory organ development. *PLoS ONE* 3, #5, e2115.

2289. Tomarev, S.I., Callaerts, P., Kos, L., Zinovieva, R., Halder, G., Gehring, W., and Piatigorsky, J. (1997). Squid Pax-6 and eye development. *PNAS* 94, 2421–2426.

2290. Tomer, R., Denes, A.S., Tessmar-Raible, K., and Arendt, D. (2010). Profiling by image registration reveals common origin of annelid mushroom bodies and vertebrate pallium. *Cell* 142, 800–809.

2291. Torgersen, J. (1947). Transposition of viscera, bronchiectasis and nasal polyps. *Acta Radiol.* 28, 17–24.

2292. Torres, M. (2016). Limb regrowth takes two. *Nature* 533, 328–330.

2293. Torres, M., Gómez-Pardo, E., and Gruss, P. (1996). *Pax2* contributes to inner ear patterning and optic nerve trajectory. *Development* 122, 3381–3391.

2294. Tosches, M.A., and Arendt, D. (2013). The bilaterian forebrain: an evolutionary chimaera. *Curr. Opin. Neurobiol.* 23, 1080–1089.

2295. Touhara, K., and Vosshall, L.B. (2009). Sensing odorants and pheromones with chemosensory receptors. *Annu. Rev. Physiol.* 71, 307–332.

2296. Towbin, B.D., Meister, P., and Gasser, S.M. (2009). The nuclear envelope: a scaffold for silencing? *Curr. Opin. Genet. Dev.* 19, 180–186.

2297. Travis, J. (1995). The ghost of Geoffroy Saint-Hilaire: frog and fly genes revive the ridiculed idea that vertebrates resemble upside-down insects. *Sci. News* 148, 216–218.

2298. Treisman, J.E. (2004). Coming to our senses. *BioEssays* 26, 825–828.

2299. Treisman, J.E. (2004). How to make an eye. *Development* 131, 3823–3827.

2300. Tromelin, A. (2016). Odour perception: a review of an intricate signalling pathway. *Flavour Fragr. J.* 31, 107–119.

2301. Troost, T., and Klein, T. (2012). Sequential Notch signalling at the boundary of Fringe expressing and non-expressing cells. *PLoS ONE* 7, #11, e49007.

2302. Troost, T., Schneider, M., and Klein, T. (2015). A re-examination of the selection of the sensory organ precursor of the bristle sensilla of *Drosophila melanogaster*. *PLoS Genet*. 11, #1, e1004911.

2303. Trotier, D. (2011). Vomeronasal organ and human pheromones. *Eur. Ann. Otorhinolaryngol. Head Neck Dis.* 128, 184–190.

2304. Trujillo-Cenóz, O. (1985). The eye: development, structure and neural connections. *In* G.A. Kerkut and L.I. Gilbert (eds.), *Comprehensive Insect Physiology, Biochemistry, and Pharmacology*. Pergamon Press, Oxford, pp. 171–223.

2305. Tsachaki, M., and Sprecher, S.G. (2012). Genetic and developmental mechanisms underlying the formation of the *Drosophila* compound eye. *Dev. Dyn.* 241, 40–56.

2306. Tschopp, P., and Duboule, D. (2011). A genetic approach to the transcriptional regulation of *Hox* gene clusters. *Annu. Rev. Genet.* 45, 145–166.

2307. Tschopp, P., Tarchini, B., Spitz, F., Zakany, J., and Duboule, D. (2009). Uncoupling time and space in the collinear regulation of *Hox* genes. *PLoS Genet.* 5, #3, e1000398.

2308. Tsigankov, D., and Koulakov, A.A. (2010). Sperry versus Hebb: topographic mapping in Isl2/EphA3 mutant mice. *BMC Neurosci.* 11, Article 155.

2309. Tuthill, J.C., and Wilson, R.I. (2016). Parallel transformation of tactile signals in central circuits of *Drosophila*. *Cell* 164, 1046–1059.

2310. Tweedt, S.M., and Erwin, D.H. (2015). Origin of metazoan developmental toolkits and their expression in the fossil record. *In* I. Ruiz-Trillo and A.M. Nedelcu (eds.), *Evolutionary Transitions to Multicellular Life: Principles and Mechanisms*. Springer, Netherlands, pp. 47–77.

2311. Uemura, T., and Shimada, Y. (2003). Breaking cellular symmetry along planar axes in *Drosophila* and vertebrates. *J. Biochem.* 134, 625–630.

2312. Ugur, B., Chen, K., and Bellen, H.J. (2016). *Drosophila* tools and assays for the study of human diseases. *Dis. Model. Mech.* 9, 235–244.

2313. Ullrich, B., Ushkaryov, Y.A., and Südhof, T.C. (1995). Cartography of neurexins: more than 1000 isoforms generated by alternative splicing and expressed in distinct subsets of neurons. *Neuron* 14, 497–507.

2314. Ullrich-Lüter, E.M., Dupont, S., Arboleda, E., Hausen, H., and Arnone, M.I. (2011). Unique system of photoreceptors in sea urchin tube feet. *PNAS* 108, #20, 8367–8372.

2315. Umesono, Y., Watanabe, K., and Agata, K. (1999). Distinct structural domains in the planarian brain defined by the expression of evolutionarily conserved homeobox genes. *Dev. Genes Evol.* 209, 31–39.

2316. Underwood, E. (2015). Plugged pores may underlie some ALS, dementia cases. *Science* 349, 911–912.

2317. Urata, M., Tsuchimoto, J., Yasui, K., and Yamaguchi, M. (2009). The *Hox8* of the hemichordate *Balanoglossus misakiensis*. *Dev. Genes Evol.* 219, 377–382.

2318. Urbach, R. (2007). A procephalic territory in *Drosophila* exhibiting similarities and dissimilarities compared to the vertebrate midbrain/hindbrain boundary region. *Neural Dev.* 2, Article 23.

2319. Urbach, R., Volland, D., Seibert, J., and Technau, G.M. (2006). Segment-specific requirements for dorsoventral patterning genes during early brain development in *Drosophila*. *Development* 133, 4315–4330.

2320. Usukura, J., and Obata, S. (1995). Morphogenesis of photoreceptor outer segments in retinal development. *Prog. Retin. Eye Res.* 15, 113–125.

2321. van Alphen, B., and van Swinderen, B. (2013). *Drosophila* strategies to study psychiatric disorders. *Brain Res. Bull.* 92, 1–11.

2322. van Amerongen, R., and Nusse, R. (2009). Towards an integrated view of Wnt signaling in development. *Development* 136, 3205–3214.

2323. Van Battum, E.Y., Brignani, S., and Pasterkamp, R.J. (2015). Axon guidance proteins in neurological disorders. *Lancet* 14, 532–546.

2324. Van de Peer, Y., Maere, S., and Meyer, A. (2009). The evolutionary significance of ancient genome duplications. *Nat. Rev. Genet.* 10, 725–732.

2325. van der Knaap, L.J., and van der Ham, J.M. (2011). How does the corpus callosum mediate interhemispheric transfer? A review. *Behav. Brain Res.* 223, 211–221.

2326. van der Linde, D., Konings, E.E.M., Slager, M.A., Witsenburg, M., Helbing, W.A., Takkenberg, J.J.M., and Roos-Hesselink, J.W. (2011). Birth prevalence of congenital heart disease worldwide: a systematic review and meta-analysis. *J. Am. Coll. Cardiol.* 58, #21, 2241–2247.

2327. Van Essen, D.C. (1997). A tension-based theory of morphogenesis and compact wiring in the central nervous system. *Nature* 385, 313–318.

2328. van Hateren, J.H., Hardie, R.C., Rudolph, A., Laughlin, S.B., and Stavenga, D.G. (1989). The bright zone, a specialized dorsal eye region in the male blowfly *Chrysomyia megacephala*. *J. Comp. Physiol. A* 164, 297–308. [*See also* Perry, M.W., and Desplan, C. (2016). Love spots. *Curr. Biol.* 26, R484–R485.]

2329. van Holde, K.E., Miller, K.I., and Decker, H. (2001). Hemocyanins and invertebrate evolution. *J. Biol. Chem.* 276, #19, 15563–15566.

2330. van Staaden, M.J., and Römer, H. (1998). Evolutionary transition from stretch to hearing organs in ancient grasshoppers. *Nature* 394, 773–776.

2331. Vandenberg, L.N., and Levin, M. (2010). Consistent left-right asymmetry cannot be established by late organizers in *Xenopus* unless the late organizer is a conjoined twin. *Development* 137, 1095–1105.

2332. Vandenberg, L.N., and Levin, M. (2010). Far from solved: a perspective on what we know about early mechanisms of left-right asymmetry. *Dev. Dyn.* 239, 3131–3146.

2333. Vandenberg, L.N., and Levin, M. (2013). A unified model for left-right asymmetry? Comparison and synthesis of molecular models of embryonic laterality. *Dev. Biol.* 379, 1–15.

2334. Vasiliauskas, D., Mazzoni, E.O., Sprecher, S.G., Brodetskiy, K., Johnston, R.J., Jr., Lidder, P., Vogt, N., Celik, A., and Desplan, C. (2011). Feedback from rhodopsin controls *rhodopsin* exclusion in *Drosophila* photoreceptors. *Nature* 479, 108–112.

2335. Vasudevan, D., and Ryoo, H.D. (2015). Regulation of cell death by IAPs and their antagonists. *Curr. Top. Dev. Biol.* 114, 185–208.

2336. Vavouri, T., and Lehner, B. (2009). Conserved noncoding elements and the evolution of animal body plans. *BioEssays* 31, 727–735.

2337. Videnovic, A., Lazar, A.S., Barker, R.A., and Overeem, S. (2014). "The clocks that time us": circadian rhythms in neurodegenerative disorders. *Nat. Rev. Neurol.* 10, 683–693.

2338. Villar, D., Flicek, P., and Odom, D.T. (2014). Evolution of transcription factor binding in metazoans: mechanisms and functional implications. *Nat. Rev. Genet.* 15, 221–233.

2339. Visel, A., Minovitsky, S., Dubchak, I., and Pennacchio, L.A. (2007). VISTA enhancer browser: a database of tissue-specific human enhancers. *Nucleic Acids Res.* 35, D88–D92.

2340. Vitruvius (1960). *The Ten Books on Architecture.* Dover Publications, New York, NY. M.H. Morgan, translator.

2341. Vogel, A., Rodriguez, C., Warnken, W., and Izpisua Belmonte, J.C. (1995). Dorsal cell fate specified by chick *Lmx1* during vertebrate limb development. *Nature* 378, 716–720.

2342. Vogt, K., Schnaitmann, C., Dylla, K.V., Knapek, S., Aso, Y., Rubin, G.M., and Tanimoto, H. (2014). Shared mushroom body circuits underlie visual and olfactory memories in *Drosophila. eLife* 3, e02395.

2343. Vogt, N., and Desplan, C. (2014). Vision. *In* J. Dubnau (ed.), *Behavioral Genetics of the Fly (Drosophila melanogaster)*. Cambridge University Press, New York, NY, pp. 37–48.

2344. Vogt, T.F., and Duboule, D. (1999). Antagonists go out on a limb. *Cell* 99, 563–566.

2345. Von Allmen, G., Hogga, I., Spierer, A., Karch, F., Bender, W., Gyurkovics, H., and Lewis, E. (1996). Splits in the fruitfly Hox gene complexes. *Nature* 380, 116.

2346. von Frisch, K. (1967). *The Dance Language and Orientation of Bees*. Harvard University Press, Cambridge, MA.

2347. von Lintig, J. (2012). Metabolism of carotenoids and retinoids related to vision. *J. Biol. Chem.* 287, #3, 1627–1634.

2348. von Schantz, M., Jenkins, A., and Archer, S.N. (2006). Evolutionary history of the vertebrate *period* genes. *J. Mol. Evol.* 62, 701–707.

2349. Vopalensky, P., and Kozmik, Z. (2009). Eye evolution: common use and independent recruitment of genetic components. *Philos. Trans. R. Soc. Lond. B* 364, 2819–2832.

2350. Vopalensky, P., Pergner, J., Liegertova, M., Benito-Gutierrez, E., Arendt, D., and Kozmik, Z. (2013). Molecular analysis of the amphioxus frontal eye unravels the evolutionary origin of the retina and pigment cells of the vertebrate eye. *PNAS* 109, #38, 15383–15388.

2351. Vora, S., and Phillips, B.T. (2015). Centrosome-associated degradation limits b-catenin inheritance by daughter cells after asymmetric division. *Curr. Biol.* 25, 1005–1016.

2352. Voss, A.K., Collin, C., Dixon, M.P., and Thomas, T. (2009). Moz and retinoic acid coordinately regulate H3K9 acetylation, *Hox* gene expression, and segment identity. *Dev. Cell* 17, 674–686.

2353. Vosshall, L.B., and Stensmyr, M.C. (2005). Wake up and smell the pheromones. *Neuron* 45, 179–187.

2354. Vosshall, L.B., and Stocker, R.F. (2007). Molecular architecture of smell and taste in *Drosophila. Annu. Rev. Neurosci.* 30, 505–533.

2355. Vosshall, L.B., Wong, A.M., and Axel, R. (2000). An olfactory sensory map in the fly brain. *Cell* 102, 147–159.

2356. Vulliemoz, S., Raineteau, O., and Jabaudon, D. (2005). Reaching beyond the midline: why are human brains cross wired? *Lancet Neurol.* 4, 87–99.

2357. Wada, H., Escriva, H., Zhang, S., and Laudet, V. (2006). Conserved RARE localization in amphioxus *Hox* clusters and implications for *Hox* code evolution in the vertebrate neural crest. *Dev. Dyn.* 235, 1522–1531.

2358. Wada, H., Garcia-Fernàndez, J., and Holland, P.W.H. (1999). Colinear and segmental expression of amphioxus Hox genes. *Dev. Biol.* 213, 131–141.

2359. Wadsworth, W.G. (2005). Axon pruning: *C. elegans* makes the cut. *Curr. Biol.* 15, R796–R798.

2360. Wagner, A. (2011). Genotype networks shed light on evolutionary constraints. *Trends Ecol. Evol.* 26, 577–584.

2361. Wagner, E.F., and Petruzzelli, M. (2015). A waste of insulin interference. *Nature* 521, 430–431.

2362. Wagner, G.P. (2007). The developmental genetics of homology. *Nat. Rev. Genet.* 8, 473–479.

2363. Wagner, G.P. (2014). *Homology, Genes, and Evolutionary Innovation.* Princeton University Press, Princeton, NJ.

2364. Wagner, G.P. (2016). What is "homology thinking" and what is it for? *J. Exp. Zool. B. Mol. Dev. Evol.* 326, 3–8.

2365. Wagner, R.A., Tabibiazar, R., Liao, A., and Quertermous, T. (2005). Genome-wide expression dynamics during mouse embryonic development reveal similarities to *Drosophila* development. *Dev. Biol.* 288, 595–611.

2366. Wake, D.B., Wake, M.H., and Specht, C.D. (2011). Homoplasy: from detecting pattern to determining process and mechanism of evolution. *Science* 331, 1032–1035.

2367. Wallace, M.T. (2015). Multisensory perception: the building of flavor representations. *Curr. Biol.* 25, R980–R1001.

2368. Wallace, V.A. (2008). Proliferative and cell fate effects of Hedgehog signaling in the vertebrate retina. *Brain Res.* 1192, 61–75.

2369. Wallingford, J.B., and Mitchell, B. (2011). Strange as it may seem: the many links between Wnt signaling, planar cell polarity, and cilia. *Genes Dev.* 25, 201–213.

2370. Wan, G., Corfas, G., and Stone, J.S. (2013). Inner ear supporting cells: rethinking the silent majority. *Semin. Cell Dev. Biol.* 24, 448–459.

2371. Wang, G.-Z., Marini, S., Ma, X., Yang, Q., Zhang, X., and Zhu, Y. (2014). Improvement of *Dscam* homophilic binding affinity throughout *Drosophila* evolution. *BMC Evol. Biol.* 14, Article 186.

2372. Wang, J., Zugates, C.T., Liang, I.H., Lee, C.-H.J., and Lee, T. (2002). *Drosophila* Dscam is required for divergent segregation of sister branches and suppresses ectopic bifurcation of axons. *Neuron* 33, 559–571.

2373. Wang, R., Chen, C.-C., Hara, E., Rivas, M.V., Roulhac, P.L., Howard, J.T., Chakraborty, M., Audet, J.-N., and Jarvis, E.D. (2014). Convergent differential regulation of SLIT-ROBO axon guidance genes in the brains of vocal learners. *J. Comp. Neurol.* 523, 892–906.

2374. Wang, S., and Samakovlis, C. (2012). Grainy head and its target genes in epithelial morphogenesis and wound healing. *Curr. Top. Dev. Biol.* 98, 35–63.

2375. Wang, W., Nossoni, Z., Berbasova, T., Watson, C.T., Yapici, I., Lee, K.S.S., Vasileiou, C., Geiger, J.H., and Borhan, B. (2012). Tuning the electronic absorption of protein-embedded all-*trans*-retinal. *Science* 338, 1340–1343.

2376. Wang, X., Wang, T., Jiao, Y., von Lintig, J., and Montell, C. (2010). Requirement for an enzymatic visual cycle in *Drosophila*. *Curr. Biol.* 20, 93–102.

2377. Wang, Y., Gao, Y., Imsland, F., Gu, X., Feng, C., Liu, R., Song, C., Tixier-Boichard, M., Gourichon, D., Li, Q., Chen, K., Li, H., Andersson, L., Hu, X., and Li, N. (2012). The Crest phenotype in chicken is associated with ectopic expression of *Hoxc8* in cranial skin. *PLoS ONE* 7, #4, e34012.

2378. Wang, Z., Nudelman, A., and Storm, D.R. (2007). Are pheromones detected through the main olfactory epithelium? *Mol. Neurobiol.* 35, 317–323.

2379. Wang, Z., Singhvi, A., Kong, P., and Scott, K. (2004). Taste representations in the *Drosophila* brain. *Cell* 117, 981–991.

2380. Wangler, M.F., Yamamoto, S., and Bellen, H.J. (2015). Fruit flies in biomedical research. *Genetics* 199, 639–653.

2381. Wanninger, A., Kristof, A., and Brinkmann, N. (2009). Sipunculans and segmentation. *Commun. Integr. Biol.* 2, 56–59.

2382. Warchol, M.E., and Montcouquiol, M. (2010). Maintained expression of the planar cell polarity molecule Vangl2 and reformation of hair cell orientation in the regenerating inner ear. *J. Assoc. Res. Otolaryngol.* 11, 395–406.

2383. Ward, E.J., Zhou, X., Riddiford, L.M., Berg, C.A., and Ruohola-Baker, H. (2006). Border of Notch activity establishes a boundary between the two dorsal appendage tube cell types. *Dev. Biol.* 297, 461–470.

2384. Wardill, T.J., List, O., Li, X., Dongre, S., McCulloch, M., Ting, C.-Y., O'Kane, C.J., Tang, S., Lee, C.-H., Hardie, R.C., and Juusola, M. (2012). Multiple spectral inputs improve motion discrimination in the *Drosophila* visual system. *Science* 336, 925–931.

2385. Ware, M., Dupé, V., and Schubert, F.R. (2015). Evolutionary conservation of the early axon scaffold in the vertebrate brain. *Dev. Dyn.* 244, 1202–1214.

2386. Warner, J.F., and McClay, D.R. (2014). Left-right asymmetry in the sea urchin. *Genesis* 52, 481–487.

2387. Warrant, E.J. (2015). Photoreceptor evolution: ancient "cones" turn out to be rods. *Curr. Biol.* 25, R148–R151.

2388. Warrant, E.J., and Johnsen, S. (2013). Vision and the light environment. *Curr. Biol.* 23, R990–R994.

2389. Warrant, E.J., and McIntyre, P.D. (1993). Arthropod eye design and the physical limits to spatial resolving power. *Prog. Neurobiol.* 40, 413–461.

2390. Wartlick, O., Jülicher, F., and Gonzalez Gaitan, M. (2014). Growth control by a moving morphogen gradient during *Drosophila* eye development. *Development* 141, 1884–1893.

2391. Washington, I., Zhou, J., Jockusch, S., Turro, N.J., Nakanishi, K., and Sparrow, J.R. (2007). Chlorophyll derivatives as visual pigments for super vision in the red. *Photochem. Photobiol. Sci.* 6, 775–779.

2392. Washington, N.L., Haendel, M.A., Mungall, C.J., Ashburner, M., Westerfield, M., and Lewis, S.E. (2009). Linking human diseases to animal models using ontology-based phenotype annotation. *PLoS Biol.* 7, #11, e1000247.

2393. Wässle, H. (1999). A patchwork of cones. *Nature* 397, 473–475.

2394. Watanabe, D., Saijoh, Y., Nonaka, S., Sasaki, G., Ikawa, Y., Yokoyama, T., and Hamada, H. (2003). The left-right determinant Inversin is a component of node monocilia and other 9+0 cilia. *Development* 130, 1725–1734.

2395. Watanabe, H., Schmidt, H.A., Kuhn, A., Höger, S.K., Kocagöz, Y., Laumann-Lipp, N., Özbek, S., and Holstein, T.W. (2014). Nodal signalling determines biradial symmetry in Hydra. *Nature* 515, 112–115.

2396. Watanabe, S., Umeki, N., Ikebe, R., and Ikebe, M. (2008). Impacts of Usher syndrome type 1B mutations on human myosin VIIa motor function. *Biochemistry* 47, 9505–9513.

2397. Watson, F.L., Püttmann-Holgado, R., Thomas, F., Lamar, D.L., Hughes, M., Kondo, M., Rebel, V.I., and Schmucker, D. (2005). Extensive diversity of Ig-superfamily proteins in the immune system of insects. *Science* 309, 1874–1878.

2398. Wawersik, S., and Maas, R.L. (2000). Vertebrate eye development as modeled in *Drosophila*. *Hum. Mol. Genet.* 9, 917–925.

2399. Weaver, J. (2012). Striking similarities in fly and vertebrate olfactory network formation. *PLoS Biol.* 10, #10, e1001401.

2400. Weavers, H., Prieto-Sánchez, S., Grawe, F., Garcia-López, A., Artero, R., Wilsch-Bräuninger, M., Ruiz-Gómez, M., Skaer, H., and Denholm, B. (2009). The insect nephrocyte is a podocyte-like cell with a filtration slit diaphragm. *Nature* 457, 322–326.

2401. Weber, T., Göpfert, M.C., Winter, H., Zimmermann, U., Kohler, H., Meier, A., Hendrich, O., Rohbock, K., Robert, D., and Knipper, M. (2003). Expression of prestin-homologous solute carrier (SLC26) in auditory organs of nonmammalian vertebrates and insects. *PNAS* 100, #13, 7690–7695.

2402. Weddell, T.D., Mellado-Lagarde, M., Lukashkina, V.A., Lukashkin, A.N., Zuo, J., and Russell, I.J. (2011). Prestin links extrinsic tuning to neural excitation in the mammalian cochlea. *Curr. Biol.* 21, R682–R683.

2403. Wehner, R. (2005). Brainless eyes. *Nature* 435, 157–159.

2404. Weiner, J. (1999). *Time, Love, Memory*. Random House, New York, NY.

2405. Weir, P.T., and Dickinson, M.H. (2012). Flying *Drosophila* orient to sky polarization. *Curr. Biol.* 22, 21–27.

2406. Weirauch, M.T., and Hughes, T.R. (2011). A catalogue of eukaryotic transcription factor types, their evolutionary origin, and species distribution. *In* T.R. Hughes (ed.), *A Handbook of Transcription Factors*. Springer, New York, NY.

2407. Weisblat, D.A., and Kuo, D.-H. (2014). Developmental biology of the leech *Helobdella*. *Int. J. Dev. Biol.* 58, 429–443.

2408. Wellik, D.M., and Capecchi, M.R. (2003). *Hox10* and *Hox11* genes are required to globally pattern the mammalian skeleton. *Science* 301, 363–367.

2409. Welniarz, Q., Dusart, I., Gallea, C., and Roze, E. (2015). One hand clapping: lateralization of motor control. *Front. Neuroanat.* 9, Article 75.

2410. Welsh, I.C., Thomsen, M., Gludish, D.W., Alfonso-Parra, C., Bai, Y., Martin, J.F., and Kurpios, N.A. (2013). Integration of left-right *Pitx2* transcription and Wnt signaling drives asymmetric gut morphogenesis via Daam2. *Dev. Cell* 26, 629–644.

2411. Werner, H.B. (2013). Do we have to reconsider the evolutionary emergence of myelin? *Front. Cell. Neurosci.* 7, Article 217.

2412. Werner, M.E., Ward, H.H., Phillips, C.L., Miller, C., Gattone, V.H., and Bacallao, R.L. (2013). Inversin modulates the cortical actin network during mitosis. *Am. J. Physiol. Cell Physiol.* 305, C36–C47.

2413. Wernet, M.F., Celik, A., Mikeladze-Dvali, T., and Desplan, C. (2007). Generation of uniform fly retinas. *Curr. Biol.* 17, R1002–R1003.

2414. Wernet, M.F., and Desplan, C. (2004). Building a retinal mosaic: cell-fate decision in the fly eye. *Trends Cell Biol.* 14, 576–584.

2415. Wernet, M.F., Huberman, A.D., and Desplan, C. (2014). So many pieces, one puzzle: cell type specification and visual circuitry in flies and mice. *Genes Dev.* 28, 2565–2584.

2416. Wernet, M.F., Mazzoni, E.O., Celik, A., Duncan, D.M., Duncan, I., and Desplan, C. (2006). Stochastic *spineless* expression creates the retinal mosaic for color vision. *Nature* 440, 174–180.

2417. Wernet, M.F., Meier, K.M., Baumann-Klausener, F., Dorfman, R., Weihe, U., Labhart, T., and Desplan, C. (2014). Genetic dissection of photoreceptor subtype specification by the *Drosophila melanogaster* zinc finger proteins Elbow and No ocelli. *PLoS Genet.* 10, #3, e1004210.

2418. Wernet, M.F., Perry, M.W., and Desplan, C. (2015). The evolutionary diversity of insect retinal mosaics: common design principles and emerging molecular logic. *Trends Genet.* 31, 316–328.

2419. Wernet, M.F., Velez, M.M., Clark, D.A., Baumann-Klausener, F., Brown, J.R., Klovstad, M., Labhart, T., and Clandinin, T.R. (2012). Genetic dissection reveals two separate retinal substrates for polarization vision in *Drosophila*. *Curr. Biol.* 22, 12–20.

2420. Wexler, J.R., Plachetzki, D.C., and Kopp, A. (2014). Pan-metazoan phylogeny of the DMRT gene family: a framework for functional studies. *Dev. Genes Evol.* 224, 175–181.

2421. Wheeler, S.R., Stagg, S.B., and Crews, S.T. (2008). Multiple *Notch* signaling events control *Drosophila* CNS midline neurogenesis, gliogenesis and neuronal identity. *Development* 135, 3071–3079.

2422. White, L.E., Lucas, G., Richards, A., and Purves, D. (1994). Cerebral asymmetry and handedness. *Nature* 368, 197–198.

2423. White, R.A.H., and Lehmann, R. (1986). A gap gene, *hunchback*, regulates the spatial expression of *Ultrabithorax*. *Cell* 47, 311–321.

2424. Whitehouse, D. (2009). *Renaissance Genius: Galileo Galilei and His Legacy to Modern Science*. Sterling, New York, NY.

2425. Whitesides, G.M., and Grzybowski, B. (2002). Self-assembly at all scales. *Science* 295, 2418–2421.

2426. Whitfield, T.T. (2015). Development of the inner ear. *Curr. Opin. Genet. Dev.* 32, 112–118.

2427. Whitfield, T.T., and Monk, N. (2014). Julian Hart Lewis, F.R.S. (1946–2014). *Dev. Cell* 29, 507–509.

2428. Wibowo, I., Pinto-Teixeira, F., Satou, C., Higashijima, S.-i., and López-Schier, H. (2011). Compartmentalized Notch signaling sustains epithelial mirror symmetry. *Development* 138, 1143–1152.

2429. Wicher, D. (2012). Functional and evolutionary aspects of chemoreceptors. *Front. Cell. Neurosci.* 6, Article 48.

2430. Wicher, D., Schäfer, R., Bauernfeind, R., Stensmyr, M.C., Heller, R., Heinemann, S.H., and Hansson, B.S. (2008). *Drosophila* odorant receptors are both ligand-gated and cyclic-nucleotide-activated cation channels. *Nature* 452, 1007–1011.

2431. Wieschaus, E., and Nüsslein-Volhard, C. (2014). Walter Gehring (1939–2014). *Curr. Biol.* 24, R632–R634.

2432. Wikler, K.C., and Rakic, P. (1991). Relation of an array of early-differentiating cones to the photoreceptor mosaic in the primate retina. *Nature* 351, 397–400.

2433. Wilkins, A.S. (1989). Organizing the *Drosophila* posterior pattern: why has the fruit fly made life so complicated for itself? *BioEssays* 11, 67–69.

2434. Wilkins, A.S. (2002). *The Evolution of Developmental Pathways*. Sinauer, Sunderland, MA.

2435. Wilkins, A.S. (2014). "The genetic tool-kit": the life history of an important metaphor. *In Advances in Evolutionary Developmental Biology*. Wiley-Blackwell, New York, NY, pp. 1–14.

2436. Williams, D.S. (1991). Actin filaments and photoreceptor membrane turnover. *BioEssays* 13, 171–178.

2437. Williams, R.J. (1971). *You Are Extraordinary*. Pyramid Books, New York, NY.

2438. Williams, T.A., Nulsen, C., and Nagy, L.M. (2002). A complex role for Distal-less in crustacean appendage development. *Dev. Biol.* 241, 302–312.

2439. Wilmer, P. (2003). Convergence and homoplasy in the evolution of organismal form. *In* G.B. Müller and S.A. Newman (eds.), *Origination of Organismal Form: Beyond the Gene in Developmental and Evolutionary Biology*. MIT Press, Cambridge, MA., pp. 33–49.

2440. Wilson, M.J., and Dearden, P.K. (2011). Diversity in insect axis formation: two *orthodenticle* genes and *hunchback* act in anterior patterning and influence dorsoventral organization in the honeybee (*Apis mellifera*). *Development* 138, 3497–3507.

2441. Wilson, R.I. (2013). Early olfactory processing in *Drosophila*: mechanisms and principles. *Annu. Rev. Neurosci.* 36, 217–241.

2442. Wilson, R.I., and Mainen, Z.F. (2006). Early events in olfactory processing. *Annu. Rev. Neurosci.* 29, 163–201.

2443. Wilson, S.I., Shafer, B., Lee, K.J., and Dodd, J. (2008). A molecular program for contralateral trajectory: Rig-1 control by LIM homeodomain transcription factors. *Neuron* 59, 413–424.

2444. Wilson-Sanders, S.E. (2011). Invertebrate models for biomedical research, testing, and education. *ILAR J.* 52, 126–152.

2445. Wimmer, E.A., Carleton, A., Harjes, P., Turner, T., and Desplan, C. (2000). *bicoid*-independent formation of thoracic segments in *Drosophila*. *Science* 287, 2476–2479.

2446. Winchell, C.J., and Jacobs, D.K. (2013). Expression of the Lhx genes apterous and lim1 in an errant polychaete: implications for bilaterian appendage evolution, neural development, and muscle diversification. *EvoDevo* 4, Article 4.

2447. Winchell, C.J., Valencia, J.E., and Jacobs, D.K. (2010). Expression of *Distal-less*, *dachshund*, and *optomotor blind* in *Neanthes arenaceodentata* (Annelida, Nereididae) does not support homology of appendage-forming mechanisms across the Bilateria. *Dev. Genes Evol.* 220, 275–295.

2448. Winston, R., and Wilson, D.E. (2006). *Human*. Dorling Kindersley, New York, NY.

2449. Winterbottom, E.F., Illes, J.C., Faas, L., and Isaacs, H.V. (2010). Conserved and novel roles for the Gsh2 transcription factor in primary neurogenesis. *Development* 137, 2623–2631.

2450. Wodarz, A., and Huttner, W.B. (2003). Asymmetric cell division during neurogenesis in *Drosophila* and vertebrates. *Mech. Dev.* 120, 1297–1309.

2451. Wojtowicz, W.M., Wu, W., Andre, I., Qian, B., Baker, D., and Zipursky, S.L. (2007). A vast repertoire of Dscam binding specificities arises from modular interactions of variable Ig domains. *Cell* 130, 1134–1145.

2452. Wolf, F.W., and Heberlein, U. (2003). Invertebrate models of drug abuse. *J. Neurobiol.* 54, 161–178.

2453. Wolff, G.H., and Strausfeld, N.J. (2015). Genealogical correspondence of mushroom bodies across invertebrate phyla. *Curr. Biol.* 25, 38–44.

2454. Wolff, G.H., and Strausfeld, N.J. (2016). Genealogical correspondence of a forebrain centre implies an executive brain in the protostome–deuterostome bilaterian ancestor. *Philos. Trans. R. Soc. Lond. B* 371, 20150055. [*See also* Esmaeeli-Nieh, S., *et al.* (2016). BOD1 is required for cognitive function in humans and *Drosophila*. *PLoS Genet.* 12, #5, e1006022.]

2455. Wollesen, T., Monje, S.V.R., Todt, C., Degnan, B.M., and Wanninger, A. (2015). Ancestral role of *Pax2/5/8* in molluscan brain and multimodal sensory system development. *BMC Evol. Biol.* 15, Article 231.

2456. Wolpert, L. (1968). The French Flag Problem: a contribution to the discussion on pattern development and regulation. *In* C.H. Waddington (ed.), *Towards a Theoretical Biology. I. Prolegomena*. Aldine, Chicago, IL, pp. 125–133.

2457. Wolpert, L. (1969). Positional information and the spatial pattern of cellular differentiation. *J. Theor. Biol.* 25, 1–47.

2458. Wolpert, L. (2014). Revisiting the F-shaped molecule: is its identity solved? *Genesis* 52, 455–457.

2459. Wootton, R.J. (1976). The fossil record and insect flight. *In* R.C. Rainey (ed.), *Insect Flight*. Wiley, New York, NY, pp. 235–254.

2460. Wotton, K.R., Weierud, F.K., Juárez-Morales, J.L., Alvares, L.E., Dietrich, S., and Lewis, K.E. (2009). Conservation of gene linkage in dispersed vertebrate NK homeobox clusters. *Dev. Genes Evol.* 219, 481–496.

2461. Wray, G.A. (1998). Promoter logic. *Science* 279, 1871–1872.

2462. Wray, G.A. (2001). Resolving the *Hox* paradox. *Science* 292, 2256–2257.

2463. Wray, G.A., and Abouheif, E. (1998). When is homology not homology? *Curr. Opin. Genet. Dev.* 8, 675–680.

2464. Wright, G.A. (2015). Olfaction: smells like fly food. *Curr. Biol.* 25, R144–R146.

2465. Wu, J., and Cohen, S.M. (2000). Proximal distal axis formation in the *Drosophila* leg: distinct functions of Teashirt and Homothorax in the proximal leg. *Mech. Dev.* 94, 47–56.

2466. Wu, J., and Mlodzik, M. (2009). A quest for the mechanism regulating global planar cell polarity of tissues. *Trends Cell Biol.* 19, 295–305.

2467. Wu, J.Y., and Rao, Y. (1999). Fringe: defining borders by regulating the Notch pathway. *Curr. Opin. Neurobiol.* 9, 537–543.

2468. Wu, P., Hou, L., Plikus, M., Hughes, M., Scehnet, J., Suksaweang, S., Widelitz, R.B., Jiang, T.-X., and Chuong, C.-M. (2004). Evo-devo of amniote integuments and appendages. *Int. J. Dev. Biol.* 48, 249–270.

2469. Wu, X. (2010). Wg signaling in *Drosophila* heart development as a pioneering model. *J. Genet. Genomics* 37, 593–603.

2470. Wurst, W., and Bally-Cuif, L. (2001). Neural plate patterning: upstream and downstream of the isthmic organizer. *Nat. Rev. Neurosci.* 2, 99–108.

2471. Wyart, C., Webster, W.W., Chen, J.H., Wilson, S.R., McClary, A., Khan, R.M., and Sobel, N. (2007). Smelling a single component of male sweat alters levels of cortisol in women. *J. Neurosci.* 27, #6, 1261–1265.

2472. Wyatt, T.D. (2015). The search for human pheromones: the lost decades and the necessity of returning to first principles. *Proc. R. Soc. Lond. B* 282, 20142994.

2473. Xavier-Neto, J., Castro, R.A., Sampaio, A.C., Azambuja, A.P., Castillo, H.A., Cravo, R.M., and Simões-Costa, M.S. (2007). Parallel avenues in the evolution of hearts and pumping organs. *Cell. Mol. Life Sci.* 64, 719–734.

2474. Xiang, Y., Yuan, Q., Vogt, N., Looger, L.L., Jan, L.Y., and Jan, Y.N. (2010). Light-avoidance-mediating photoreceptors tile the *Drosophila* larval body wall. *Nature* 468, 921–926.

2475. Xie, B., Charlton-Perkins, M., McDonald, E., Gebelein, B., and Cook, T. (2007). Senseless functions as a molecular switch for color photoreceptor differentiation in *Drosophila*. *Development* 134, 4243–4253.

2476. Xu, P.-X., Zhang, X., Heaney, S., Yoon, A., Michelson, A.M., and Maas, R.L. (1999). Regulation of *Pax6* expression is conserved between mice and flies. *Development* 126, 383–395.

2477. Xue, T., Do, M.T.H., Riccio, A., Jiang, Z., Hsieh, J., Wang, H.C., Merbs, S.L., Welsbie, D.S., Yoshioka, T., Weissgerber, P., Stolz, S., Flockerzi, V., Freichel, M., Simon, M.I., Clapham, D.E., and Yau, K.-W. (2011). Melanopsin signalling in mammalian iris and retina. *Nature* 479, 67–73.

2478. Yack, J.E. (2004). The structure and function of auditory chordotonal organs in insects. *Microsc. Res. Tech.* 63, 315–337.

2479. Yager, D., and Hoy, R.R. (1986). The cyclopean ear: a new sense for the praying mantis. *Science* 231, 727–729.

2480. Yager, D., and May, M. (1993). Coming in on a wing and an ear. *Nat. Hist.* 102, #1, 28–33.

2481. Yamagata, M., and Sanes, J.R. (2008). Dscam and Sidekick proteins direct lamina-specific synaptic connections in vertebrate retina. *Nature* 451, 465–469.

2482. Yamaguchi, S., Desplan, C., and Heisenberg, M. (2010). Contribution of photoreceptor subtypes to spectral wavelength preference in *Drosophila*. *PNAS* 107, #12, 5634–5639.

2483. Yamaguchi, S., Wolf, R., Desplan, C., and Heisenberg, M. (2008). Motion vision is independent of color in *Drosophila*. *PNAS* 105, #12, 4910–4915.

2484. Yamaguchi, Y., and Miura, M. (2015). Programmed cell death and caspase functions during neural development. *Curr. Top. Dev. Biol.* 114, 159–184.

2485. Yamaguchi, Y., and Miura, M. (2015). Programmed cell death in neurodevelopment. *Dev. Cell* 32, 478–490.

2486. Yamamoto, S., Jaiswal, M., Charng, W.-L., Gambin, T., Karaca, E., Mirzaa, G., Wiszniewski, W., Sandoval, H., Haelterman, N.A., Xiong, B., Zhang, K., Bayat, V., David, G., Li, T., Chen, K., Gala, U., Harel, T., Pehlivan, D., Penney, S., Vissers, L.E.L.M., de Ligt, J., Jhangiani, S.N., Xie, Y., Tsang, S.H., Parman, Y., Sivaci, M., Battaloglu, E., Muzny, D., Wan, Y.-W., Liu, Z., Lin-Moore, A.T., Clark, R.D., Curry, C.J., Link, N., Schulze, K.L., Boerwinkle, E., Dobyns, W.B., Allikmets, R., Gibbs, R.A., Chen, R., Lupski, J.R., Wangler, M.F., and Bellen, H.J. (2014). A *Drosophila* genetic resource of mutants to study mechanisms underlying human genetic diseases. *Cell* 159, 200–214.

2487. Yan, B. (2010). Numb: from flies to humans. *Brain Dev.* 32, 293–298.

2488. Yang, H.H., and Clandinin, T.R. (2014). What can fruit flies teach us about karate? *eLife* 3, e04040.

2489. Yang, J., Ortega-Hernández, J., Butterfield, N.J., Liu, Y., Boyan, G.S., Hou, J.-b., Lan, T., and Zhang, X.-g. (2016). Fuxianhuiid ventral nerve cord and early nervous system evolution in Panarthropoda. *PNAS* 113, #11, 2988–2993.

2490. Yang, L., Li, R., Kaneko, T., Takle, K., Morikawa, R.K., Essex, L., Wang, X., Zhou, J., Emoto, K., Xiang, Y., and Ye, B. (2014). Trim9 regulates activity-dependent fine-scale topography in *Drosophila*. *Curr. Biol.* 24, 1024–1030.

2491. Yang, M., and Meyer-Rochow, V.B. (2004). Fine-structural details of the photoreceptor membranes in the ocellus of the scale-insect parasite *Centrodora* sp. (Hymenoptera; Aphenelidae): a case of gene transfer between host and parasite? *Biocell* 28, 151–154.

2492. Yang, Y., Kovács, M., Sakamoto, T., Zhang, F., Kiehart, D.P., and Sellers, J.R. (2006). Dimerized *Drosophila* myosin VIIa: a processive motor. *PNAS* 103, #15, 5746–5751.

2493. Yang, Y., and Mlodzik, M. (2015). Wnt-Frizzled/Planar cell polarity signaling: cellular orientation by facing the wind (Wnt). *Annu. Rev. Cell Dev. Biol.* 31, 623–646.

2494. Yang, Z., Bertolucci, F., Wolf, R., and Heisenberg, M. (2013). Flies cope with uncontrollable stress by learned helplessness. *Curr. Biol.* 23, 799–803.

2495. Yao, Z., and Shafer, O.T. (2014). The *Drosophila* circadian clock is a variably coupled network of multiple peptidergic units. *Science* 343, 1516–1520.

2496. Yarmolinsky, D.A., Zuker, C.S., and Ryba, N.J.P. (2009). Common sense about taste: from mammals to insects. *Cell* 139, 234–244.

2497. Yau, K.-W., and Hardie, R.C. (2009). Phototransduction motifs and variations. *Cell* 139, 246–264.

2498. Yekta, S., Tabin, C.J., and Bartel, D.P. (2008). MicroRNAs in the Hox network: an apparent link to posterior prevalence. *Nat. Rev. Genet.* 9, 789–796.

2499. Yeo, R.A., and Gangestad, S.W. (1994). Developmental origins of variation in human hand preference. *In* T.A. Markow (ed.), *Developmental Instability: Its Origins and Evolutionary Implications*. Kluwer, London, pp. 283–298.

2500. Yi, C.H., and Yuan, J. (2009). The Jekyll and Hyde functions of caspases. *Dev. Cell* 16, 21–34.

2501. Yi, H., and Norell, M.A. (2015). The burrowing origin of modern snakes. *Sci. Adv.* 1, e1500743.

2502. Yogev, S., and Shen, K. (2014). Cellular and molecular mechanisms of synaptic specificity. *Annu. Rev. Cell Dev. Biol.* 30, 417–437.

2503. Yokoyama, S., and Radlwimmer, F.B. (2001). The molecular genetics and evolution of red and green color vision in vertebrates. *Genetics* 158, 1697–1710.

2504. Yokoyama, S., Xing, J., Liu, Y., Faggionato, D., Altun, A., and Starmer, W.T. (2014). Epistatic adaptive evolution of human color vision. *PLoS Genet.* 10, #12, e1004884.

2505. Yokoyama, T., Copeland, N.G., Jenkins, N.A., Montgomery, C.A., Elder, F.F.B., and Overbeek, P.A. (1993). Reversal of left-right asymmetry: a situs inversus mutation. *Science* 260, 679–682.

2506. Yong, E. (2015). Seeing the light. *Natl. Geogr.* 229, #2, 30–57.

2507. Yoshiba, S., and Hamada, H. (2014). Roles of cilia, fluid flow, and Ca^{2+} signaling in breaking of left-right symmetry. *Trends Genet.* 30, 10–17.

2508. Yoshida, M.-a., Shigeno, S., Tsuneki, K., and Furuya, H. (2010). Squid vascular endothelial growth factor receptor: a shared molecular signature in the convergent evolution of closed circulatory systems. *Evol. Dev.* 12, 25–33.

2509. Yoshida, M.-a., Yura, K., and Ogura, A. (2014). Cephalopod eye evolution was modulated by the acquisition of Pax-6 splicing variants. *Sci. Rep.* 4, Article 4256.

2510. Yost, H.J. (1999). Diverse initiaiton in a conserved left-right pathway? *Curr. Opin. Genet. Dev.* 9, 422–426.

2511. Yost, H.J. (2003). Left-right asymmetry: nodal cilia make and catch a wave. *Curr. Biol.* 13, R808–R809.

2512. Young, M.W. (2000). The tick-tock of the biological clock. *Sci. Am.* 282, #3, 64–71.

2513. Young, M.W., and Kay, S.A. (2001). Time zones: a comparative genetics of circadian clocks. *Nat. Rev. Genet.* 2, 702–715.

2514. Yu, J.-K., Holland, L.Z., and Holland, N.D. (2002). An amphioxus *nodal* gene (*AmphiNodal*) with early symmetrical expression in the organizer and mesoderm and later asymmetrical expression associated with left-right axis formation. *Evol. Dev.* 4, 418–425.

2515. Yu, J.-K., Satou, Y., Holland, N.D., Shin-I, T., Kohara, Y., Satoh, N., Bronner-Fraser, M., and Holland, L.Z. (2007). Axial patterning in cephalochordates and the evolution of the organizer. *Nature* 445, 613–617.

2516. Yu, W., and Hardin, P.E. (2006). Circadian oscillators of *Drosophila* and mammals. *J. Cell Sci.* 119, 4793–4795.

2517. Yuan, S., Yu, X., Asara, J.M., Heuser, J.E., Ludtke, S.J., and Akey, C.W. (2011). The holo-apoptosome: activation of procaspase-9 and interactions with caspase-3. *Structure* 19, 1084–1096.

2518. Yuan, S., Yu, X., Topf, M., Dorstyn, L., Kumar, S., Ludtke, S.J., and Akey, C.W. (2011). Structure of the *Drosophila* apoptosome at 6.9 Å resolution. *Structure* 19, 128–140.

2519. Zaccardi, G., Kelber, A., Sison-Mangus, M.P., and Briscoe, A.D. (2006). Color discrimination in the red range with only one long-wavelength sensitive opsin. *J. Exp. Biol.* 209, 1944–1955.

2520. Zaffran, S., El Robrini, N., and Bertrand, N. (2014). Retinoids and cardiac development. *J. Dev. Biol.* 2, 50–71.
2521. Zaffran, S., Reim, I., Qian, L., Lo, P.C., Bodmer, R., and Frasch, M. (2006). Cardioblast-intrinsic Tinman activity controls proper diversification and differentiation of myocardial cells in *Drosophila. Development* 133, 4073–4083.
2522. Zagozewski, J.L., Zhang, Q., Pinto, V.I., Wigle, J.T., and Eisenstat, D.D. (2014). The role of homeobox genes in retinal development and disease. *Dev. Biol.* 393, 195–208.
2523. Zak, M., Klis, S.F.L., and Grolman, W. (2015). The Wnt and Notch signalling pathways in the developing cochlea: formation of hair cells and induction of regenerative potential. *Int. J. Dev. Neurosci.* 47, 247–258.
2524. Zakaria, S., Mao, Y., Kuta, A., de Sousa, C.F., Gaufo, G.O., McNeill, H., Hindges, R., Guthrie, S., Irvine, K.D., and Francis-West, P.H. (2014). Regulation of neuronal migration by Dchs1-Fat4 planar cell polarity. *Curr. Biol.* 24, 1620–1627.
2525. Zakin, L., and De Robertis, E.M. (2010). Extracellular regulation of BMP signaling. *Curr. Biol.* 20, R89–R92.
2526. Zamora, S., Rahman, I.A., and Smith, A.B. (2012). Plated Cambrian bilaterians reveal the earliest stages of echinoderm evolution. *PLoS ONE* 7, #6, e38296.
2527. Zampini, V., Rüttiger, L., Johnson, S.L., Franz, C., Furness, D.N., Waldhaus, J., Xiong, H., Hackney, C.M., Holley, M.C., Offenhauser, N., Di Fiore, P.P., Knipper, M., Masetto, S., and Marcotti, W. (2011). Eps8 regulates hair bundle length and functional maturation of mammalian auditory hair cells. *PLoS Biol.* 9, #4, e1001048.
2528. Zarin, A.A., Asadzadeh, J., Hokamp, K., McCartney, D., Yang, L., Bashaw, G.J., and Labrador, J.-P. (2014). A transcription factor network coordinates attraction, repulsion, and adhesion combinatorially to control motor axon pathway selection. *Neuron* 81, 1297–1311.
2529. Zarin, A.A., Asadzadeh, J., and Labrador, J.-P. (2014). Transcriptional regulation of guidance at the midline and in motor circuits. *Cell. Mol. Life Sci.* 71, 419–432.
2530. Zdobnov, E.M., and Bork, P. (2007). Quantification of insect genome divergence. *Trends Genet.* 23, 16–20.
2531. Zeitouni, B., Sénatore, S., Séverac, D., Aknin, C., Sémériva, M., and Perrin, L. (2007). Signalling pathways involved in adult heart formation revealed by gene expression profiling in *Drosophila. PLoS Genet.* 3, #10, e174.
2532. Zhang, J. (2003). Evolution by gene duplication: an update. *Trends Ecol. Evol.* 18, 292–298.
2533. Zhang, J., and Webb, D.M. (2003). Evolutionary deterioration of the vomeronasal pheromone transduction pathway in catarrhine primates. *PNAS* 100, #14, 8337–8341.
2534. Zhang, W., Cheng, L.E., Kittelmann, M., Li, J., Petkovic, M., Cheng, T., Jin, P., Guo, Z., Göpfert, M.C., Jan, L.Y., and Jan, Y.N. (2015). Ankyrin repeats convey force to gate the NOMPC mechanotransduction channel. *Cell* 162, 1391–1402.
2535. Zhang, Y., Andl, T., Yang, S.H., Teta, M., Liu, F., Seykora, J.T., Tobias, J.W., Piccolo, S., Schmidt-Ullrich, R., Nagy, A., Taketo, M.M., Dlugosz, A.A., and Millar, S.E. (2008). Activation of β-catenin signaling programs embryonic epidermis to hair follicle fate. *Development* 135, 2161–2172.
2536. Zhang, Y., Tomann, P., Andl, T., Gallant, N.M., Huelsken, J., Jerchow, B., Birchmeier, W., Paus, R., Piccolo, S., Mikkola, M.L., Morrisey, E.E., Overbeek, P.A., Scheidereit, C., Millar, S.E., and Schmidt-Ullrich, R. (2009). Reciprocal requirements for EDA/EDAR/NF-κB and Wnt/β-catenin signaling pathways in hair follicle induction. *Dev. Cell* 17, 49–61.

2537. Zhang, Y., Yang, Y., Trujillo, C., Zhong, W., and Leung, Y.F. (2012). The expression of *irx7* in the inner nuclear layer of zebrafish retina is essential for a proper retinal development and lamination. *PLoS ONE* 7, #4, e36145.

2538. Zhao, L., Svingen, T., Ng, E.T., and Koopman, P. (2015). Female-to-male sex reversal in mice caused by transgenic overexpression of *Dmrt1*. *Development* 142, 1083–1088.

2539. Zheng, J., Shen, W., He, D.Z.Z., Long, K.B., Madison, L.D., and Dallos, P. (2000). Prestin is the motor protein of cochlear outer hair cells. *Nature* 405, 149–155.

2540. Zheng, L., Zheng, J., Whitlon, D.S., García-Añoveros, J., and Bartles, J.R. (2010). Targeting of the hair cell proteins cadherin 23, harmonin, myosin XVa, espin, and prestin in an epithelial cell model. *J. Neurosci.* 30, #21, 7187–7201.

2541. Zhong, Y.-f., and Holland, P.W.H. (2011). The dynamics of vertebrate homeobox gene evolution: gain and loss of genes in mouse and human lineages. *BMC Evol. Biol.* 11, Article 169.

2542. Zhou, F., and Roy, S. (2015). Snapshot: motile cilia. *Cell* 162, 224.

2543. Zhu, B., Pennack, J.A., McQuilton, P., Forero, M.G., Mizuguchi, K., Sutcliffe, B., Gu, C.-J., Fenton, J.C., and Hidalgo, A. (2008). *Drosophila* neurotrophins reveal a common mechanism for nervous system formation. *PLoS Biol.* 6, #11, e284.

2544. Zhukovsky, E.A., and Oprian, D.D. (1989). Effect of carboxylic acid side chains on the absorption maximum of visual pigments. *Science* 246, 928–930.

2545. Zhukovsky, E.A., Robinson, P.R., and Oprian, D.D. (1991). Transducin activation by rhodopsin without a covalent bond to the 11-*cis*-retinal chromophore. *Science* 251, 558–560.

2546. Ziauddin, J., and Schneider, D.S. (2012). Where does innate immunity stop and adaptive immunity begin? *Cell Host Microbe* 12, 394–395.

2547. Ziegler, A.B., Berthelot-Grosjean, M., and Grosjean, Y. (2013). The smell of love in *Drosophila. Front. Physiol.* 4, Article 72.

2548. Zimmer, C. (2001). *Evolution: The Triumph of an Idea*. HarperCollins, New York, NY.

2549. Zimmer, C. (2013). Genes are us. And them. *Natl. Geogr.* 224, #1, 102–103.

2550. Zimmerman, A., Bai, L., and Ginty, D.D. (2014). The gentle touch receptors of mammalian skin. *Science* 346, 950–954.

2551. Zinn, K. (2007). Dscam and neuronal uniqueness. *Cell* 129, 455–456.

2552. Zinn, K., and Sun, Q. (1999). Slit branches out: a secreted protein mediates both attractive and repulsive axon guidance. *Cell* 97, 1–4.

2553. Zinzen, R.P., Cande, J., Ronshaugen, M., Papatsenko, D., and Levine, M. (2006). Evolution of the ventral midline in insect embryos. *Dev. Cell* 11, 895–902.

2554. Zipursky, S.L., and Sanes, J.R. (2010). Chemoaffinity revisited: Dscams, protocadherins, and neural circuit assembly. *Cell* 143, 343–353.

2555. Zollinger, D.R., Baalman, K.L., and Rasband, M.N. (2015). The ins and outs of polarized axonal domains. *Annu. Rev. Cell Dev. Biol.* 31, 647–667.

2556. Zordan, M.A., and Sandrelli, F. (2015). Circadian clock dysfunction and psychiatric disease: could fruit flies have a say? *Front. Neurol.* 6, Article 80.

2557. Zou, Z., and Buck, L.B. (2006). Combinatorial effects of odorant mixes in olfactory cortex. *Science* 311, 1477–1481.

2558. Zuber, M.E., Gestri, G., Viczian, A.S., Barsacchi, G., and Harris, W.A. (2003). Specification of the vertebrate eye by a network of eye field transcription factors. *Development* 130, 5155–5167.

2559. Zuker, C.S. (1994). On the evolution of eyes: would you like it simple or compound? *Science* 265, 742–743.

2560. Zumdahl, S.S. (1986). *Chemistry*. D.C. Heath, Lexington, MA.

Index

Printed in the United States
By Bookmasters